Plant Life and Environment Dynamics

This book series aims to publish volumes focusing on plant responses to changing environmental conditions. The series is centered on the theme of plant-environment interaction, ecosystem changes, plant physiology, various physio-biochemical and molecular mechanism of adaptation working within the plants. The book series also provides knowledge and understanding of plant's stress tolerance strategies such as ionomic transcriptomic, proteomic, phytohormones, cellular redox regulation, energy metabolism, nanomaterial in plant protection and genomics of plants under environmental stress metabolomic approaches which are currently being used in order to protect plant life against adverse effects of changing environment. This series will cover book ideas focusing on all aspects of plant's responses to the change in the respective environment. The books published in this series are a useful source of information for academicians and researchers working in the field of agriculture, botany, agronomy, biotechnology, plant science, crop physiology, plant stress physiology, plant and environment interactions related courses.

Series plans to cover dynamics of plant-environment interaction via following broad themes:

(a) Abiotic stress and plant development
(b) Photosynthesis and energy metabolism
(c) Bio-stimulants and plant life
(d) Phytohormones in stress tolerance
(e) Cellular redox regulation (ROS, RNS and RSS), oxidative stress and antioxidant defense system
(f) Transcriptomics, Proteomics and Functional Genomics
(g) Ionomics: Uptake and Assimilation
(h) Nanomaterials and plant life

Anket Sharma
Devendra Kumar Chauhan
Durgesh Kumar Tripathi
Editors

Hormones and Heavy Metals

Springer

Editors
Anket Sharma
Texas Tech University
Lubbock, TX, USA

Zhejiang A&F University
Hangzhou, China

Durgesh Kumar Tripathi
Amity Institute of Organic Agriculture
Amity University
Noida, India

Devendra Kumar Chauhan
University of Allahabad
Prayagraj, India

ISSN 2730-6755 ISSN 2730-6763 (electronic)
Plant Life and Environment Dynamics
ISBN 978-981-96-9621-5 ISBN 978-981-96-9622-2 (eBook)
https://doi.org/10.1007/978-981-96-9622-2

This Springer imprint is published by the registered company Springer Nature Singapore Pte Ltd.
The registered company address is: 152 Beach Road, #21-01/04 Gateway East, Singapore 189721, Singapore

Contents

About the Editors and Contributors

About the Editors

Anket Sharma is currently working as a Visiting Scientist at Texas Tech University, Lubbock, USA. Additionally, he is also working as an Adjunct Professor at Zhejiang A&F University, China. Prior to that, he worked as a postdoctoral scientist at Zhejiang A&F University, China. He did his PhD in Plant Science at Guru Nanak Dev University, India. He got DST-INSPIRE fellowship from the Department of Science and Technology, Government of India. His research interests are to study plant hormone-mediated regulation of plant physiology under abiotic stress conditions. He has authored over 100 publications in internationally reputed journals. He has been listed in the "Highly Cited Researchers 2022" by Clarivate Analytics.

Devendra Kumar Chauhan is working as a professor, and he was former head of the Department of Botany, University of Allahabad. He has authored 250 scientific publications. His research interest includes the role of nutrient including silicon, zinc, etc. in management of abiotic stress of plants. He is also working as editor and reviewer of several reputed international journals and has published more than 10 international books.

Durgesh Kumar Tripathi is currently working as an Associate Professor and Joint Coordinator (R&D) at Amity Institute of Organic Agriculture, Amity University Uttar Pradesh, India. He has received his DPhil in Science from University of Allahabad, India. During this period, he worked extensively on Silicon analysis, crop stress physiology, agro-nanotechnology, and molecular stress biology. His research interests encompass stress tolerance mechanisms in plants. Presently, he is working with nutrient and nano-nutrient regulation in plants under various stresses. He is also working on nitric oxide and hormonal crosstalk against abiotic stress in plants. He has published more than 200 research papers and review articles in internationally reputed journals and several book chapters. He has also edited 15 books for Elsevier, Wiley, CRC Press, and Springer Nature. He is Associate Editor in very high impact journals and the Chief Editor of *Plant Nanobiology* published with Elsevier. He is recognized as top 2% researcher by Stanford University, USA, from 2019 to 2024, and Top 1% highly cited researchers all over the globe by Clarivate web of Science, USA.

Contributors

S. Maqbool Ahmed Botany Section, School of Sciences, Maulana Azad National Urdu University, Hyderabad, Telangana, India

Savita Bhardwaj Department of Botany, MCM DAV College, Kangra, Himachal Pradesh, India

Romila Rawat Bisht Department of Botany, Hansraj College, University of Delhi, Delhi, India

Ambika Choudhary School of Biological and Environmental Sciences, Shoolini University, Solan, Himachal Pradesh, India

Anuj Choudhary Department of Biology and Environmental Sciences, CSK Himachal Pradesh Agricultural University, Palampur, Himachal Pradesh, India

Manik Devgan Department of Plant Breeding and Genetics, Punjab Agricultural University, Ludhiana, Punjab, India

Yashika Dhingra Department of Plant Molecular Biology, South Campus, University of Delhi, Delhi, India

Nahla El-Sherif Botany Department, Faculty of Science, Ain Shams University, Cairo, Egypt

Biology Department, Faculty of Science, Taibah University, Madinah, Kingdom of Saudi Arabia

Mohammad Faizan Botany Section, School of Sciences, Maulana Azad National Urdu University, Hyderabad, Telangana, India

Aurosha Fayaz Botany Section, School of Sciences, Maulana Azad National Urdu University, Hyderabad, Telangana, India

Suresh Gali Department of Botany and Plant Physiology, Chaudhary Charan Singh Haryana Agricultural University, Hisar, Haryana, India

Shruti Gupta Department of Botany, Hansraj College, University of Delhi, Delhi, India

Sadaf Jan Department of Biotechnology, School of Bioengineering and Biosciences, Lovely Professional University, Phagwara, Punjab, India

Shivam Jasrotia Department of Biosciences, Chandigarh University, Mohali, Punjab, India

Kaveri Umesh Kadam School of Engineering Sciences and Technology, Jamia Hamdard, New Delhi, India

Dhriti Kapoor School of Biological and Environmental Sciences, Shoolini University, Solan, Himachal Pradesh, India

Arshdeep Kaur Department of Plant Breeding and Genetics, Punjab Agricultural University, Ludhiana, Punjab, India

Harleen Kaur Department of Botany and Environment Studies, Jalandhar, Punjab, India

Harmanjot Kaur Department of Botany, Punjab Agricultural University, Ludhiana, Punjab, India

Ravneet Kaur Department of Botany and Environment Studies, DAV University, Jalandhar, Punjab, India

Sukhmanjot Kaur Department of Botany, Punjab Agricultural University, Ludhiana, Punjab, India

Ekhlaque A. Khan Department of Biotechnology, Chaudhary Bansi Lal University, Bhiwani, Haryana, India

Ira Khan Botany Section, School of Sciences, Maulana Azad National Urdu University, Hyderabad, Telangana, India

Lisha Khungar Department of Plant Molecular Biology, South Campus, University of Delhi, Delhi, India

Monika Koul Department of Botany, Hansraj College, University of Delhi, Delhi, India

Asha Kumari School of Crop Science, ICAR-Indian Agricultural Research Institute, Hazaribagh, Jharkhand, India

Krishna Kumar Department of Plant Physiology, Institute of Agricultural Sciences, Varanasi, Uttar Pradesh, India

Lalit Kumar Department of Botany, Dharam Samaj College, Aligarh, Uttar Pradesh, India

Rahul Kumar Faculty of Agricultural Sciences, DAV University, Jalandhar, Punjab, India

Simar Malhotra Department of Botany, Hansraj College, University of Delhi, Delhi, India

Vasudha Maurya Department of Biotechnology, DAV University, Jalandhar, Punjab, India

Faculty of Agricultural Sciences, DAV University, Jalandhar, Punjab, India

Sahil Mehta Department of Botany, Hansraj College, University of Delhi, Delhi, India

Amarendra N. Misra Office of the Vice Chancellor, Jagdguru Kripalu University, Banara, Cuttack, Odisha, India

Slimane Mokrani Laboratoire de Maitrise des Energies Renouvelables, Faculté des Sciences de la Nature et de la Vie, Université de Bejaia, Bejaia, Algeria

El-hafid Nabti Laboratoire de Maitrise des Energies Renouvelables, Faculté des Sciences de la Nature et de la Vie, Université de Bejaia, Bejaia, Algeria

Om Prakash Narayan Division of Chemical Biology and Medicinal Chemistry, University of Texas at Austin, Austin, TX, USA

Puja Panda Department of Biotechnology, School of Bioengineering and Biosciences, Lovely Professional University, Phagwara, Punjab, India

Vijay Rani Rajpal Department of Botany, Hansraj College, University of Delhi, Delhi, India

Amisha Rani Department of Forest Products, Dr YS Parmar University of Horticulture & Forestry, Nauni, Solan, Himachal Pradesh, India

Rajeev Rathore Department of Botany, Hansraj College, University of Delhi, Delhi, India

Ankush Sharma ICAR-Indian Institute of Maize Research, Ludhiana, Punjab, India

Ashish Sharma Department of Botany and Environment Studies, DAV University, Jalandhar, Punjab, India

Ashutosh Sharma Faculty of Agricultural Sciences, DAV University, Jalandhar, Punjab, India

Binny Sharma Department of Plant Physiology, Institute of Agricultural Sciences, Varanasi, Uttar Pradesh, India

Dhriti Sharma Department of Botany, School of Bioengineering and Biosciences, Lovely Professional University, Phagwara, Punjab, India

Indu Sharma Department of Life Sciences, University Institute of Sciences, Sant Baba Bhag Singh University, Jalandhar, Punjab, India

Priya Sharma Department of Botany, University of Delhi, Delhi, India

Priyanka Sharma Department of Crop Physiology, Assam Agricultural University, Jorhat, Assam, India

Radhika Sharma Department of Soil Science, Punjab Agricultural University, Ludhiana, Punjab, India

Shivam Sharma Department of Vegetable Sciences, CSK Himachal Pradesh Agricultural University, Palampur, Himachal Pradesh, India

Sonika Sharma Faculty of Agricultural Sciences, DAV University, Jalandhar, Punjab, India

Sadaf Shehzad University of Texas Rio Grande Valley, Edinburg, TX, USA

Amanpreet Singh Department of Agronomy, Punjab Agricultural University, Ludhiana, Punjab, India

Harjot Singh Faculty of Agricultural Sciences, DAV University, Jalandhar, Punjab, India

Mohar Singh Division of Agrotechnology, CSIR – Institute of Himalayan Bioresource Technology, Palampur, Himachal Pradesh, India

Narayan Singh Department of Biotechnology, DAV University, Jalandhar, Punjab, India

Faculty of Agricultural Sciences, DAV University, Jalandhar, Punjab, India

Kiran Soni Department of Forest Products, Dr YS Parmar University of Horticulture & Forestry, Nauni, Solan, Himachal Pradesh, India

Usha Thakur Department of Botany, MCM DAV College, Kangra, Himachal Pradesh, India

Sheen Tickoo Department of Botany, Hansraj College, University of Delhi, Delhi, India

Mohit Uppal Department of Botany, Hansraj College, University of Delhi, Delhi, India

Bindu Yadav School of Environmental Sciences, Jawaharlal Nehru University, New Delhi, India

Sudha Yadav Department of Botany, Chaudhary Bansi Lal University, Bhiwani, Haryana, India

Chapter 1
Regulation of Plant Biology by Phytohormones

Harmanjot Kaur, Amanpreet Singh, Anuj Choudhary, Ankush Sharma, Sukhmanjot Kaur, Shruti Gupta, Vijay Rani Rajpal, Sahil Mehta, Kaveri Umesh Kadam, and Om Prakash Narayan

Abstract Plants, which are sessile life forms, are frequently subjected to a variety of abiotic and biotic stressors. Drought, heat, cold, and salinity are examples of abiotic stress conditions, whereas bacteria, fungi, viruses, nematodes, and insects are the principal sources of biotic stress. Plants have developed sophisticated systems that assist them in recognizing the stress signal and permit the best possible growth response to adapt to such unfavorable conditions. Phytohormones are essential for the plant's ability to adapt to challenging environmental circumstances. The complex signaling networks and the sophisticated interaction between the various hormone-signaling pathways have been clarified by recent study discoveries. With a particular emphasis on the relevance of crosstalk between various hormones in producing a complex and effective stress response, we summarize the roles of the key plant hormones in regulating abiotic and biotic stress responses in this study. The explanation of seed dormancy and germination serves to show the delicate bal-

---

H. Kaur · S. Kaur
Department of Botany, Punjab Agricultural University, Ludhiana, Punjab, India

A. Singh
Department of Agronomy, Punjab Agricultural University, Ludhiana, Punjab, India

A. Choudhary
Department of Biology and Environmental Sciences, CSK Himachal Pradesh Agricultural University, Palampur, Himachal Pradesh, India

A. Sharma
ICAR-Indian Institute of Maize Research, Ludhiana, Punjab, India

S. Gupta · V. R. Rajpal · S. Mehta
Department of Botany, Hansraj College, University of Delhi, Delhi, India

K. U. Kadam
School of Engineering Sciences and Technology, Jamia Hamdard, New Delhi, India

O. P. Narayan (✉)
Division of Chemical Biology and Medicinal Chemistry, University of Texas at Austin, Austin, TX, USA

A. Sharma et al. (eds.), *Hormones and Heavy Metals*, Plant Life and Environment Dynamics, https://doi.org/10.1007/978-981-96-9622-2_1

ance that may be imposed by the phytohormones in controlling plant responses to environmental cues.

Keywords Plants · Stresses · Hormones · Physiology · Supplementation · Effect · Genes

1.1 Introduction

Phytoregulators, commonly referred to as plant growth regulators, are chemical substances that act as bioregulators and are derived from plants. They can have effects like those generated by phytohormones (Yakhin et al. 2017). They come from an external source, and their use is usually targeted and in line with certain productivity goals. The synthesis of phytohormones in plants can take place anywhere throughout the plant, in contrast to the origin of animal hormones, which normally come from specialized places or organs. Nutrition, absorption, circulation, transpiration, photosynthesis, respiration, growth, blooming, fruit ripening, and rooting are only a few examples of the numerous physiological processes in plants that are complex and diverse collections of chemical reactions and physical occurrences. To ensure their life and sustainability, plants constantly adapt to their environment. Chemical messengers are essential in these complex routes and gradients for stimulating development, coping with stress, and preserving homeostasis (Jaillais and Chory 2010).

The traditional phytohormones—auxin, gibberellic acid, cytokinin, ethylene, and abscisic acid—are now recognized as different categories of phytohormones that control plant growth (Davies 2010) (Figs. 1.1, 1.2, and 1.3). Additionally, it has been shown that there are additional signaling molecules with comparable activities to those of phytohormones in plants. These molecules are essential for disease resistance and herbivore defense. Brassinosteroids, jasmonates, salicylates, strigolactones, systemin peptides, and nitric oxide are some of them (Smith et al. 2017) (Table 1.1). Therefore, the current chapter will deal with both traditional as well as newly established hormones.

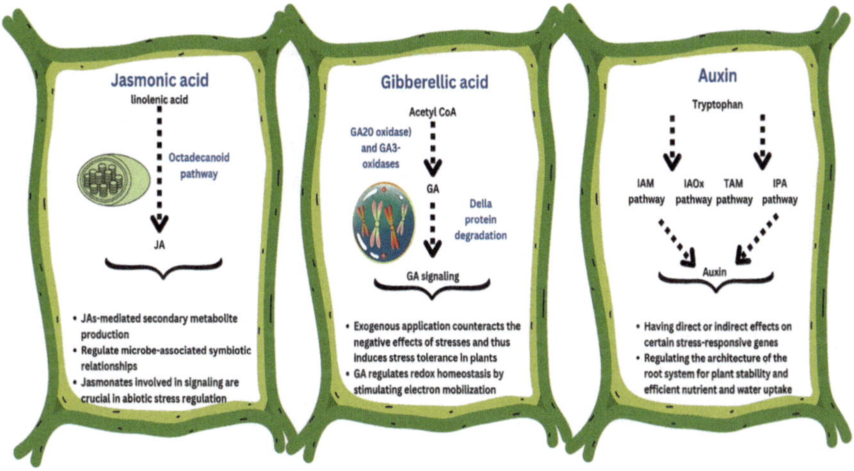

Fig. 1.1 Diagrammatic representation of biosynthesis as well as functions of three different acidic phytohormones (Auxin, Gibberellic acid, and Jasmonic acid)

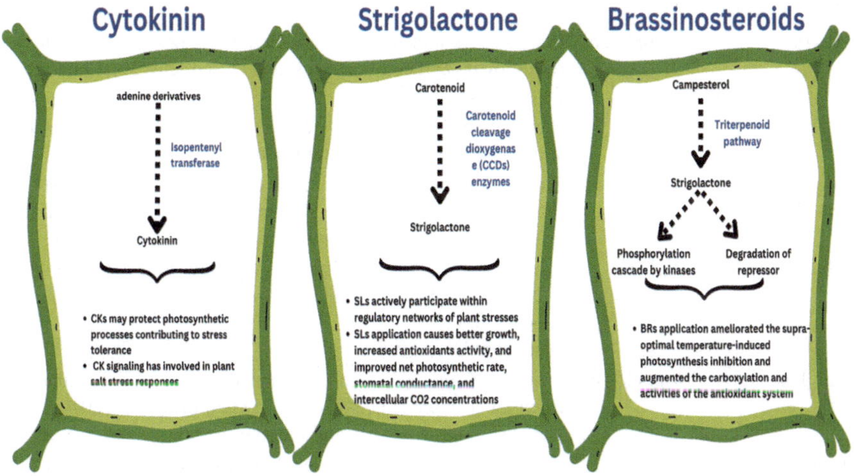

Fig. 1.2 Diagrammatic representation of biosynthesis as well as functions of cytokinin, strigolactones, and brassinosteroids

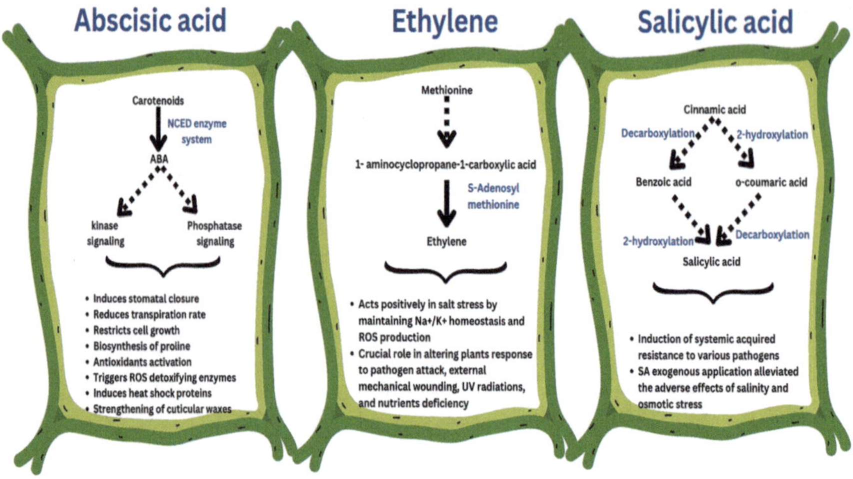

Fig. 1.3 Neck-to-neck comparison of biosynthesis and functions of three stress-related hormones in plant biology

Table 1.1 Comparative account of different plant hormones, their functions, and effects of their overproduction

Plant hormone	Plant	Function	Overproduction effect	References
Auxins	Arabidopsis	Promotes stem growth, root growth, and cell division	Stunted growth, thick stems, and curled leaves	Qingguo et al. (2014)
Cytokinins	Cotton	Promotes cell division and leaf growth	Delayed flowering and fruit ripening	Zhao et al. (2015)
Gibberellins	Apricot	Promotes stem growth, flowering, and fruit growth	Gigantic plants, but weak stems and poor fruit quality	Chauhan et al. (2020)
Abscisic acid	Barley	Promotes seed dormancy, stomatal closure, and leaf abscission	Stunted growth, wilting, and early leaf drop	Gietler et al. (2020)
Ethylene	Arabidopsis	Promotes fruit ripening, senescence, and abscission	Overripe and mushy fruits, early leaf drop, and lodging	Tatsuki (2010)
Brassinosteroids	Rice	Promote cell division, elongation, and differentiation	Stunted growth, thick leaves, and abnormal flower development	Tsuda et al. (2014)
Salicylic acid	Chickpea	Induces systemic acquired resistance (SAR) to diseases	Reduced plant growth and yield	Kaur et al. (2022)
Jasmonic acid	Rice	Induces systemic acquired resistance (SAR) to pests and diseases	Reduced plant growth and yield	Bhavanam and Stout (2021)

(continued)

Table 1.1 (continued)

Plant hormone	Plant	Function	Overproduction effect	References
Strigolactones	Arabidopsis	Promote root architecture and shoot branching	Stunted growth, reduced branching, and increased susceptibility to pests and diseases	Santoro et al. (2021)
Peptide hormones	Peach	Regulate a variety of processes, including growth, development, and stress responses	Effects of overproduction vary depending on the specific peptide hormone	Tadiello et al. (2016)

1.2 Major Morphological Events Controlled by Plant Hormones

1.2.1 Seed Germination

Controlling and coordinating cell division, growth, and differentiation are some of the plant hormones' most significant roles. Plant hormones can influence a variety of plant processes, including seed germination and dormancy (Graeber et al. 2012). Many physiological and biochemical processes in plants are regulated by plant hormones, which include abscisic acid, ethylene, gibberellins, auxin, cytokinins, and brassinosteroids.

As embryonic organs arise as a result of cell division and differentiation during seed development, the embryo body is also formed. This stage includes the development of the embryo's size and weight as well as the maturity of the seed, which is followed by the development of desiccation tolerance and dormancy. As seeds mature, the cell cycle is inhibited, seed moisture decreases, ABA levels rise, storage reservoirs are produced, and dormancy is formed (Bewley et al. 2012).

For seed germination and the beginning of dormancy, respectively, gibberellins and ABA are required. The processes required for seed maturity or the capacity of seeds to germinate are determined by the gibberellins/ABA balance (Finch-Savage and Leubner-Metzger 2006). Gibberellins are required for seed germination, whereas ABA governs seed dormancy and prevents seeds from germinating (Matilla and Matilla-Vázquez 2008).

Gibberellins are the most significant hormone; however, ethylene and gibberellins both have an impact on radicle development. Ethylene is not required for the mannanase synthesis that is required for seed germination, but gibberellins are required. However, because the seeds may germinate fully in gibberellin-deficient mutants, ethylene can behave similarly to gibberellins in these situations (Wang et al. 2009).

It has been suggested that ABA and the genes that respond to it have a role in the germination of seeds. According to Graeber et al. (2012), ABA inhibits seed germination by weakening endosperm, delaying radicle growth, and increasing transcription factor expression. These actions may have a negative impact on the process of seed germination. By increasing the synthesis of ABA and the associated transcription factors, the gibberellin repressor RGL2 can prevent seed germination (Piskurewicz et al. 2008). During many plant growth phases, including seed germination, the H component of the chloroplast enzyme Mg-chelatase can function as an ABA receptor (Shen et al. 2006). Recent research suggests that GPROTEIN COUPLED RECEPTOR 2 may function as an additional ABA receptor, modulating ABA's many effects, including those on seed germination (Liu et al. 2007).

Ethylene production can be stimulated by brassinosteroids (BR) and IAA (Arteca and Arteca 2008). By rupturing the testa and endosperm, gibberellins, ethylene, and BR may stimulate seed germination while counteracting the inhibitory effects of ABA on seed germination (Finkelstein 2013). Dormant seeds can sprout when exposed to ethylene. According to some research (Borghetti et al. 2002), ethylene can break the dormancy of seeds by controlling the expression of the cysteine-proteinase genes and the proteasome, which is a protein complex that contains these genes. During the early phases of germination, these enzymes can break down seed proteins.

The growth and development of the embryo, which leads to the eventual generation of a new plant, depends on the seed germination process, which is governed by a variety of processes. Seed germination is influenced by interactions between plant hormones and genes. Seed dormancy is influenced by variables such as linked genes, chromatin-associated elements, and non-enzymatic processes. The maturing genes, hormonal and epigenetic regulatory genes, and the genes that govern emergence from dormancy are among the genes that control dormancy (Graeber et al. 2012).

1.2.2 Shoot and Root Growth

The first significant development of widely applicable, commercially viable tree growth retardants (TGRs) was made in the late 1970s by the cell elongation inhibitors paclobutrazol (PBZ), uniconazole (UCZ), and flurprimidol (Kaur et al. 2021). Compared to other TGRs, PBZ is more effective and only has to be present at relatively small concentrations to restrict shoot development. To prevent internode elongation and promote greener, more compact growth, PBZ primarily affects trees by lowering tree height. The leaves get greener as a result of PBZ's compact growth, which increases chlorophyll content per unit leaf area. Additional photosynthetic activity can result from the higher chlorophyll content per unit leaf area. Fruit trees, including mango, cashew, peach, apricot, and pecan trees, all experienced considerably slower development after the PBZ treatment (Kaur et al. 2021).

An evolutionarily conserved developmental system in terrestrial plants, adaptive modulation of the shoot/root growth ratio provides maximum progeny generation under changing environmental conditions. Shoot growth is preferred in growth conditions where water and mineral nutrients are not limited because it embodies traits that support the species' ability to reproduce and survive, and root development is constrained to a level that allows shoot development to continue without unnecessarily depleting photosynthetic capacity. Thus, it is projected that plants will have a high shoot/root ratio under ideal water and nutritional circumstances (Fageria and Moreira 2011). The growth of the shoot—the primary location for water loss and mineral nutrient consumption—must be slowed when water or mineral nutrient availability declines in favor of the development of a bigger root system.

Auxin and cytokinin are essential for the development of the vascular system and, in part, use vascular transport to influence plant growth. The amount of nitrogen in a soil affects auxin synthesis and auxin-dependent root development. A slight nitrogen shortage enhances auxin content and signaling, which promotes lateral root development, lateral root growth, and primary root elongation, as opposed to a severe nitrogen deficiency, which restricts the growth of both the primary and lateral roots (Pélissier et al. 2021). With ammonium encouraging lateral root branching and nitrate encouraging lateral root elongation, the auxin-dependent stimulation of lateral root development in Arabidopsis functions in an ion-dependent manner. The absorption of minerals and nutrients is often negatively regulated by cytokinin. It has been demonstrated that cytokinin content decreases in response to nitrogen deprivation, which simultaneously promotes cytokinin breakdown and increases root development and mineral nutrient accumulation (Ramireddy et al. 2021).

1.2.3 Flowering

The differentiation and development of floral buds and flowers include several physiological and biochemical processes. Three steps make up flowering: induction, initiation, and differentiation. Auxins are essential for the formation of floral primordia; without them, a plant cannot produce flowers. In addition to controlling cell division, apical dominance, food metabolism, chloroplast formation, senescence, blooming, nodulation, and circadian rhythms, cytokinin also regulates growth and differentiation. The postponement of senescence is one of cytokinin's key functions in blooming (Wu et al. 2017). Additionally, they affect the floral meristem's activity by contributing to cell differentiation in the floral meristem. GA controls several pathways that govern photoperiodism, stress reactions, flower induction, and autonomous pathways. Signaling 'to stimulate conception' is the main function of gibberellic acid in the flowering pathway (Dong et al. 2017). Gibberellins also stimulate development at lower temperatures and modify the rate of cell division and dormancy breaking in seeds and buds. In Arabidopsis, ethylene slows flowering, which plays a significant role in the control of flower initiation (Dar et al. 2021). Due to its function in the processes of drought stress and its capacity to postpone blooming, ABA is referred to as a "floral repressor" (Martignago et al. 2020). In response to abiotic stress, the hormone SA instructs proteins to generate floral buds. Floral maturation is influenced by the JA hormone (Chandler 2011).

Transcriptome investigations often reveal the genes that are differently expressed under various circumstances, providing fresh insight into the genes or pathways related to the particular organ or situation. The genes for blooming LOCUS C (FLC), CONSTANS (CO), and FLOWERING LOCUS T (FT) are regulated by hormonal signaling during blooming (Kim et al. 2008). For instance, salicylic acid (SA) controls CO, FLC, FT, and SOC1 (SUPPRESSOR OF OVEREXPRESSION OF CO1) transcription in Arabidopsis. SA facilitates the change from the vegetative to the reproductive phase in Arabidopsis under light stress. It controls the time of flowering in unstressed Arabidopsis plants and functions as a negative regulator of the FLC floral repressor gene and other elements of the autonomous flowering pathway (Martínez et al. 2004). Under stress, the short-day plant *Pharbitis nil* showed a comparable impact of SA on flower induction. The expression of PnFT2 (a flowering gene FT ortholog) under stressful circumstances shows that it plays a part in stress-induced blooming (Yamada and Takeno 2014). High levels of GA4, ABA, and specific CTK may have a beneficial impact on olive flower development, while high levels of GA3 have an inhibiting effect during the induction and initiation stages (Zhang et al. 2016).

1.3 Regulation of Biotic Stress Responses by Plant Hormones

To defend themselves against invasive infections, plants have created defense-signaling systems. Salicylic acid, jasmonates, and ethylene are examples of plant hormones that operate as signals to initiate and mediate a wide range of defensive responses (Yang et al. 2015). Other hormones that have been linked to developmental and abiotic stress responses, including auxin, gibberellic acids, cytokinins, and brassinosteroids, are also crucial in defense signaling against pathogens (Verma et al. 2016). The antagonistic or synergistic interactions between these hormone-signaling pathways provide plants with a large regulatory ability to respond quickly to biotic stress. Pathogens, on the other hand, have created techniques to influence the signaling network and boost their virulence.

Pattern recognition receptors (PRRs) detect pathogen-associated molecular patterns (PAMPs) and start PAMP-triggered immunity (PTI) to stop pathogen colonization (Boller and Felix 2009). Resistance (R) proteins in plants recognize the effector proteins released by microbial pathogens as a second layer of induced defense and strongly stimulate disease-resistance responses (effector-triggered immunity, ETI) (Wu et al. 2014). The ETI is typically accompanied by hypersensitivity reactions (HR), which are marked by rapid programmed cell death at the infection sites. Both PTI and ETI depend heavily on the immunological responses mediated by SA (Yuan et al. 2021). Systemic acquired resistance is produced as a result of secondary immune responses being triggered by both PTI and ETI in distant, uninfected regions. It has been demonstrated that the application of exogenous SA or its functional analogs, such as 2,6-dichloro isonicotinic acid (INA), benzothiadiazole S-methyl ester (BTH), and probenazole, increases the expression of the PR genes and confers resistance to a variety of oomycete, viral, bacterial, fungal, and oomycete pathogens (Nakashita et al. 2002). When ethylene is present, plant defenses against necrotrophic diseases are strengthened.

Jasmonic acid alone triggers plant responses to injury and herbivory (De Bruxelles and Roberts 2001). Increased resistance to pathogen assaults may result from the interplay of the SA and JA signaling pathways. Numerous biotrophic infections use the JA-SA antagonistic relationship to weaken host defenses (Huang et al. 2020). According to Boutrot et al. (2010), plants with mutations in the essential ET-signaling protein EIN2 displayed decreased FLS2-mediated responses, highlighting the critical function of ET in plant immune responses. Ethylene is often believed to collaborate with JA in the resistance responses to herbivore pests and necrotrophic diseases (Yang et al. 2015).

Along with its functions in the transmission of abiotic signals and the control of developmental processes, abscisic acid has become a crucial signaling molecule in interactions between plants and pathogens. The sensitivity of different plant species to bacterial and fungal diseases is raised by exogenous ABA application or increased endogenous ABA accumulation caused by genetic abnormalities (Song et al. 2011). Numerous studies have demonstrated the beneficial effects of ABA on plant defense, including how it may control stomatal closure and prime callose deposition (Flors et al. 2009). As a result, depending on the lifestyles and stages of infection of the pathogens, ABA appears to play distinct roles in defense.

1.4 Regulation of Abiotic Stress by Plant Hormones

1.4.1 Heat Stress

The phytohormone ABA, one of the hormones that regulate heat stress responses that have been widely investigated, is necessary to generate thermotolerance during the majority of embryonic stages (Hemantaranjan et al. 2014). The PYR/PYL receptors sense the experience of heat stress by releasing ABA, which interacts with the PYR/PYL complex to inhibit the activity of protein phosphatase 2Cs (PP2Cs). The positive regulator of the ABA response (SnRK2/Open Stomata 1; OST1) is released as a result of this interaction. To cause heat stress reactions, the released SnRK2s autophosphorylate and activate a variety of transcription factors (TFs), including MYB1, ABA-Responsive Promoter Elements (ABREs)-Binding Factors (ABFs), and heat shock transcription factors (HSF) (Devireddy et al. 2021). To promote thermotolerance, ABA is necessary for the majority of the embryonic phases. ABA has been demonstrated to stimulate the SPL transcription factors that provide thermotolerance during the reproductive stage in Arabidopsis (Chao et al. 2017).

The active BR1 interacts with a coreceptor called BRI1-ASSOCIATED KINASE1 (BAK1) thanks to the recognition of BR by its receptor, BRASSINOSTEROID INSENSITIVE1 (BRI1) (Kour et al. 2021). To increase the expression of heat-responsive genes, this interaction positively regulates the BR-signaling proteins BRI1-EMS SUPPRESSOR1 (BES1) and BZR1

(BRASSINAZOLERESISTANT1). It was shown that BRs play a crucial role in achieving plant acclimatization to high temperatures in the barley (*Hordeum vulgare*) mutants lacking in BR production and signaling, which adversely affected the accumulation of HSP mRNA and heat-shock proteins (Sadura and Janeczko 2021). By controlling ABA production, it has been demonstrated that BRs interact to increase the endogenous level of ABA in *Chlorella vulgaris*, improving the plant's ability to withstand heat (Bajguz 2019).

According to reports, salicylic acid is an important signaling molecule in plants growing in hot climates. According to earlier research, SA largely promotes basal thermotolerance by activating several HSPs. Plants that are under heat stress have greater SA levels, which have been proven to enhance photosynthetic performance by shielding the photosystem II (PSII) complex from elevated ROS levels (Mathur et al. 2014). The primary roles of CKs in plant growth and development include controlling organ size, branching, and meristem activity of shoots and roots (Yan et al. 2020). Although the precise nature of CKs in high temperatures has not yet been determined, certain research points to their function in heat stress adaptation processes.

1.4.2 Cold Stress

By reducing membrane integrity, antioxidant defense enzyme activity, protein and secondary metabolite synthesis, and phytohormonal signaling, cold stress (chilling 0–15 °C or freezing 0 °C) slows down or stops plant development owing to cellular damage. Phytohormones are essential for increasing plant growth and development under cold stress via intricate mechanisms like physiological and biochemical control (Raza et al. 2023).

Low temperatures lead to osmotic stress and decreased water availability in plant cells, which in turn increases ABA production and biosynthesis. Stomatal closure, a reduction in transpiration rate, and slowed cell proliferation are all caused by endogenous ABA being transferred from roots to leaves and building up in guard cells. Plants normally require very small amounts of ABA for growth, and any excess is removed by controlling its endogenous production (LeNoble et al. 2004). As a result of increased ABA levels, cold-tolerant plants produce new proteins as a result of acclimatization to the cold. Exogenous ABA can also improve CS tolerance by inducing the transcription of antioxidant enzymes for ROS scavenging, which lessens the symptoms of CS damage in many plants (Shi and Yang 2014).

Salicylic acid (SA) is a unique phenolic compound that alters the expression of proteins involved in pathogenesis. Additionally, SA increases a plant's ability to withstand abiotic stresses, including cold stress, as well as growth, ripening, and expansion. In muskmelon (*Cucumis melo* L.), SA treatment (0.1 mM) improved seed germination and many antioxidant enzymes under cold stress (Kaur and Gupta 2017). At low temperatures, cytokinins play important roles in a variety of growth and developmental processes. Cytokinin has been employed by several studies to increase plant output and resistance to cold stress. Exogenous cytokinins boosted the relative growth rate of wild-type plants because of the increased cell counts, which are higher in amp1 mutant plants than in wild-type plants at low temperatures.

1.4.3 *Salt Stress*

The endogenous signal molecules known as ABA are widely recognized for helping plants withstand extremely harsh environmental circumstances, including salt and drought stressors (Raghavendra et al. 2010). A calcium-dependent phosphorylation pathway has been well characterized as the mechanism by which the salinity stress rapidly activates genes related to ABA biosynthesis, including zeaxanthin oxidase, 9-cisepoxycarotenoid dioxygenase, ABA-aldehyde oxidase, and molybdenum cofactor sulfurase (Chinnusamy et al. 2006). By shutting stomata and storing many proteins and osmoprotectants for osmotic adjustment, the increased ABA hormone aids plants in acclimating to a decreasing water supply.

Numerous studies have been done on the essential functions of the plant hormone cytokinins (CKs) in the growth and developmental processes of plants. Additionally, it has been claimed that cytokinins help plants better withstand abiotic conditions, including excessive salt and temperature (Javid et al. 2011). By antagonistically interacting with the stress hormone ABA, CKs may be able to completely eradicate plants' salt tolerance (Javid et al. 2011). In addition, JA is a key signaling molecule for a variety of plant developmental processes and defense reactions (Kazan and Manners 2012). The biological relevance of JA in salt stress tolerance has been shown by recent molecular investigations.

1.5 Conclusion and Future Perspectives

Plant hormones are signaling molecules that control crucial aspects of development, growth, and reactions to environmental stress. Crucial new insights into how plants perceive and react to abiotic challenges are being offered by recent developments in plant biology. The identification of the genes, processes, and mechanisms governing these features might open up new avenues for translational research and applications aimed at improving yields against abiotic and biotic stresses. Even though significant progress has been made recently in understanding the molecular basis of phytohormone activity, many essential issues still need to be answered. In the majority of phytohormones, receptors and other elements of the upstream signaling pathway have yet to be discovered.

References

Arteca RN, Arteca JM (2008) Effects of brassinosteroid, auxin, and cytokinin on ethylene production in Arabidopsis thaliana plants. J Exp Bot 59(11):3019–3026

Bajguz A (2019) Brassinosteroids in microalgae: application for growth improvement and protection against abiotic stresses. In: Brassinosteroids: plant growth and development. Springer, Singapore, pp 45–58

Bewley JD, Bradford K, Hilhorst H (2012) Seeds: physiology of development, germination, and dormancy. Springer, New York

Bhavanam S, Stout M (2021) Seed treatment with jasmonic acid and methyl jasmonate induces resistance to insects but reduces plant growth and yield in rice, Oryza sativa. Front Plant Sci 12:691768

Boller T, Felix G (2009) A renaissance of elicitors: perception of microbe–associated molecular patterns and danger signals by pattern–recognition receptors. Annu Rev Plant Biol 60:379–406

Borghetti F, Noda FN, Sá CMD (2002) Possible involvement of proteasome activity in ethylene–induced germination of dormant sunflower embryos. Braz J Plant Physiol 14:125–131

Boutrot F, Segonzac C, Chang KN, Qiao H, Ecker JR, Zipfel C, Rathjen JP (2010) Direct transcriptional control of the Arabidopsis immune receptor FLS2 by the ethylene–dependent transcription factors EIN3 and EIL1. Proc Natl Acad Sci 107(32):14502–14507

Chandler JW (2011) The hormonal regulation of flower development. J Plant Growth Regul 30(2):242–254

Chao LM, Liu YQ, Chen DY, Xue XY, Mao YB, Chen XY (2017) Arabidopsis transcription factors SPL1 and SPL12 confer plant thermotolerance at the reproductive stage. Mol Plant 10(5):735–748

Chauhan N, Sharma JB, Rana K, Mir W, Bakshi M (2020) Effects of gibberellins and Promalin on the growth and development of fruit crops: a review. J Pharmacogn Phytochem 9(6):1284–1289

Chinnusamy V, Zhu J, Zhu JK (2006) Salt stress signaling and mechanisms of plant salt tolerance. Genet Eng 27:141–177

Dar RA, Nisar S, Tahir I (2021) Ethylene: a key player in ethylene sensitive flower senescence: a review. Sci Hortic 290:110491

Davies PJ (2010) Their nature, occurrence, and functions. Plant hormones. Springer, New York, pp 1–15

De Bruxelles GL, Roberts MR (2001) Signals regulating multiple responses to wounding and herbivores. Crit Rev Plant Sci 20(5):487–521

Devireddy AR, Tschaplinski TJ, Tuskan GA, Muchero W, Chen JG (2021) Role of reactive oxygen species and hormones in plant responses to temperature changes. Int J Mol Sci 22(16):8843

Dong B, Deng Y, Wang H, Gao R, Stephen GUK, Chen S, Chen F (2017) Gibberellic acid signaling is required to induce flowering of chrysanthemums grown under both short and long days. Int J Mol Sci 18(6):1259

Fageria NK, Moreira A (2011) The role of mineral nutrition on root growth of crop plants. Adv Agron 110:251–331

Finch-Savage WE, Leubner-Metzger G (2006) Seed dormancy and the control of germination. New Phytol 171(3):501–523

Finkelstein R (2013) Abscisic acid synthesis and response. In: The Arabidopsis book, vol 11, p e0166. [PubMed: 24273463]

Flors V, Ton J, Mauch-Mani B (2009) Role of abscisic acid in disease resistance. In: Signal crosstalk in plant stress responses. Wiley, Chichester, pp 1–22

Gietler M, Fidler J, Labudda M, Nykiel M (2020) Abscisic acid—enemy or savior in the response of cereals to abiotic and biotic stresses? Int J Mol Sci 21(13):4607

Graeber KAI, Nakabayashi K, Miatton E, Leubner-Metzger G, Soppe WJJ (2012) Molecular mechanisms of seed dormancy. Plant Cell Environ 35(10):1769–1786

Hemantaranjan A, Bhanu AN, Singh MN, Yadav DK, Patel PK, Singh R, Katiyar D (2014) Heat stress responses and thermotolerance. Adv Plants Agric Res 1(3):1–10

Huang S, Zhang X, Fernando WD (2020) Directing trophic divergence in plant–pathogen interactions: antagonistic phytohormones with NO doubt? Front Plant Sci 11:600063

Jaillais Y, Chory J (2010) Unraveling the paradoxes of plant hormone Signaling integration. Nat Struct Mol Biol 17:642–645

Javid MG, Sorooshzadeh A, Moradi F, Sanavy SAMM, Allahdadi I (2011) The role of phytohormones in alleviating salt stress in crop plants. Aust J Crop Sci 5:726–734

Kaur S, Gupta N (2017) Effect of proline and salicylic acid on germination and antioxidant enzymes at different temperatures in muskmelon (*Cucumis melo* L.) seeds. J Appl Nat Sci 9:2165–2169

Kaur A, Maness N, Ferguson L, Deng W, Zhang L (2021) Role of plant hormones in flowering and exogenous hormone application in fruit/nut trees: a review of pecans. Fruit Res 1(1):1–9

Kaur H, Hussain SJ, Kaur G, Poor P, Alamri S, Siddiqui MH, Khan MIR (2022) Salicylic acid improves nitrogen fixation, growth, yield, and antioxidant defense mechanisms in chickpea genotypes under salt stress. J Plant Growth Regul 41(5):2034–2047

Kazan K, Manners JM (2012) JAZ repressors and the orchestration of phytohormone crosstalk. Trends Plant Sci 17:22–31

Kim SY, Yu X, Michaels SD (2008) Regulation of CONSTANS and FLOWERING LOCUS T expression in response to changing light quality. Plant Physiol 148(1):269–279

Kour J, Kohli SK, Khanna K, Bakshi P, Sharma P, Singh AD, Sharma A (2021) Brassinosteroid signaling, crosstalk, and, physiological functions in plants under heavy metal stress. Front Plant Sci 12:608061

LeNoble ME, Spollen WG, Sharp RE (2004) Maintenance of shoot growth by endogenous ABA: genetic assessment of the involvement of ethylene suppression. J Exp Bot 55(395):237–245

Liu JX, Srivastava R, Che P, Howell SH (2007) An endoplasmic reticulum stress response in Arabidopsis is mediated by proteolytic processing and nuclear relocation of a membrane–associated transcription factor, bZIP28. Plant Cell 19(12):4111–4119

Martignago D, Siemiatkowska B, Lombardi A, Conti L (2020) Abscisic acid and flowering regulation: many targets, different places. Int J Mol Sci 21(24):9700

Martínez C, Pons E, Prats G, León J (2004) Salicylic acid regulates flowering time and links defense responses and reproductive development. Plant J 37(2):209–217

Mathur S, Agrawal D, Jajoo A (2014) Photosynthesis: response to high–temperature stress. J Photochem Photobiol B Biol 137:116–126

Matilla AJ, Matilla-Vázquez MA (2008) Involvement of ethylene in seed physiology. Plant Sci 175(1–2):87–97

Nakashita H, Yoshioka K, Yasuda M, Nitta T, Arai Y, Yoshida S, Yamaguchi I (2002) Probenazole induces systemic acquired resistance in tobacco through salicylic acid accumulation. Physiol Mol Plant Pathol 61(4):197–203

Pélissier PM, Motte H, Beeckman T (2021) Lateral root formation and nutrients: nitrogen in the spotlight. Plant Physiol 187(3):1104–1116

Piskurewicz U, Jikumaru Y, Kinoshita N, Nambara E, Kamiya Y, Lopez-Molina L (2008) The gibberellic acid signaling repressor RGL2 inhibits Arabidopsis seed germination by stimulating abscisic acid synthesis and ABI5 activity. Plant Cell 20(10):2729–2745

Qingguo C, Dai X, De Paoli H, Cheng Y, Takebayashi Y, Kasahara H, Kamiya Y, Zhao Y (2014) Auxin overproduction in shoots cannot rescue auxin deficiencies in Arabidopsis roots. Plant Cell Physiol 55(6):1072–1079

Raghavendra AS, Gonugunta VK, Christmann A, Grill E (2010) ABA perception and signaling. Trends Plant Sci 15:395–401

Ramireddy E, Nelissen H, Leuendorf JE, Van Lijsebettens M, Inzé D, Schmülling T (2021) Root engineering in maize by increasing cytokinin degradation causes enhanced root growth and leaf mineral enrichment. Plant Mol Biol 106:555–567

Raza A, Charagh S, Najafi-Kakavand S, Abbas S, Shoaib Y, Anwar S, Siddique KH (2023) Role of phytohormones in regulating cold stress tolerance: physiological and molecular approaches for developing cold–smart crop plants. Plant Stress 8:100152

Sadura I, Janeczko A (2021) Brassinosteroids and the tolerance of cereals to low and high–temperature stress: photosynthesis and the physicochemical properties of cell membranes. Int J Mol Sci 23(1):342
Santoro V, Schiavon M, Visentin I, Constán-Aguilar C, Cardinale F, Celi L (2021) Strigolactones affect phosphorus acquisition strategies in tomato plants. Plant Cell Environ 44(11):3628–3642
Shen A, Kamp HD, Gründling A, Higgins DE (2006) A bifunctional O–GlcNAc transferase governs flagellar motility through anti–repression. Genes Dev 20(23):3283–3295
Shi Y, Yang S (2014) ABA regulation of the cold stress response in plants. In: Abscisic acid: metabolism, transport, and Signaling. Springer, Netherlands, pp 337–363
Smith S, Li C, Li J (2017) Hormone function in plants. In: Hormone metabolism and signaling in plants. Elsevier Inc, pp 1–38
Song W, Ma X, Tan H, Zhou J (2011) Abscisic acid enhances resistance to Alternaria solani in tomato seedlings. Plant Physiol Biochem 49(7):693–700
Tadiello A, Ziosi V, Negri AS, Noferini M, Fiori G, Busatto N, Trainotti L (2016) On the role of ethylene, auxin, and a GOLVEN–like peptide hormone in the regulation of peach ripening. BMC Plant Biol 16:1–17
Tatsuki M (2010) Ethylene biosynthesis and perception in fruit. J Jpn Soc Hortic Sci 79(4):315–326
Tsuda K, Kurata N, Ohyanagi H, Hake S (2014) Genome–wide study of KNOX regulatory network reveals brassinosteroid catabolic genes important for shoot meristem function in rice. Plant Cell 26(9):3488–3500
Verma V, Ravindran P, Kumar PP (2016) Plant hormone–mediated regulation of stress responses. BMC Plant Biol 16:1–10
Wang L, Wang Z, Xu Y, Joo SH, Kim SK, Xue Z, Chong K (2009) OsGSR1 is involved in crosstalk between gibberellins and brassinosteroids in rice. Plant J 57(3):498–510
Wu L, Chen H, Curtis C, Fu ZQ (2014) Go in for the kill: how plants deploy effector–triggered immunity to combat pathogens. Virulence 5(7):710–721
Wu L, Ma N, Jia Y, Zhang Y, Feng M, Jiang CZ, Gao J (2017) An ethylene–induced regulatory module delays flower senescence by regulating cytokinin content. Plant Physiol 173(1):853–862
Yakhin OI, Lubyanov AA, Yakhin IA, Brown PH (2017) Biostimulants in plant science: a global perspective. Front Plant Sci 7:2049
Yamada M, Takeno K (2014) Stress and salicylic acid induce the expression of PnFT2 in the regulation of the stress–induced flowering of Pharbitis nil. J Plant Physiol 171(3–4):205–212
Yan Y, Zhao N, Tang H, Gong B, Shi Q (2020) Shoot branching regulation and signaling. Plant Growth Regul 92(2):131–140
Yang YX, Ahammed GJ, Wu C, Fan S, Zhou YH (2015) Crosstalk among jasmonate, salicylate, and ethylene signaling pathways in plant disease and immune responses. Curr Protein Pept Sci 16(5):450–461
Yuan M, Ngou BPM, Ding P, Xin XF (2021) PTI–ETI crosstalk: an integrative view of plant immunity. Curr Opin Plant Biol 62:102030
Zhang S, Zhang D, Fan S, Du L, Shen Y, Xing L, Han M (2016) Effect of exogenous GA3 and its inhibitor paclobutrazol on floral formation, endogenous hormones, and flowering–associated genes in 'Fuji'apple (Malus domestica Borkh.). Plant Physiol Biochem 107:178–186
Zhao J, Bai W, Zeng Q (2015) Moderately enhancing cytokinin level by down–regulation of *GhCKX* expression in cotton concurrently increases fiber and seed yield. Mol Breed 35:60

Chapter 2
Impact of Heavy Metals on Plant Biology

Slimane Mokrani and El-hafid Nabti

Abstract Actually, heavy metals cause major problems in terms of the contamination of the environment (soil, air, and water), as well as humans, animals, and plants. They have two main sources: the natural origin and the ever-increasing anthropological activities. Particularly, plants are affected by those toxic contaminants at various levels, starting with the cells, organs, metabolism, and physiological processes like photosynthesis, plant water balance, nutrient accumulation, gaseous exchanges, enzymatic activities, growth regulation, plant water balance, and reproduction. Purposely, various processes are involved in the interaction of plants, heavy metals, and the environment, including toxicity, uptake, sequestration, detoxification, tolerance, and hyperaccumulation. The main objective of this review is to deal with the impact of heavy metals on plant biology, discuss, and explain a number of advances, and the various processes involved.

Keywords Adverse effects · Heavy metals · Plant biology · Mechanisms · Toxicity

2.1 Introduction

Although it is sometimes not strictly defined, the expression "heavy metal" refers to metals with specific weights exceeding 5 g/cm^{-3} (Holleman and Wiberg 1985; Khanna et al. 2018). Chemically, the notion of "heavy metal" only applies to transition metals having an atomic mass greater than 20 and a specific gravity greater than 5. In the field of biology, the term "heavy" describes a number of metals and metalloids that, even in very small quantities, may be poisonous to both plants and animals. These extremely phytotoxic substances will be referred to as "heavy metals" in this context (Rascio and Navari-Izzo 2011). This group includes around 40

S. Mokrani · El-h. Nabti (✉)
Laboratoire de Maitrise des Energies Renouvelables, Faculté des Sciences de la Nature et de la Vie, Université de Bejaia, Bejaia, Algeria
e-mail: el-hafid.nabti@univ-bejaia.dz

A. Sharma et al. (eds.), *Hormones and Heavy Metals*, Plant Life and Environment Dynamics, https://doi.org/10.1007/978-981-96-9622-2_2

different components. In formations of rock, heavy metals are often present in dispersed form (Sharma and Agrawal 2005). Nickel, cobalt, thallium, copper, zinc, silver, antimony, cadmium, lead, mercury, arsenic, and selenium are some examples of heavy metals. Because of their high levels of toxicity and ranking as preceding elements, lead, cadmium, arsenic, chromium, and mercury are considered the most important components (Rubiya et al. 2018).

Heavy metal toxicity has drawn a lot of attention among abiotic stresses as a result of increased environmental contamination. Metals usually have negative biological impacts because they are not biodegradable (Jaleel et al. 2009). Heavy metals are a worry for the environment in many regions of the world and may have negative impacts on agriculture in particular (Gaur and Adholeya 2004). A number of these heavy metals, such as As, Hg, Cd, Se, or Pb, are not necessary because there is no evidence that they have any physiological effects on plants (Rascio and Navari-Izzo 2011). As they are necessary for a plant's regular growth, several additional heavy metals play a significant part in nature. Significant heavy metals, including Zn, Cu, Fe, Mn, Ni, and Mo, have an important effect on the biochemistry and physiology of plants. (Zhuang et al. 2009). Generally speaking, plants may be divided into three groups depending on their mechanisms for surviving in stressful environments: indicators, accumulators, and excluders, as proposed by Baker (1981). The biochemical processes of the plants growing on soils contaminated with effluent (heavy metals) exhibit many physiological alterations, such as gaseous exchanges, respiration, nutrient accumulation, etc. The substantial metal has a toxic nature as a result of overpowering plant stress, which varies depending on a few factors like the species of plants, the convergence of the substantive metal and its concoct structure, and soil formation (Nagajyoti et al. 2010). The aim of this chapter is to deal with different aspects related to the impact of heavy metals on plant biology.

2.2 Heavy Metal Source and Bioavailability in Soils

In spite of the fact that heavy metals are already present in soil, human and geological processes have increased their concentration to levels that are detrimental to both animals and plants. Burning fossil fuels, smelting and metal mining, manufacturing batteries and other metal goods in factories, and disposing of municipal garbage are a few of these operations (Alloway 1990; Shen et al. 2002). Transport (aircraft, diesel-powered vehicles, and automobiles) along with landfills and waste incineration are other sources of heavy metals (Sandeep et al. 2019). Liming, pesticides, and sewage sludge are among the most significant sources of heavy metals in agricultural soils, along with other inorganic and organic fertilizers that are chemicals added to soil to promote plants' development and productivity. Depending on their origins, other substances, such as inorganic fertilizers, fungicides, and phosphate fertilizers, can include varying amounts of Cr, Cd, Ni, Zn, and Pb (Sandeep et al. 2019). One of the main causes of the soil's high heavy metal concentration is wastewater irrigation (Mapanda et al. 2005). Vegetables grown near untreated wastewater irrigation systems have been shown to contain high quantities of heavy metals (Sinha and Gupta 2005; Sharma et al. 2006).

When heavy metals accumulate in the environment, they are carried by air and water before being deposited in soil and sediments (Ozturk et al. 2008). The chemical characteristics of the soil, such as its pH, redox potential, loading rate, soil texture, cation exchange capacity, organic matter, and clay content, largely determine the metals' solubility and bioavailability (Verloo and Eeckhout 1990). The charge type and density in soil colloids, the level of association with ligands, and the relative surface area of the soil are other parameters that impact the metal availability in soil (Norvell 1984; Marques et al. 2009). One heavy metal's presence may have an impact on another's plant and soil availability. In other words, there are behaviors that are both synergistic and antagonistic among heavy metals (Chibuike and Obiora 2014).

2.3 Mechanistic Insights into the Accumulation of Toxic Elements in Plants

2.3.1 Heavy Metal Uptake and Chelation

As immobile organisms, plants' roots serve as the primary point of interaction for heavy metal ions. Plants absorb heavy metals largely via their roots as well as through the surfaces of their leaves as a result of this interaction and the deposition of particles containing these metals (Kumar et al. 2019). Heavy metals go toward the rhizosphere during the chelation process. It occurs both external to the plant's body and inside the cell of the plant. Different peptides, amino acids, and organic acids are in charge of intracellular heavy metal chelation within the plant cell, whereas certain organic acids found in roots exude the extracellular chelation of heavy metals. Preferably, only harmful metals (such as Hg, Cd, Cr, and Pb) are appropriated by metal-binding proteins and peptides, while essential metals like Cu, Zn, etc. are disregarded (Ryu et al. 2003; Yang et al. 2005). The majority of heavy metals that do get into the plant are subsequently maintained in the root cells, where they are detoxified by complexing with organic acids, amino acids, or metal-binding peptides and/or being stored in vacuoles (Hall 2002).

Due to particles deposited on the foliar surfaces, heavy metals can also be directly adsorbed to the leaves (Sharma and Agrawal 2005). Despite significant advancements in recent years in our understanding of how plants absorb heavy metals via their leaves, there are still fascinating scientific problems to be solved, especially those relating to the mechanisms by which plants absorb metals through their leaves. Metal that is deposited on plant leaves enters the plants via the cuticle and stomatal openings. Heavy metal absorption by plant foliar components is influenced by a number of parameters, including the physico-chemical characteristics of the metals and the cuticle, the texture, morphology, and surface area of the leaves, the period of exposure, and ambient conditions (Shahid et al. 2017).

2.3.2 Root-to-Shoot Translocation

Translocation to the shoot depends on the rate of trapping in various root cell sections, movement within the root symplast, and over obstacles like the loading into the xylem, endodermis, and upward mobility in the xylem. Uptake from the xylem and the apoplast, transportation to use sites and storage sections, and the presence of binding molecules are all necessary for element accumulation in leaves. The mobility and loading of phloem, the effectiveness of xylem-to-phloem exchange, the rates of different symplast-to-apoplast and apoplast-to-symplast transporting mechanisms, and the availability of storage sites all influence distribution among leaves and seed loading (Clemens and Ma 2016). The most common instances include the uptake of Cd by the Fe(II) transporter iron-regulated transporter 1 (IRT1) in *A. thaliana* (27) and the absorption of Cd by the divalent metal transporters 1 (DMT1) and ZRT- and IRT-like protein 8 (ZIP8) in mammalian cells (Gunshin et al. 1997; Dalton et al. 2005). This significantly limits translocation to the aboveground organs, preventing heavy metal injury to the leaf tissues, especially the metabolically active photosynthetic cells.

2.3.3 Sequestration, Detoxification, and Accumulation

The primary method of removing harmful metals from plants is vacuolar (Dutta et al. 2018). Initially discovered in Arabidopsis, the process is responsible for As vacuolar sequestration. As(III)-PC complexes or apoPCs are carried into the vacuoles by two C-type ABC transporters, AtABCC1 and AtABCC2, and are essential for As resistance (Song et al. 2010). Later, it was discovered that OsABCC1 was also crucial for the sequestration of As(III)- PC complexes in rice vacuoles (Song et al. 2014).

In general, plant responses to HM toxicity comprise immobilization, exclusion, chelation, and compartmentalization of metal ions inside the cellular organelle as part of the detoxifying process (Cobbett 2000). Some of these measures limit the binding of heavy metal ions to the cell wall, vascular sequestration, and the generation of different biochemical substances, such as phyto-chelators and organic acids, to reduce the toxicity (Ejaz et al. 2023). The enzyme phytochelatin synthase (PC_S) produces short-chain thiol-rich repetitions of low-molecular weight peptides with the general structure of (γ-glutamyl-cysteinyl) n-glycine ($n = 2$–11), which have a high affinity to bind to HMs when they are present at toxic levels. This process is a key method used by plants to detoxify HMs (Lee et al. 2003; Gupta et al. 2013). PCs are shown to be involved in plants' defensive mechanisms not only in response to challenges associated with metals but also upon exposure to other stressors such as high temperatures, salt, UV-B, and herbicide (Zagorchev et al. 2013). In addition to complexation with PCs and vacuole sequestration, toxic metals like Cd, Pb, and Hg are detoxified (Cobbett and Goldsbrough 2002; Park et al. 2012; Fischer et al. 2014).

Metal accumulators include plants from the Chenopodiaceae, Solanaceae, Asteraceae, and Brassicaceae families (Thiebeauld et al. 2005). Some halophytes and metal hyperaccumulator plants can reduce metal toxicity by limiting their uptake, compartmentalizing the metals into vacuoles, and producing phytochelatins, hormones, metallothioneins, non-enzymatic and enzymatic antioxidants, etc. (Ghuge et al. 2023). The distribution of heavy metal accumulation was largest in the root, then in the stem, branch, leaf, and finally in the fruit (Moreno-Caselles et al. 2000; Aery and Rana 2003). Varied plant sections have varied concentrations of various heavy metals. For instance, Pb accumulates more in *Silene vulgaris* leaves than in *Armeria maritima* or *Thalaspi alpestre* roots (Baker 1981). For instance, Gopal et al. (2003) reported that the accumulation of cobalt in tomato (*Lycopersicon esculentum*) cv. Pusa ruby was largest in the old leaves and roots and lowest in the stem. Numerous elements, including a plant's structure, life cycle, vigor, soil pH, depth of the root system, temperature, partial oxygen pressure, level of carbohydrates, rate of respiration, microbial presence, nutrient interface, etc., have a substantial impact on a plant's ability to accumulate metals (Chen et al. 2006).

2.4 Heavy Metal Toxicity to Plants

Toxic heavy metals' impacts on plants have been the subject of numerous studies around the globe (Reeves and Baker 2000). These scientists looked at variations in the internal architecture of the plant's roots, stems, and leaves in response to heavy metals (Gomes et al. 2012; Noreen et al. 2019; El-Shabasy 2021). The aforementioned information leads to the conclusion that there are three negative causes of heavy metal toxicity: (a) MG and ROS generation is stimulated by the Fenton reaction, auto-oxidation, or by the glyoxalase pathway and modifying the system of antioxidant defense, (b) through direct protein interaction, heavy metals target cellular structures, catalytic, and transport sites due to their propensities for histidyl-, carboxyl-, and thioyl-groups, and (c) elimination of important metal ions from specific binding sites, causing the function to collapse (Sharma and Dietz 2009).

Like all living organisms, plants frequently face the conundrum that while some heavy metal ions are necessary micronutrients, others, like Cd, As, and Hg, are highly toxic to metabolic processes at higher levels (Fernandes and Henariques 1991; Reeves and Baker 2000). If the plant's metal concentration is higher than usual, heavy metal stress will negatively affect it. High metal amounts can directly affect plants by inhibiting cytoplasmic enzymes and breaking down cell assemblies due to oxidative stress (Rubiya et al. 2018). Elevated heavy metal levels also lead to an increase in reactive oxygen species (ROS) generation, like hydroxyl free radicals, superoxide free radicals, or non-free radicals such as hydrogen peroxide, which disrupt the balance between antioxidant and pro-oxidant in plant cells and result in oxidative stress (Zengin and Munzuroglu 2005; Sytar et al. 2013). The majority of ROS species are produced under stressful conditions and have strong oxidizing properties that can target biomolecules like nucleic acids, carbohydrates, proteins, lipids, etc. (Wojtaszek 1997). Figure 2.1 indicates the effects and toxicity of heavy metals on plants.

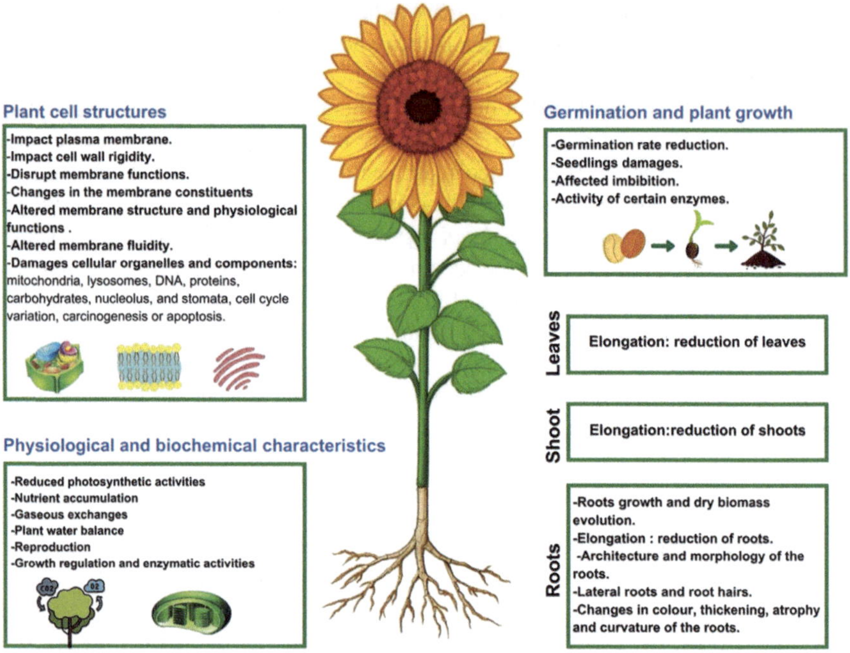

Fig. 2.1 Effects and toxicity of heavy metals on plants

2.5 Effects of Heavy Metals on Plants

2.5.1 Impact on Germination and Plant Growth

For plants, an excessive amount of heavy metals in soils may have a detrimental effect on seed germination, evolution of dry biomass, and root growth (Pavel et al. 2013). Numerous studies have been conducted on the effects of heavy metals on seed germination (Moosavi et al. 2012; Márquez-García et al. 2013), with some of the most frequent consequences being a decrease in the germination rate or harm to seedlings, such as a reduction in the growth and length of leaves, shoots, or roots, which could result in the seedlings' death (Singh et al. 2007; Peralta et al. 2001).

The majority of seeds and seedlings respond to the presence of heavy metals by showing a reduction in germination and vigor; mechanisms impacted include imbibition (Kranner and Colville 2011), or the activity of specific enzymes, such as proteases, acid phosphatases, amylases, and proteolytic enzymes, involved in reserve mobilization (Ko et al. 2012). Reduced photosynthetic activities, decreased enzyme activity, and plant mineral nutrition are mostly responsible for the decline in growth characteristics of plants growing in polluted soils (Kabata-Pendias and Pendias 2001). The germination and seedling growth of wheat (*Triticum aestivum* L.), pea (*Pisum sativum*), and tomato (*Solanum lycopersicum* L.) were investigated during lead (Pb), cadmium (Cd), and copper (Cu) exposure. Five different metal concentrations (20, 65, 110, 175, and 220 ppm) were used. Among the examined agricultural seeds, copper had the greatest impact (51.2%), followed by lead (47.5%) and cadmium (35.3%) on germination (Baruah et al. 2019). Studies on the resistance of plants to heavy metals reveal that the growth of roots is particularly susceptible to heavy metals (Punz and Sieghardt 1993). The morphology and architecture of the roots change in tandem with the suppression of root extension. Common signs include a decrease in the growth of the lateral roots and root hairs, color changes, thickness, atrophy, and curvature of the roots (Čiamporová 2000).

2.5.2 Impact on Physiological and Biochemical Characteristics of Plants

Various physiological modifications in the biochemical process are present in plants cultivated on soils polluted with effluent (heavy metals) like photosynthesis (Souri et al. 2019), nutrient accumulation, plant water balance, gaseous exchanges, and reproduction (Kumar et al. 2019), and growth regulation and enzymatic activities (Tuna et al. 2002). Also, low biomass output, chlorosis, senescence, necrosis, wilting, stunted growth, few seeds, and eventual death are all evident signs of heavy metal toxicity (Kumar et al. 2019).

The impact of heavy metals on photosynthesis in plants has been documented in a number of research studies. Toxic metals disrupt many photosynthetic mechanisms and result in metal accumulation in plant foliage (Gill et al. 2012), a modification in the chloroplast membrane's function and the amount of metals present in the mesophyll, stomata, and bundle sheath of leaves (Romanowska et al. 2012), and decrease in the production of pigments generated through photosynthesis (Chandra and Kang 2016; Shah et al. 2017).

The plant's ability to absorb macro- and micronutrients can be affected by heavy metals (Singh et al. 2020b). For instance, mineral nutrient absorption can increase, decrease, or remain constant in the presence of Ni. The similar ionic ratio of Ni and other cations' shared binding sites determines how much less macro- and micronutrients are absorbed. Due to its chemical affinity for Fe, Mg, and Zn, high concentrations of Ni bind to organic macromolecules, denature them, interfere with their metabolism, and quickly replace them (Philip et al. 1998).

It has been demonstrated that the considerable decline in the net photosynthetic rate of plants during heavy metal stress is related to stomatal restriction for gas exchange characteristics (De Matos et al. 2020; Rios et al. 2020). The sensitivity of gas exchange parameters to heavy metal stress was greater. Intercellular carbon dioxide (CO_2) level, stomatal conductance, net photosynthetic rate, and transpiration rate were all impacted (Yang et al. 2020).

The water balance in plants may be also impacted by heavy metals. Abscisic acid (ABA) levels in plants have been observed to rise in response to heavy metal stress, which leads to water loss in guard cells and stimulates stomatal closure (Poschenrieder et al. 1989). Metals have the ability to slow down short-distance water transfer in both apoplast and symplast, which in turn decreases the flow of water into the vascular system and affects the supply of water to the shoot. The root, stem, and leaf midrib's reduced hydraulic conductivity, which is brought on by a decline in the size of the vessels and tracheids, and the partial obstruction of xylem elements by cellular debris or gums are further factors that impede long-distance transport. Due to their restriction of transpiration, heavy metals lower the size and thickness of the leaf lamina, intercellular spaces, stomata density, and aperture, which all impair water transport to the shoot (Rucińska-Sobkowiak 2016).

It has been claimed that heavy metals have a negative effect on plant reproduction. Metals can accumulate in the nectar of plants that are grown in heavy metal-rich soils. Nectar's chemical composition can influence the foraging habits of floral visitors, such as pollinators and floral adversaries, and may further impact the reproductive health of plants (Xun et al. 2018). However, it is unclear if heavy metals from the soil may be transferred by non-hyperaccumulator plants into their flowering rewards, such as pollen and nectar, and if this happens, how this would affect the reproduction of plants. Hyperaccumulator plants' floral rewards and reproductive organs may absorb metalloids and metals from the soil, which might affect the plants' reproductive ability (Xun et al. 2017).

Additionally, heavy metals can affect how plants regulate their growth. Heavy metals can interfere with metabolism, growth, and development once they enter biological systems, having a variety of short- and long-term impacts (Wang and Shi 2001; Tchounwou et al. 2012). Macrophytes require phytohormones at certain sites and periods during their growth and development to maximize performance. External exposure to heavy metals triggers defensive mechanisms in plants, which include biochemical processes that amplify signaling pathways and encourage the manufacture and storage of phytohormones that are connected to defense (Mithöfer et al. 2004; Maksymiec 2007). Increased detoxification and/or accumulation can result from phytohormone signaling, which can support heavy metal tolerance (Yadav 2010; Bücker-Neto et al. 2017).

Stimulating the activities of both enzymatic (SOD, POD, APX, GR, and CAT) and non-enzymatic (GSH and ASA) antioxidants is the initial step in countering heavy metals in plants (Ali and Gill 2022). While several additional enzymes may be inhibited by higher plants absorbing phytotoxic quantities of metal. Some enzymes, such as peroxidase, are thought to play a substantial part in the stress metabolism brought on by metal phytotoxicity (Van Assche and Clijsters 1990). In general, key enzymes involved in the metabolism and absorption of nitrate and ammonia are inhibited by heavy metals, like nitrate reductase, glutamine synthetase, glutamine dehydrogenase, nitrite reductase, and glutamine oxoglutarate aminotransferase (Chaffei et al. 2003). Heavy metals can prevent enzymes from functioning properly by disrupting the enzyme-substrate complex, denaturing the enzyme protein, interfering with the active sites, or preventing microbial organisms from producing enzymes. Enzyme activity may also be impacted by modifications in group structure brought on by metals (Yeboah et al. 2021).

2.5.3 Effects of Heavy Metals on Plant Cell Structures

Plasma membranes and cell walls may be affected by heavy metals. Another location where any metal may bind and impair membrane activities is the plasma membrane (Llamas et al. 2008). After interaction, the metal promotes changes in the membrane's contents, particularly its lipids, changing the membrane's structure and physiological activities as well as those of other processes in the cell (Rizvi et al. 2020). Heavy metals affect plant plasma membranes in a number of different ways, such as the oxidation and cross-linking of protein thiols, the inhibition of important membrane proteins like H+-ATPase, or the alteration of the composition and fluidity of membrane lipids (Meharg 1993). Metal concentration and exposure time may affect cell wall stiffness, leading to the rupture of the meristem's outer cortex and rhizodermis, which may prevent the growth of root tips (Jones and Dangl 2006; Kopittke et al. 2008).

There are several other impacts of heavy metals on plant cell structure. Heavy metals harm mitochondria, DNA, lysosomes, proteins, nuclei, and carbohydrates, among other cellular constituents and organelles in biological systems, causing structural alterations that may lead to cell cycle variation, apoptosis, or carcinogenesis (Chang et al. 1996; Beyersmann and Hartwig 2008). For instance, Peng and Wang (1991) investigated the impacts of Cd on the maize cellular structure and demonstrated that under low amounts of Cd stress, the grana cascade of chloroplast mitochondria was reduced and/or eliminated. In addition, high metal concentrations have a number of direct harmful consequences, such as cytoplasmic enzyme inhibition and oxidative stress-related cell structural destruction (Van Assche and Clijsters 1990; Jadia and Fulekar 2009). Plants' ability to divide their cells is impacted by heavy metals, and the consequences vary depending on the concentration (Cheng 2003). Zhang (1997) has evaluated the impact of Cd, Pb, and Hg on the division of barley (*Hordeum vulgate* Linn.) cells and demonstrated the tendency of cell cycle prolongation under the 0.01 mol/L concentration treatment. Following a 24-hour treatment with a dose of 0.005–0.0005 mol/L, Cd, Pb, and Hg also interacted with nuclear acid to disrupt the nucleolus' structure and impede the DNase and RNase activities (Duan and Wang 1992). Cd-treated barley demonstrated that the combination of Cd and nuclear acid caused chromosomal fragmentation, conglutination, liquefaction, and aberration (Zhang 1997). Under heavy metal stress, the grana cascade of chloroplast mitochondria diminishes, enlarges, or disappears, and the polypeptide compositions of the thylakoid membrane degenerate (Cheng 2003). To carry out processes like photosynthesis, respiration, and transpiration, stomata, which are specialized pores in the epidermis of plant cells, are necessary (Guo et al. 2023). A growing body of research has demonstrated that exposure to heavy metals may alter the structure and function of plant stomata, which in turn can affect plant ecology and physiology (Khudsar et al. 2008; Alkhatib et al. 2019; Farooq et al. 2020). Table 2.1 shows the adverse impacts of heavy metals on different crops.

Table 2.1 Adverse impacts of heavy metals on different crops

Heavy Metals	Crops	Adverse impacts	References
Mercury (Hg)	Rice (*Oryza sativa* L.)	Decreased biomass of the roots and shoots	Du et al. (2005)
Arsenic (As)	Wheat (*Triticum aestivum* L. cv. PBW-343)	Hill activity and the amount of chlorophyll decreased, the amount of starch phosphorylase increased, the amount of starch decreased, the amount of sugar increased as a result of the starch's decomposition, the amount of putrescine significantly increased, and the amount of spermidine and spermine decreased.	Sil et al. (2019)
Cadmium (Cd)	Rice (*Oryzaesativum*)	Affected nutrient transport and leaf growth	Huybrechts et al. (2021)
Cobalt (Co)	Tomato (*Lycopersicon esculentum* cv. Pusaruby)	Reduction in biomass, concentrations of phosphorus (P), sulfur (S), and iron (Fe), DNA and RNA, chlorophyll a and b, starch, total soluble proteins, protein and non-protein nitrogen, and elevated phenol and Co contents. Catalase activity dropped in leaves, but peroxidase, ribonuclease, and acid phosphatase activity increased	Gopal et al. (2003)
Zinc (Zn)	Rice (*Oryzae sativum*)	Decreased nitrate reductase activity	Singh et al. (2019)
Lead (Pb) and Zinc (Zn)	Alfalfa (*Medicago sativa* L.)	Influenced seed germination and seedling development	Yahaghi et al. (2019)
Chromium (Cr)	Chickpea (*Cicer arietinum* L.)	Malondialdehyde and hydrogen peroxide concentration increased and electrolyte leaks showed elevated oxidative damage Enhanced proline content, superoxide dismutase activity, and peroxide content in leaves Reduced plant growth, chlorophyll content, protein in grains, pod number, and grain production per plant	Singh et al. (2020a)
Nikle (Ni)	Maize (*Zea mays* L.)	The amount of total proteins as well as photosynthetic pigments (chl-a, chl-b, carotenoids, and total chlorophyll) showed a declining trend	Tipu et al. (2021)
Chromium (Cr)	Radish (*Raphanus sativus*)	Impaired nutrition absorption and metabolism	Tiwari et al. (2013)

2.6 Plant Tolerance to Heavy Metals

It's critical to note that some plants can survive in environments with high concentrations of heavy metals (Chibuike and Obiora 2014). There are three processes through which plants may withstand certain metals, namely, (i) exclusion: limiting metal transfer and maintaining a consistent metal amount in the shoot over a wide range of soil amounts; (ii) inclusion: metal amounts in the shoot linearly correlated with those in the soil solution; and (iii) bioaccumulation: metal accumulation in plant shoots and roots, both at low- and high soil amounts (Baker 1981). For example, Cd toxicity can be brought on by difficulties with plant metabolism, mineral nutrient absorption and translocation, and the development and growth of plants. The strategies used by plants to tolerate Cd include constrained Cd absorption, reduced Cd root-to-shoot translocation, improved antioxidant enzyme activity, and enhanced phytochelatin synthesis (Shiyu et al. 2020). Toxic heavy metals may accumulate in the trichomes of plants that can tolerate heavy metals as a part of their elemental defense detoxifying mechanism. In addition to their traditional roles, trichomes assist ion and metal homeostasis in a variety of ways, including the sequestration, compartmentalization, accumulation, and excretion of heavy metals (Koul et al. 2021). Enhancing cell antioxidant systems to combat oxidative stress is a key defensive strategy commonly used by heavy metal-exposed plants (Navari-Izzo et al. 1998; Sgherri et al. 2003).

Omics approaches have demonstrated that HM stress controls a number of genes, stress metabolites or small substances, microRNAs, and phytohormones to develop resistance to HM stress and to effectively control the metabolic processes for survival (Ghuge et al. 2023). There is a wealth of knowledge on the plant molecular and metabolic regulators involved in metal resistance, thanks to the development of omics strategies, including proteomics, genome, transcriptomics, miRNAomics, and metabolomics (Ghuge et al. 2023). The particular genes are being recognized as essential plant metal resistance pathway regulators. Using RNA-Seq, five genes unique to Cucumis melo's metal ion content were discovered (Shen et al. 2022). Heavy metals and plant resistance to salt appear to be strongly correlated. To counteract the high HM concentration in soil, halophytes have a variety of mechanisms, including phytostabilization, phytoextraction, and phytodegradation (Huang and Wang 2009).

2.7 Perspectives and Challenges

Recently, heavy metal contamination in agricultural fields has drawn a lot of attention due to its negative effects on the agroecosystem (Pande et al. 2022). The circumstances (pollution and population) are placing an ever-increasing load on global agriculture yield; hence, there is a need for crop types that are flexible and resilient to varied stresses (Singh et al. 2016). Significant efforts are being undertaken to support plant health and production under difficult abiotic stress conditions, but little progress has been gained as a result of our inadequate comprehension of stress resistance processes (Tiwari and Bae 2023).

The majority of research investigations to date have concentrated on the study of HM's impact on adult plant growth stages, but it is crucial to learn about how plants react to HM stress starting with seed germination and seedling growth because these are the plant lifecycle stages that are most susceptible to external stress since the plants that are used for phytoremediation must endure stress from the very beginning of growth (Pulford and Watson 2003; Thakur and Sharma 2016). Targeting the first stage of a plant's heavy metal absorption is one method for reducing heavy metal toxicity in plants. To comprehend the molecular process behind the transport of heavy metals across cell membranes and into the cells, research on membrane proteins, including ion channels and pumps, should be encouraged (Goyal et al. 2020). Thus, understanding the process by which the target plant responds to various metal stresses is also crucial (Thakur et al. 2016).

The transcription of one or more genes whose products are either engaged in regulatory and signaling pathways or in the production of structural and functional proteins and metabolites that confer heavy metal stress resistance is necessary for heavy metal tolerance to be achieved in plants. Current attempts to increase heavy metal stress resistance capacity by genetic engineering have yielded a number of successes; nonetheless, it is challenging to transfer innovation to actual field circumstances due to the genetically complicated processes of heavy metal stress resistance (Singh et al. 2016). Understanding the detrimental effects of heavy metals on plant development is crucial for achieving increased plant growth and production, as well as a pollution-free environment and ecological balance (Goyal et al. 2020). There are significant differences between the foliar and root metal routes in terms of the processes of metal absorption, transport, compartmentation, detoxification, and toxicity. However, in the framework of risk assessment studies, there was much that remained to be investigated about heavy metal foliar absorption, particularly in locations close to mining activities. A thorough understanding of the physical, chemical, environmental, biological, and physiological mechanisms and variables that regulate the absorption of foliarly deposited heavy metals is necessary for this field's advancements (Shahid et al. 2017).

2.8 Conclusion

The interaction of plants with various heavy metals is linked to different cytotoxicity of plant structures and organs, which are in turn linked to modifications and alterations of plant metabolism, functions, and physiology, besides various essential mechanisms like photosynthesis, respiration, reproduction, and nutrient uptake. Despite the advances elaborated in the field of plant and heavy metal interaction, there are still many areas to be clarified. They will undoubtedly contribute to our understanding of the mechanisms involved and to better exploitation in ecological agriculture, bioremediation, and environmental preservation.

References

Aery NC, Rana DK (2003) Growth and cadmium uptake in barley under cadmium stress. J Environ Biol 24:117–123

Ali B, Gill RA (2022) Heavy metal toxicity in plants: recent insights on physiological and molecular aspects, volume II. Front Plant Sci 13:1–3. https://doi.org/10.3389/fpls.2022.1016257

Alkhatib R, Mheidat M, Abdo N, Tadros M, Al-Eitan L, Al-Hadid K (2019) Effect of lead on the physiological, biochemical and ultrastructural properties of *Leucaena leucocephala*. Plant Biol 21:1132–1139. https://doi.org/10.1111/plb.13021

Alloway BJ (1990) Heavy metals in soils. Wiley, NY

Baker AJ (1981) Accumulators and excluders-strategies in the response of plants to heavy metals. J Plant Nutr 3(1–4):643–654. https://doi.org/10.1080/01904168109362867

Baruah N, Mondal SC, Farooq M, Gogoi N (2019) Influence of heavy metals on seed germination and seedling growth of wheat, pea, and tomato. Water Air Soil Pollut 230:1–15. https://doi.org/10.1007/s11270-019-4329-0

Beyersmann D, Hartwig A (2008) Carcinogenic metal compounds: recent insight into molecular and cellular mechanisms. Arch Toxicol 82(8):493–512. https://doi.org/10.1007/s00204-008-0313-y

Bücker-Neto L, Paiva ALS, Machado RD, Arenhart RA, Margis-Pinheiro M (2017) Interactions between plant hormones and heavy metals responses. Genet Mol Biol 40(Suppl. 1):373–386. https://doi.org/10.1590/1678-4685-gmb-2016-0087

Chaffei C, Gouia H, Ghorbel MH (2003) Nitrogen metabolism in tomato plants under cadmium stress. J Plant Nutr 26:1617–1634. https://doi.org/10.1081/PLN-120022372

Chandra R, Kang H (2016) Mixed heavy metal stress on photosynthesis, transpiration rate, and chlorophyll content in poplar hybrids. For Sci Technol 12(2):55–61. https://doi.org/10.1080/21580103.2015.1044024

Chang LM, Magos L, Suzuki T (1996) Toxicology of metals. CRC Press, Boca Raton

Chen Y, Wang Y, Wu W, Lin Q, Xue S (2006) Impacts of chelate-assisted phytoremediation on microbial community composition in the rhizosphere of a copper accumulator and non-accumulator. Sci Total Environ 356(1–3):247–255. https://doi.org/10.1016/j.scitotenv.2005.04.028

Cheng S (2003) Effects of heavy metals on plants and resistance mechanisms. Environ Sci Pollut Res 10:256–264. https://doi.org/10.1065/espr2002.11.141.2

Chibuike GU, Obiora SC (2014) Heavy metal polluted soils: effect on plants and bioremediation methods. Appl Environ Soil Sci 2014:12. https://doi.org/10.1155/2014/752708

Čiamporová M (2000) Diverse responses of root cell structure to aluminium stress. Plant Soil 226(1):113–116

Clemens S, Ma JF (2016) Toxic heavy metal and metalloid accumulation in crop plants and foods. Annu Rev Plant Biol 67:489–512. https://doi.org/10.1146/annurev-arplant-043015-112301

Cobbett CS (2000) Phytochelatins and their roles in heavy metal detoxification. Plant Physiol 123(3):825–832. https://doi.org/10.1104/pp.123.3.825

Cobbett C, Goldsbrough P (2002) Phytochelatins and metallothioneins: roles in heavy metal detoxification and homeostasis. Annu Rev Plant Biol 53:159–182

Dalton TP, He L, Wang B, Miller ML, Jin L, Stringer KF, Chang X, Stuart Baxter C, Nebert DW (2005) Identification of mouse SLC39a8 as the transporter responsible for cadmium-induced toxicity in the testis. PNAS 102:3401–3406. https://doi.org/10.1073/pnas.040608510

De Matos LP, de Andrade HM, Marinato CS, Prado IGDO, Coelho DG, Montoya SG, Kasuya MCM, de Oliveira JA (2020) Limitations to use of *Cassia grandis* L. in the revegetation of the areas impacted with mining tailings from Fundão Dam. Water Air Soil Pollut 231:127. https://doi.org/10.1007/s11270-020-04479-0

Du X, Zhu YG, Liu WJ, Zhao XS (2005) Uptake of mercury (Hg) by seedlings of rice (*Oryza sativa* L.) grown in solution culture and interactions with arsenate uptake. Environ Exp Bot 54(1):1–7. https://doi.org/10.1016/j.envexpbot.2004.05.001

Duan C, Wang H (1992) Effects of heavy metals on the content of nuclear acids and nuclear enzymatic activities in the root tips of bean (*Vicia faba*). Environ Sci 13(5):31–35. (In Chinese)

Dutta S, Mitra M, Agarwal P, Mahapatra K, De S, Sett U, Roy S (2018) Oxidative and genotoxic damages in plants in response to heavy metal stress and maintenance of genome stability. Plant Signal Behav 13(8):1–49. https://doi.org/10.1080/15592324.2018.1460048

Ejaz U, Khan SM, Khalid N, Ahmad Z, Jehangir S, Fatima Rizvi Z, HeejungLho L, Han H, Raposo A (2023) Detoxifying the heavy metals: a multipronged study of tolerance strategies against heavy metals toxicity in plants. Front Plant Sci 14:1–17. https://doi.org/10.3389/fpls.2023.1154571

El-Shabasy A (2021) Effect of heavy metals in the cement dust pollution on morphological and anatomical characteristics of *Cenchrus ciliaris* L. Saudi. J Biol Sci 28:1069–1079. https://doi.org/10.1016/j.sjbs.2020.11.015

Farooq A, Nadeem M, Abbas G, Shabbir A, Khalid MS, Javeed HMR, Saeed MF, Akram A, Younis A, Akhtar G (2020) Cadmium partitioning, physiological and oxidative stress responses in Marigold (*Calendula calypso*) grown on contaminated soil: implications for phytoremediation. Bull Environ Contam Toxicol 105:270–276. https://doi.org/10.1007/s00128-020-02934-6

Fernandes JC, Henariques FS (1991) Biochemical, physiological and structural effects of excess copper in plants. Bot Rev 57:246–273. https://doi.org/10.1007/bf02858564

Fischer S, Kuehnlenz T, Thieme M, Schmidt H, Clemens S (2014) Analysis of plant pb tolerance at realistic submicromolar concentrations demonstrates the role of phytochelatin synthesis for pb detoxification. Environ Sci Technol 48:7552–7559. https://doi.org/10.1021/es405234p

Gaur A, Adholeya A (2004) Prospects of arbuscular mycorrhizal fungi in phytoremediation of heavy metal contaminated soils. Curr Sci 86:528–534

Ghuge SA, Nikalje GC, Kadam US, Suprasanna P, Hong JC (2023) Comprehensive mechanisms of heavy metal toxicity in plants, detoxification, and remediation. J Hazard Mater 450:131039. https://doi.org/10.1016/j.jhazmat.2023.131039

Gill SS, Khan NA, Tuteja N (2012) Cadmium at high dose perturbs growth, photosynthesis and nitrogen metabolism while at low dose it up regulates sulfur assimilation and antioxidant machinery in garden cress (*Lepidium sativum* L.). Plant Sci 182:112–120. https://doi.org/10.1016/j.plantsci.2011.04.018

Gomes M, Marques T, Carneiro M, Soares ÂM (2012) Anatomical characteristics and nutrient uptake and distribution associated with the cd phytoremediation capacity of *eucalyptus camaldulenses*dehnh. J Soil Sci Plant Nutr 12:481–496. https://doi.org/10.4067/S0718-95162012005000010

Gopal R, Dube BK, Sinha P, Chatterjee C (2003) Cobalt toxicity effects on growth and metabolism of tomato. Commun Soil Sci Plant Anal 34(5–6):619–628. https://doi.org/10.1081/css-120018963

Goyal D, Yadav A, Prasad M, Singh TB, Shrivastav P, Ali A, Dantu PK, Mishra S (2020) Effect of heavy metals on plant growth: An overview. In: Naeem M, Ansari A, Gill S (eds) Contaminants in agriculture: sources, impacts and management. Springer, Cham, pp 79–101

Gunshin H, Mackenzie B, Berger UV, Gunshin Y, Romero MF, Boron WF, Nussberger S, Gollan JL, Hediger MA (1997) Cloning and characterization of a mammalian proton-coupled metal-ion transporter. Nature 388:482–488. https://doi.org/10.1038/41343

Guo Z, Gao Y, Yuan X, Yuan M, Huang L, Wang S, Liu C, Duan C (2023) Effects of heavy metals on stomata in plants: a review. Int J Mol Sci 24(11):1–16. https://doi.org/10.3390/ijms24119302

Gupta D, Vandenhove H, Inouhe M (2013) Role of phytochelatins in heavy metal stress and detoxification mechanisms in plants. In: Gupta D, Corpas F, Palma J (eds) Heavy metal stress in plants. Springer, Berlin, pp 73–94

Hall JÁ (2002) Cellular mechanisms for heavy metal detoxification and tolerance. J Exp Bot 53(366):1–11. https://doi.org/10.1093/jexbot/53.366.1

Holleman AF, Wiberg E (1985) Lehebuch du Anoranischenchemie. Water de Gruyter, Berlin, p 868

Huang GY, Wang YS (2009) Expression analysis of type 2 metallothionein gene in mangrove species (*Bruguieragymnorrhiza*) under heavy metal stress. Chemosphere 77(7):1026–1029. https://doi.org/10.1016/j.chemosphere.2009.07.073

Huybrechts M, Hendrix S, Kyndt T, Demeestere K, Vandamme D, Cuypers A (2021) Short-term effects of cadmium on leaf growth and nutrient transport in rice plants. Plant Sci 313:111054. https://doi.org/10.1016/j.plantsci.2021.111054

Jadia CD, Fulekar MH (2009) Phytoremediation of heavy metals: recent techniques. Afr J Biotechnol 8(6):921–928

Jaleel CA, Jayakumar K, Xing ZC, Azooz MM (2009) Antioxidant potentials protect *Vigna radiata* (L.) Wilczek plants from soil cobalt stress and improve growth and pigment composition. Plant Omics 2:120–126

Jones JD, Dangl JL (2006) The plant immune system. Nature 444(7117):323–329. https://doi.org/10.1038/nature05286

Kabata-Pendias A, Pendias H (2001) Trace elements in soils and plants. CRC Press Inc, Boca Raton

Khanna K, Kohli SK, Bali S, Kaur P, Saini P, Bakshi P, Bhardwaj R (2018) Role of micro-organisms in modulating antioxidant defence in plants exposed to metal toxicity. In: Paul S, Banerjee A, Roychoudhury A (eds) Plants under metal and metalloid stress. Springer, Singapore, pp 303–335

Khudsar T, Arshi A, Siddiqi TO, Mahmooduzzafar, Iqbal M (2008) Zinc-induced changes in growth characters, foliar properties, and Zn-accumulation capacity of pigeon pea at different stages of plant growth. J Plant Nutr 31:281–306. https://doi.org/10.1080/01904160701853894

Ko KS, Lee PK, Kong IC (2012) Evaluation of the toxic effects of arsenite chromate cadmium and copper using a battery of four bioassays. Appl Microbiol Biotechnol 95:1343–1350. https://doi.org/10.1007/s00253-011-3724-2

Kopittke PM, Blamey FPC, Menzies NW (2008) Toxicities of soluble Al, cu, and La include ruptures to rhizodermal and root cortical cells of cowpea. Plant Soil 303:217–227. https://doi.org/10.1007/s11104-007-9500-5

Koul M, Thomas L, Karmakar K (2021) Functional aspects of solanaceae trichomes in heavy metal detoxification. Nord J Bot 39(5):1–14. https://doi.org/10.1111/njb.03171

Kranner I, Colville L (2011) Metals and seeds: biochemical and molecular implications and their significance for seed germination. Environ Exp Bot 72:93–105. https://doi.org/10.1016/j.envexpbot.2010.05.005

Kumar V, Singh J, Kumar P (2019) Heavy metals accumulation in crop plants: sources, response mechanisms, stress tolerance and their effects. In: Kumar V, Kumar R, Singh J, Kumar P (eds) Contaminants in agriculture and environment: health risks and remediation. Agriculture and Environmental Science Academy, Uttarakhand, pp 38–57

Lee S, Moon JS, Ko TS, Petros D, Goldsbrough PB, Korban SS (2003) Overexpression of *Arabidopsis* phytochelatin synthase paradoxically leads to hypersensitivity to cadmium stress. Plant Physiol 131(2):656–663. https://doi.org/10.1104/pp.014118

Llamas MA, Mooij MJ, Sparrius M, Vandenbroucke-Grauls CM, Ratledge C, Bitter W (2008) Characterization of five novel *Pseudomonas aeruginosa* cell-surface signalling systems. Mol Microbiol 67(2):458–472. https://doi.org/10.1111/j.1365-2958.2007.06061.x

Maksymiec W (2007) Signaling responses in plants to heavy metal stress. Acta Physiol Plant 29:177–187. https://doi.org/10.1007/s11738-007-0036-3

Mapanda F, Mangwayana EN, Nyamangara J, Giller KE (2005) The effect of long-term irrigation using wastewater on heavy metal contents of soils under vegetables in Harare, Zimbabwe. Agric Ecosyst Environ 107(2–3):151–165. https://doi.org/10.1016/j.agee.2004.11.005

Marques AP, Rangel AO, Castro PM (2009) Remediation of heavy metal contaminated soils: phytoremediation as a potentially promising clean-up technology. Crit Rev Environ Sci Technol 39(8):622–654. https://doi.org/10.1080/10643380701798272

Márquez-García B, Márquez C, Sanjosé I, Nieva FJJ, Rodríguez-Rubio P, Muñoz-Rodríguez AF (2013) The effects of heavy metals on germination and seedling characteristics in two halophyte species in Mediterranean marshes. Mar Pollut Bull 70:119–124. https://doi.org/10.1016/j.marpolbul.2013.02.019

Meharg AA (1993) The role of plasmalemma in metal tolerance in angiosperm. Physiol Plant 88(1):191–198. https://doi.org/10.1034/j.1399-3054.1993.880126.x

Mithöfer A, Schulze B, Boland W (2004) Biotic and heavy metal stress response in plants: evidence for common signals. FEBS Lett 566(1–3):1–5. https://doi.org/10.1016/j.febslet.2004.04.011

Moosavi SE, Gharineh MH, Afshari RT, Ebrahimi A (2012) Effects of some heavy metals on seed germination characteristics of canola (*Brassica napus*) wheat (*Triticum aestivum*) and safflower (*Carthamus tinctorious*) to evaluate phytoremediation potential of these crops. J Agric Sci 4:11–19. https://doi.org/10.5539/jas.v4n9p11

Moreno-Caselles J, Moral R, Pere-Espinosa A, Marcia MD (2000) Cadmium accumulation and distribution in cucumber plants. J Plant Nutr 23:243–250. https://doi.org/10.1080/01904160009382011

Nagajyoti PC, Lee KD, Sreekanth TVM (2010) Heavy metals occurrence and toxicity for plants: a review. Environ Chem Lett 8(3):199–216. https://doi.org/10.1007/s10311-010-0297-8

Navari-Izzo F, Quartacci MF, Pinzino C, Vecchia FD, Sgherri CL (1998) Thylakoid-bound and stromal antioxidative enzymes in wheat treated with excess copper. Physiol Plant 104(4):630–638. https://doi.org/10.1034/j.1399-3054.1998.1040416.x

Noreen I, Khan SM, Ahmad Z, Rahman I, Tabassum A, Abd Allah EF (2019) Response of different plant species to pollution emitted from oil and gas plant with special reference to heavy metals accumulation. Pak J Bot 51:1231–1240. https://doi.org/10.30848/PJB2019-4(39)

Norvell WA (1984) Comparison of chelating agents as extractants for metals in diverse soil materials. Soil Sci Soc Am J 48(6):1285–1292. https://doi.org/10.2136/sssaj1984.03615995004800060017x

Ozturk M, Yucel E, Gucel S, Sakcali S, Aksoy A (2008) Plants as biomonitors of trace elements pollution in soil. In: Prasad MNV (ed) Trace elements: environmental contamination, nutritional benefits and health implications. Wiley, New York, pp 723–744

Pande A, Mun BG, Methela NJ, Rahim W, Lee DS, Lee GM, Hong JK, Hussain A, Loake G, Yun BW (2022) Heavy metal toxicity in plants and the potential NO-releasing novel techniques as the impending mitigation alternatives. Front Plant Sci 13:1–13. https://doi.org/10.3389/fpls.2022.1019647

Park J, Song W-Y, Ko D, Eom Y, Hansen TH, Schiller M, Lee TG, Martinoia E, Lee Y (2012) The phytochelatin transporters AtABCC1 and AtABCC2 mediate tolerance to cadmium and mercury. Plant J 69:278–288. https://doi.org/10.1111/j.1365-313x.2011.04789.x

Pavel VL, Sobariu DL, Diaconu M, Stătescu F, Gavrilescu M (2013) Effects of heavy metals on *Lepidium sativum* germination and growth. Environ Eng Manag J 12(4):727–733. https://doi.org/10.30638/eemj.2013.089

Peng M, Wang H (1991) The variation of cell ultrastructure of maize (*Zea mays* L.) seedlings. China Environ Sci 11(6):426–431. (In Chinese with English abstract)

Peralta JR, Gardea-Torresdey JL, Tiemann KJ, Gomez E, Arteaga S, Rascon E, Parsons JG (2001) Uptake and effects of five heavy metals on seed germination and plant growth in alfalfa (*Medicago sativa* L.). Bull Environ Contam Toxicol 66:727–734. https://doi.org/10.1007/s001280069

Philip L, Iyengar L, Venkobachar C (1998) Cr (VI) reduction by *Bacillus coagulans* isolated from contaminated soils. J Environ Eng 124(12):1165–1170. https://doi.org/10.1061/(asce)0733-9372(1998)124:12(1165)

Poschenrieder C, Gunsé B, Barceló J (1989) Influence of cadmium on water relations, stomatal resistance, and abscisic acid content in expanding bean leaves. Plant Physiol 90:1365–1371. https://doi.org/10.1104/pp.90.4.1365

Pulford ID, Watson C (2003) Phytoremediation of heavy metal-contaminated land by trees – a review. Environ Int 29(4):529–540. https://doi.org/10.1016/s0160-4120(02)00152-6

Punz WF, Sieghardt H (1993) The responses of root of herbaceous plants species to heavy metals. Environ Exp Bot 33:85–98. https://doi.org/10.1016/0098-8472(93)90058-n

Rascio N, Navari-Izzo F (2011) Heavy metal hyperaccumulating plants: how and why do they do it? And what makes them so interesting? Plant Sci 180(2):169–181. https://doi.org/10.1016/j.plantsci.2010.08.016

Reeves RD, Baker AJM (2000) Metal accumulating plants. In: Raskin I, Ensely BD (eds) Phytoremediation of toxic metals: using plant to clean up the environment. Wiley, Torento, pp 193–229

Rios CO, Siqueira-Silva AI, Pereira EG (2020) How does drought affect native grasses' photosynthesis on the revegetation of iron ore tailings? Environ Sci Pollut Res 28:14797–14811. https://doi.org/10.1007/s11356-020-11599-x

Rizvi A, Zaidi A, Ameen F, Ahmed B, AlKahtani MD, Khan MS (2020) Heavy metal induced stress on wheat: phytotoxicity and microbiological management. RSC Adv 10(63):38379–38403. https://doi.org/10.1039/d0ra05610c

Romanowska E, Wasilewska W, Fristedt R, Vener AV, Zienkiewicz M (2012) Phosphorylation of PSII proteins in maize thylakoids in the presence of pb ions. J Plant Physiol 169(4):345–352. https://doi.org/10.1016/j.jplph.2011.10.006

Rubiya S, Dhriti K, Bhat AA (2018) Heavy metal toxicity in plants: a review. Plant Arch 18(2):1229–1238

Rucińska-Sobkowiak R (2016) Water relations in plants subjected to heavy metal stresses. Acta Physiol Plant 38:1–13. https://doi.org/10.1007/s11738-016-2277-5

Ryu SK, Park JS, Lee IS (2003) Purification and characterization of a copper-binding protein from Asian periwinkle Littorina brevicula. Comp Biochem Physiol Part C Toxicol Pharmacol 134(1):101–107. https://doi.org/10.1016/s1532-0456(02)00217-x

Sandeep G, Vijayalatha KR, Anitha T (2019) Heavy metals and its impact in vegetable crops. Int J Chem Stud 7(1):1612–1621

Sgherri C, Cosi E, Navari-Izzo F (2003) Phenols and antioxidative status of *Raphanus sativus* grown in copper excess. Physiol Plant 118(1):21–28. https://doi.org/10.1034/j.1399-3054.2003.00068.x

Shah K, Mankad AU, Reddy MN (2017) Cadmium accumulation and its effects on growth and biochemical parameters in *Tagetes erecta* L. J Pharmacogn Phytochem 6(3):111–115

Shahid M, Dumat C, Khalid S, Schreck E, Xiong T, Niazi NK (2017) Foliar heavy metal uptake, toxicity and detoxification in plants: a comparison of foliar and root metal uptake. J Hazard Mater 325:36–58. https://doi.org/10.1016/j.jhazmat.2016.11.063

Sharma RK, Agrawal M (2005) Biological effects of heavy metals: an overview. J Environ Biol 26(2):301–313

Sharma SS, Dietz KJ (2009) The relationship between metal toxicity and cellular redox imbalance. Trends In Plant Sci 14(1):43–50. https://doi.org/10.1016/j.tplants.2008.10.007

Sharma RK, Agrawal M, Marshall F (2006) Heavy metal contamination in vegetables grown in wastewater irrigated areas of Varanasi. India Bull Environ Contam Toxicol 77(2):312–318. https://doi.org/10.1007/s00128-006-1065-0

Shen ZG, Li XD, Wang CC, Chen HM, Chua H (2002) Lead phytoextraction from contaminated soil with high-biomass plant species. J Environ Qual 31(6):1893–1900. https://doi.org/10.2134/jeq2002.1893

Shen Q, Wu X, Tao Y, Yan G, Wang X, Cao S, Wang C, He W (2022) Mining candidate genes related to heavy metals in mature melon (*Cucumis melo* L.) peel and pulp using WGCNA. Genes 13(10):1–16. https://doi.org/10.3390/genes13101767

Shiyu QIN, Hongen LIU, Zhaojun NIE, Rengel Z, Wei GAO, Chang LI, Peng ZHAO (2020) Toxicity of cadmium and its competition with mineral nutrients for uptake by plants: a review. Pedosphere 30(2):168–180. https://doi.org/10.1016/s1002-0160(20)60002-9

Sil P, Das P, Biswas S, Mazumdar A, Biswas AK (2019) Modulation of photosynthetic parameters, sugar metabolism, polyamine and ion contents by silicon amendments in wheat (*Triticum aestivum* L.) seedlings exposed to arsenic. Environ Sci Pollut Res 26:13630–13648. https://doi.org/10.1007/s11356-019-04896-7

Singh D, Nath K, Sharma YK (2007) Response of wheat seed germination and seedling growth under copper stress. J Environ Biol 28:409–414

Singh S, Parihar P, Singh R, Singh VP, Prasad SM (2016) Heavy metal tolerance in plants: role of transcriptomics, proteomics, metabolomics, and ionomics. Front Plant Sci 6:1–36. https://doi.org/10.3389/fpls.2015.01143

Singh P, Singh I, Shah K (2019) Reduced activity of nitrate reductase under heavy metal cadmium stress in rice: an *in silico* answer. Front Plant Sci 9:1–11. https://doi.org/10.3389/fpls.2018.01948

Singh D, Sharma NL, Singh CK, Sarkar SK, Singh I, Dotaniya ML (2020a) Effect of chromium (VI) toxicity on morpho-physiological characteristics, yield, and yield components of two chickpea (*Cicer arietinum* L.) varieties. PLoS One 15(12):e0243032. https://doi.org/10.1371/journal.pone.0243032

Singh SB, Singh K, Butola SS, Rawat S, Arunachalam K (2020b) Determination of macronutrients, micronutrients and heavy metals present in *Spilanthesacmella* Hutch and Dalz: possible health effects. Nat Prod Sci 26(1):50–58. https://doi.org/10.20307/nps.2020.26.1.50

Sinha S, Gupta AK (2005) Translocation of metals from fly ash amended soil in the plant of Sesbania cannabina L. Ritz: effect on antioxidants. Chemosphere 61(8):1204–1214. https://doi.org/10.1016/j.chemosphere.2005.02.063

Song WY, Park J, Mendoza-Cozatl DG, Suter-Grotemeyer M, Shim D, Hortensteiner S, Geisler M, Weder B, Rea PA, Rentsch D, Schroeder JI, Lee Y, Martinoia E (2010) Arsenic tolerance in *Arabidopsis* is mediated by two ABCC-type phytochelatin transporters. Proc Natl Acad Sci USA 107:21187–21192. https://doi.org/10.1073/pnas.1013964107

Song W-Y, Yamaki T, Yamaji N, Ko D, Jung K-H, Fujii-Kashino M, An G, Martinoia E, Lee Y, Ma JF (2014) A rice ABC transporter, OsABCC1, reduces arsenic accumulation in the grain. Proc Natl Acad Sci USA 111:15699–15704. https://doi.org/10.1073/pnas.1414968111

Souri Z, Cardoso AA, da Silva CJ, de Oliveira LM, Dari B, Sihi D, Karimi N (2019) Heavy metals and photosynthesis: recent developments. In: Ahmad P, Ahanger MA, Alyemeni MN, Alam P (eds) Photosynthesis, productivity and environmental stress. Wiley, New York, pp 107–134

Sytar O, Kumar A, Latowski D, Kuczynska P, Strzaka K, Prasad MNV (2013) Heavy metal-induced oxidative damage, defense reactions, and detoxification mechanisms in plants. Acta Physiol Plant 35:985–999. https://doi.org/10.1007/s11738-012-1169-6

Tchounwou PB, Yedjou CG, Patlolla AK, Sutton DJ (2012) Heavy metal toxicity and the environment. Exp Suppl 101:133–164. https://doi.org/10.1007/978-3-7643-8340-4_6

Thakur S, Sharma SS (2016) Characterization of seed germination, seedling growth, and associated metabolic responses of *Brassica juncea* L. cultivars to elevated nickel concentrations. Protoplasma 253:571–580. https://doi.org/10.1007/s00709-015-0835-0

Thakur S, Singh L, Wahid ZA, Siddiqui MF, Atnaw SM, Din MFM (2016) Plant-driven removal of heavy metals from soil: uptake, translocation, tolerance mechanism, challenges, and future perspectives. Environ Monit Assess 188:1–11. https://doi.org/10.1007/s10661-016-5211-9

Thiebeauld O, Soler S, Raigón MD, Prohens J, Nuez F (2005) Variation among Solanaceae crops in cadmium tolerance and accumulation. Agron Sustain Dev 25:237–241. https://doi.org/10.1051/agro:2005014

Tipu MI, Ashraf MY, Sarwar N, Akhtar M, Shaheen MR, Ali S, Damalas CA (2021) Growth and physiology of maize (*Zea mays* L.) in a nickel-contaminated soil and phytoremediation efficiency using EDTA. J Plant Growth Regul 40:774–786. https://doi.org/10.1007/s00344-020-10132-1

Tiwari P, Bae H (2023) Trends in harnessing plant endophytic microbiome for heavy metal mitigation in plants: a perspective. Plants 12(7):1–21. https://doi.org/10.3390/plants12071515

Tiwari KK, Singh NK, Rai UN (2013) Chromium phytotoxicity in radish (*Raphanus sativus*): effects on metabolism and nutrient uptake. Bull Environ Contam Toxicol 91:339–344. https://doi.org/10.1007/s00128-013-1047-y

Tuna LA, Burun B, Yokas I, Coban E (2002) The effects of heavy metals on pollen germination and pollen tube length in the tobacco plant. Turk J Biol 26(2):109–113

Van Assche F, Clijsters H (1990) Effects of metals on enzyme activity in plants. Plant Cell Environ 24:1–15. https://doi.org/10.1111/j.1365-3040.1990.tb01304.x

Verloo M, Eeckhout M (1990) Metal species transformations in soil: an analytical approach. Int J Environ Anal Chem 39:170–186. https://doi.org/10.1080/03067319008027695

Wang S, Shi X (2001) Molecular mechanisms of metal toxicity and carcinogenesis. Mol Cell Biochem 222:3–9. https://doi.org/10.1007/978-1-4615-0793-2_1

Wojtaszek P (1997) Oxidative burst: an early plant response to pathogen infection. Biochem J 322:681–692. https://doi.org/10.1042/bj3220681

Xun E, Zhang Y, Zhao J, Guo J (2017) Translocation of heavy metals from soils into floral organs and rewards of Cucurbita pepo: implications for plant reproductive fitness. Ecotoxicol Environ Saf 145:235–243. https://doi.org/10.1016/j.ecoenv.2017.07.045

Xun E, Zhang Y, Zhao J, Guo J (2018) Heavy metals in nectar modify behaviors of pollinators and nectar robbers: consequences for plant fitness. Environ Pollut 242:1166–1175. https://doi.org/10.1016/j.envpol.2018.07.128

Yadav SK (2010) Heavy metals toxicity in plants: An overview on the role of glutathione and phytochelatins in heavy metal stress tolerance of plants. S Afr J Bot 76:167–179. https://doi.org/10.1016/j.sajb.2009.10.007

Yahaghi Z, Shirvani M, Nourbakhsh F, Pueyo JJ (2019) Uptake and effects of lead and zinc on alfalfa (*Medicago sativa* L.) seed germination and seedling growth: role of plant growth promoting bacteria. S Afr J Bot 124:573–582. https://doi.org/10.1016/j.sajb.2019.01.006

Yang X, Feng Y, He Z, Stoffella PJ (2005) Molecular mechanisms of heavy metal hyperaccumulation and phytoremediation. J Trace Elem Med Biol 18(4):339–353. https://doi.org/10.1016/j.jtemb.2005.02.007

Yang Y, Zhang L, Huang X, Zhou Y, Quan Q, Li Y, Zhu X (2020) Response of photosynthesis to different concentrations of heavy metals in *Davidiainvolucrata*. PLoS One 15(3):e0228563. https://doi.org/10.1371/journal.pone.0228563

Yeboah JO, Shi G, Shi W (2021) Effect of heavy metal contamination on soil enzymes activities. J Geosci Environ Prot 9(6):135–154. https://doi.org/10.4236/gep.2021.96008

Zagorchev L, Seal CE, Kranner I, Odjakova M (2013) A central role for thiols in plant tolerance to abiotic stress. Int J Mol Sci 14(4):7405–7432. https://doi.org/10.3390/ijms14047405

Zengin FK, Munzuroglu O (2005) Effects of some heavy metals on content of chlorophyll, proline and some antioxidant chemicals in bean (*Phaseolus vulgaris* L.) seedlings. Acta Biol Crac Ser Bot 47:157–164

Zhang Y (1997) The toxicity of heavy metals to barley (*Hordeum vulgate*). Acta Sci Circum 17(2):199–204. (In Chinese with English abstract)

Zhuang P, McBride MB, Xia H, Li N, Li Z (2009) Health risk from heavy metals via consumption of food crops in the vicinity of Dabaoshan mine, South China. Sci Total Environ 407:1551–1561. https://doi.org/10.1016/j.scitotenv.2008.10.061

Chapter 3
Auxin-Mediated Regulation of Plant Biology Under Heavy Metal Stress: Physiological and Molecular Aspects

Ravneet Kaur, Harleen Kaur, and Ashish Sharma

Abstract Auxin is a plant growth hormone responsible for regulating growth, metabolism, apical dominance, and various other physiological processes of plants. The stress of heavy metal (HM) in plants is one of the main concerns nowadays. Excessive levels of HM in plants have severe negative effects on plants, such as a reduction in protein content and oxidative damage due to uncontrolled production of ROS as a result of HM toxicity. Auxin, a plant growth promoter, helps in mitigating negative effects of HM toxicity in plants. Auxin activates various genes or various enzymes responsible for conferring tolerance to plants under stress. Auxin promotes the stimulation of antioxidant machinery of plants, thereby restoring plant growth and development by relieving them from HM toxicity. The chapter discusses various biosynthesis pathways of auxin occurring in plants as well as the role of auxin involved in relieving plants from HM toxicity. The chapter discusses the effect of auxin on the antioxidant enzymatic machinery of plants so that the plants act against HM toxicity and develop tolerance to it.

Keywords Auxin · Abiotic stress · Heavy metal toxicity · Stress physiology · Toxicity

3.1 Auxin: A Phytohormone

In the book *The Power of Movement in Plants* (Darwin and Darwin 1880), Charles and Frances Darwin coined the term "auxin" to describe a portable plant growth regulator. It was Went in 1926 who "introduced into plant physiology the general conception of regulators or coordinators" (Peirce 1936). This is exemplified by the

R. Kaur · H. Kaur · A. Sharma (✉)
Department of Botany and Environment Studies, DAV University, Jalandhar, Punjab, India
e-mail: ashish10210@davuniversity.org

A. Sharma et al. (eds.), *Hormones and Heavy Metals*, Plant Life and Environment Dynamics, https://doi.org/10.1007/978-981-96-9622-2_3

enduring value of the Cholodny-Went hypothesis, which posits that auxin's dynamic redistribution directs tropic growth. George Peirce first mentioned auxin in Plant Physiology (Whippo and Hangarter 2006).

The exploration of "auxins" commenced with investigations into the bending responses of coleoptiles towards a light source (Simon and Petrášek 2011). It was demonstrated that the signal perceived by the coleoptile tip traveled asymmetrically downward, inducing differential elongation on one side of the coleoptile (Went 1926). Charles and Francis Darwin observed in *Phalaris canariensis* that both root and shoot perceive light intensity, with the stimuli being transmitted to various parts of the plant during its growth and development (Morffy and Strader 2020). Subsequently, in the absence of a unilateral light source, an experiment by Fritz Went and Nicolai Cholodny on *Avena sativa* coleoptile segments demonstrated curvature (Cholodny 1928; Went 1928; Du et al. 2020). They coined the term "auxin" to denote the growth-promoting factor in cotyledonous plants, a chemical inherent in plants. The term "auxin" derives from the Greek word "auxein," meaning "to grow." Auxin serves as a broad term referring to a group of substances capable of inducing elongation in shoot cells (Moore 1989).

Indole-3-acetic acid (IAA), the primary auxin found in the majority of plants, was the first identified auxin molecule. Its discovery followed extensive research into the plant molecule responsible for the tropic movement of plants in response to light and gravity (Abel and Theologis 2010). Besides IAA, numerous natural compounds exhibit auxin-like activity in bioassays, including phenylacetic acid (PAA) and the halogenated derivative of IAA, 4-chloro-indole-3-acetic acid (4-Cl-IAA). Both are recognized by the auxin signaling mechanism (Shimizu-Mitao and Kakimoto 2014; Jayasinghege et al. 2019). 4-Cl-IAA is relatively uncommon (Lam et al. 2015), and research on PAA has been limited compared to IAA due to its generally weaker effects (Cook 2019; Bandurski et al. 1995; Ludwig-Müller and Cohen 2002). Some IAA precursors, such as indole-3-acetonitrile and indole-3-pyruvic acid, also exhibit activity in bioassays because they convert into IAA within the tissue (Thimann 1977).

Auxin serves as a signaling molecule regulating various physiological processes in plant growth and development, including the development of apical dominance, root gravitropism and phototropism, and the establishment of bilateral symmetry in embryos (Du et al. 2020). Additionally, it regulates tissue and organ growth, meristem formation, cell division and expansion, polarity maintenance, and embryogenesis (Mroue et al. 2018; Woodward and Bartel 2005). Apical dominance, tropic response, organ patterning, and vascular development are also influenced by auxin (Woodward and Bartel 2005; Cao et al. 2019). It has been firmly established that auxin, a crucial endogenous plant hormone, is essential for the growth and development of plants at every stage. Numerous recent studies have demonstrated that auxins contribute to plant tolerance to abiotic stressors (Bulak et al. 2014; Zhou et al. 2020; Huang et al. 2021). Auxin-responsive genes are broadly classified into three groups: auxin/indole acetic acid (AUX/IAA); GH3; and SAUR (Guilfoyle et al. 1998; Jain et al. 2006a, b, c). Auxins have been reported to be involved in improving plant tolerance to heavy metals (Sytar et al. 2019). Auxin is produced in various plant tissues and can travel over long distances in plants owing to its mobile nature. It has been found that auxin has structural similarity with tryptophan, the first discovered plant hormone (Moore 1989), and it has been recognized that auxin is a hormone necessary for plant growth and development (Zhao 2010).

Furthermore, auxin demonstrates polar/directional transport, with its distribution occurring through the phloem elements. Because auxin is transported basipetally—that is, from the tip to the base—maintaining the gradient is crucial to the architecture of plants (Leyser 2009). Research has unveiled the involvement of auxin in various plant activities. Auxin may play a stimulatory or inhibitory role in plants depending on its concentration. A slight increase in auxin concentration may stimulate plant growth, while the opposite effect was observed in plants when the auxin concentration exceeds the optimum level. Studies have shown that auxin is responsible for cell division and oriented cell expansion, alongside other plant growth hormones (Skoog and Miller 1957; Perrot-Rechenmann 2010), influencing all aspects of plant growth (Vanneste and Friml 2009).

3.2 Biosynthesis of Auxin

The biosynthesis of IAA is a complex process involving multiple pathways (Fig. 3.1). Studies have identified L-tryptophan (Trp), an aromatic amino acid, as a crucial precursor for IAA biosynthesis in plants. Further investigation has revealed that Trp-independent IAA synthesis is facilitated by cytosolic indole synthase (INS), which converts indole-3-glycerolphosphate to indole (Zhang et al. 2008; Wang et al. 2015). The tryptophan-dependent pathways of IAA biosynthesis are classified based on the synthesis of four primary compounds: (i) indole-3-pyruvic acid (IPyA), (ii) tryptamine (TAM), (iii) indole-3-acetaloxime (IAOx), and (iv) indoleacetamide (IAM), all derived from a common precursor, tryptophan. The shikimate pathway, which also produces aromatic amino acids, is how chloroplasts make Trp (Maeda and Dudareva 2012).

(i) *Indole-3-pyruvic acid (IPyA):*

It is recognized that the IPyA pathway is the most extensively studied pathway for IAA production in plants. IAA biosynthesis via the IPyA pathway occurs in two steps (Zhao et al. 2002; Stepanova et al. 2008; Tao et al. 2008). Initially, TRYPTOPHAN AMINOTRANSFERASE OF ARABIDOPSIS 1 (TAA1) and TAA1-RELATED proteins (TARs) deaminate tryptophan to IPyA (Stepanova et al. 2008; Tao et al. 2008; Yamada et al. 2009). Then, IPyA is converted into IAA by the release of CO_2, a reaction catalyzed by YUCCA (YUC) family flavin-containing monooxygenases (Mashiguchi et al. 2011; Stepanova et al. 2011; Won et al. 2011). The YUC flavin-containing monooxygenases of the auxin biosynthetic pathway were the first to be discovered and were found to be essential for several plant developmental pathways (Zhao 2014). Homologs of TAA1 and YUC are present in both vascular and nonvascular plants (Yue et al. 2014; Eklund et al. 2015; Poulet and

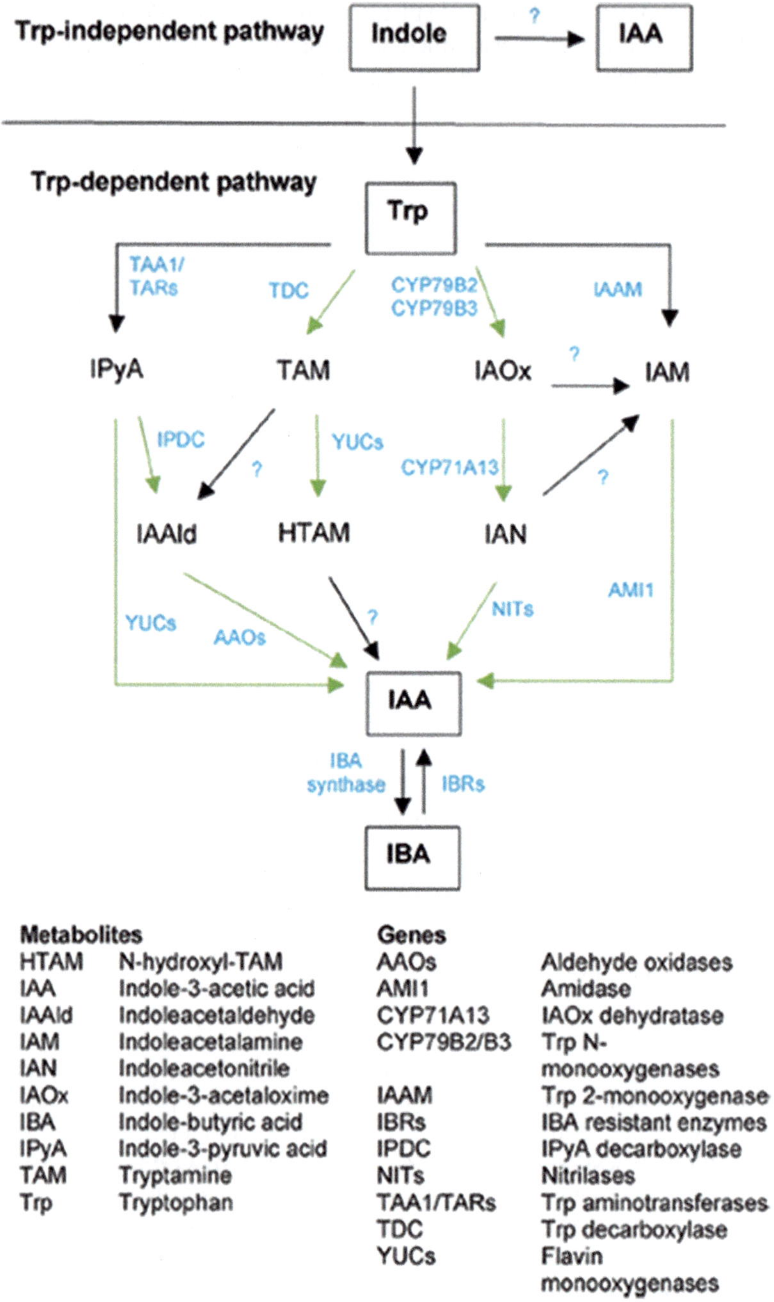

Fig. 3.1 Indole 3-acetic acid biosynthesis: tryptophan-dependent and -independent routes. (Source: Tan et al. 2021)

Kriechbaumer 2017; Matthes et al. 2019), and recent research has demonstrated the functional conservation of TAA1 and YUC homologs in *Zea mays* (Gallavotti et al. 2008; Phillips et al. 2011; Chourey et al. 2012), *Oryza sativa* (Yamamoto et al. 2007; Yoshikawa et al. 2014), and *Marchantia polymorpha* (Eklund et al. 2015). Alterations in the functional YUCCA (YUC) pathway of auxin biosynthesis may lead to changes in plant responses to the environment (Lee et al. 2012). For instance, in barley (*Hordeum vulgare*) and *Arabidopsis thaliana*, high temperatures that inhibit the production of the YUC2 and YUC6 genes result in a decrease in endogenous auxin levels in developing anthers, leading to male sterility (Cheng et al. 2006; Sakata et al. 2010).

(ii) *Tryptamine (TAM):*

Another pathway which involves enzyme flavin monooxygenase (YUC) converts TAM to N-hydroxyl-TAM and then to IAA (Tan et al. 2021).

(iii) *Indole-3-acetaloxime (IAOx):*

These pathways have been identified to play an important role in plants exposed to stress (Julkowska et al. 2017; Lehmann et al. 2017). Two isozymes of the cytochrome P450 (CYP) monooxygenase family, CYP79B2 and CYP79B3, catalyze the conversion of Trp to IAOx (Hull et al. 2000; Mikkelsen et al. 2000). However, this pathway is only restricted to the Brassicaceae family because IAOx and CYP79B2/3 genes are only present in *Brassica* species (Sugawara et al. 2009).

(iv) *Indoleacetamide (IAM):*

A well-known intermediate in auxin biosynthesis, IAM, is produced via the conversion of Trp to IAA in several bacteria associated with plants (Patten et al. 2013). Moreover, IAM has been found in non-*Brassica* species lacking IAOx (Pollmann et al. 2002; Sugawara et al. 2009). Tryptophan is converted to IAM by tryptophan-2-monoxygenase (iaaM), and then IAM is hydrolyzed into IAA by iaaH in a two-step process (Camilleri and Jouanin 1991). IAM hydrolases from tobacco and *Arabidopsis* (AtAMI1, NtAMI1) that convert IAM to IAA have been identified. Research suggests that IAM could potentially function as an intermediary in the production of IAA from indole-3-acetaldoxime (IAOx) (Sugawara et al. 2009). Research indicates that IAOx is the precursor of IAN and indole-3-acetaldehyde which nitrilases and aldehyde oxidases can employ to produce IAA (Zhao 2010).

3.3 Role of Auxin in Abiotic Stress

Various environmental stressors, such as extreme temperatures, drought, heavy metals, osmotic stress, and soil salinity, act as primary abiotic stressors that adversely affect plant growth, reproductive success, and ultimately yield. Throughout evolution, to cope with these adverse climatic factors (Fig. 3.2) plants have developed

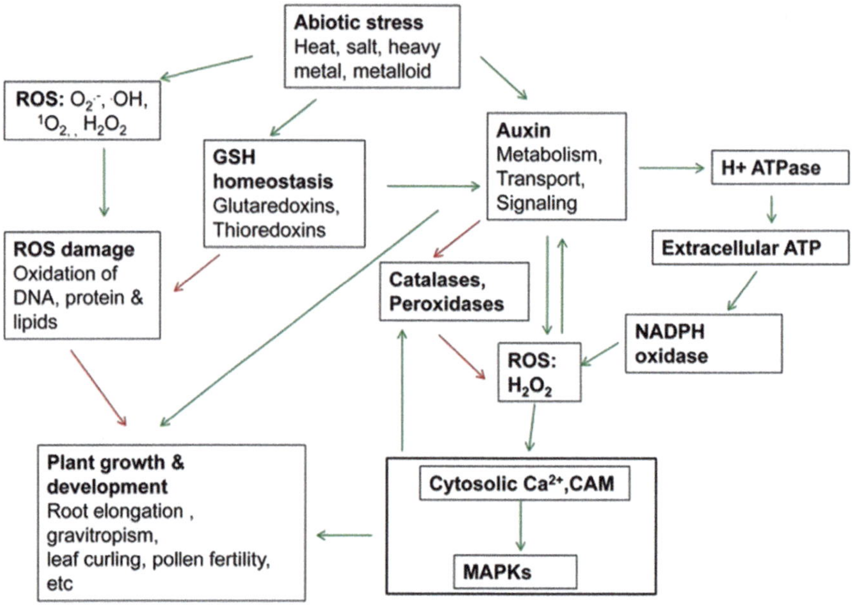

Fig. 3.2 Auxin and hydrogen peroxide's metabolic interaction regulates plant growth, development, and stress tolerance. Red arrows denote negative effects, and green arrows show favorable effects. (Modified from: Krishnamurthy and Rathinasabapathi 2013)

various strategies that eventually help them withstand such conditions (Chaves and Oliveira 2004).

Plant responses vary from species to species as well as from stress to stress. It has been observed that plants subjected to osmotic stress experience reduced plant growth and reproductive success, ultimately affecting plant yield (Hsiao et al. 1976). Under cold stress, the inflorescence gravitropism of *Arabidopsis* was greatly inhibited due to alterations in auxin activity (Fukaki et al. 1996; Wyatt et al. 2002).

In response to various abiotic stresses, the metabolic pathways of plants undergo alterations. Plants respond to abiotic stressors by simultaneously changing the concentration of several phytohormones, including auxin. Changes in auxin transport within plants affect the distribution and concentration of auxin within plant parts. Plants require the endogenous auxin pool to be maintained at a certain level to synchronize different cellular functions (Shibasaki et al. 2009).

Research has found that plants under salt stress alter their levels of auxin and genes associated with auxin. Auxin levels and auxin transporter expression have been found to decrease in plants under salt stress (Park et al. 2007; Du et al. 2012; Liu et al. 2015). Reduced or impaired auxin transport and uneven distribution along the shoot result from a reduction in auxin levels in parallel (Sun et al. 2008; Shen et al. 2010). Since auxin regulates the growth of lateral roots, studies on plants exposed to salt stress have shown that there is an antagonistic link between the development of lateral roots mediated by auxin and the arrest of growth in response to elevated salt concentrations (Julkowska and Testerink 2015).

Changes in auxin levels occur in plants subjected to different abiotic stresses. A study revealed that drought stress caused a decline in auxin levels, while extreme temperatures caused a rise in auxin levels in rice plants. One way plants adapt to drought stress is by altering the levels of auxin in the root and shoot. To reduce water loss, plants under water stress increase the concentration of auxin in their roots and decrease it in their leaves, a process known as auxin signaling. It has been documented that auxin levels decline under salt or drought stress in plants producing poplar xylem, as observed in herbaceous plants such as maize (Naqvi and Ansari 1974), *Anastatica hierochuntica*, and *Helianthus annuus* (Hartung and Witt 1968) (Junghans et al. 2006). Additionally, it has been shown that polar auxin transport is affected by osmotic stress induced by drought or elevated salinity; however, the reaction of the auxin transport machinery appears to be plant- and tissue-specific. Osmotic stress significantly increases auxin transport in *Avena* mesocotyls but significantly decreases it in *Pisum* hypocotyls (Kaldewey et al. 1974). Under salt stress, *Zea mays* coleoptiles do not exhibit any changes in auxin transport (Naqvi 1972).

3.4 Role of Auxin in Plants Under Heavy Metal Stress

A considerable group of elements with specific densities greater than 5 g/cm^3 and relative atomic masses exceeding 40 are categorized as heavy metals. Many of these metals possess a strong affinity for ligands containing sulfur, binding to them aggressively. Consequently, upon entering plant cells, heavy metals react with SH-groups, thereby inactivating various plant enzymes and disturbing numerous metabolic processes, ultimately affecting normal plant activities (Seregin and Ivanov 2001).

In response to plants subjected to heavy metal stress, auxin levels undergo alterations. The concentration of auxin may either increase or decrease depending on the level of heavy metal stress. For instance, exposure of *Glycine max* to heavy metals (Cd) inhibited auxin biosynthesis, resulting in a decline in auxin levels; however, low Cd concentrations promoted auxin biosynthesis (Małkowski et al. 2020).

Phytohormones play a significant role in nutrient uptake, root morphology, and rhizosphere activity. When denser root hairs grow on the root system of a plant or when roots cluster to enhance nutrient uptake, auxins contribute to these morphological adaptation modifications (Wittenmayer and Merbach 2005). As IAA is regarded as an important regulator for the development and growth of plants, auxin homeostasis plays a significant role in plants under heavy metal toxicity (Fig. 3.3).

The hormone inhibits the transcription of auxin-sensitive genes within the plant cell by promoting the degradation of AUX/IAA transcriptional repressors linked to auxin-responsive factors (ARFs). This process ultimately guides adjustments in plant physiology (Vanneste and Friml 2009). The Aux/IAA-SCFTIR1 complex, formed when an auxin receptor (TIR1) binds to it, is broken down by the ubiquitin/

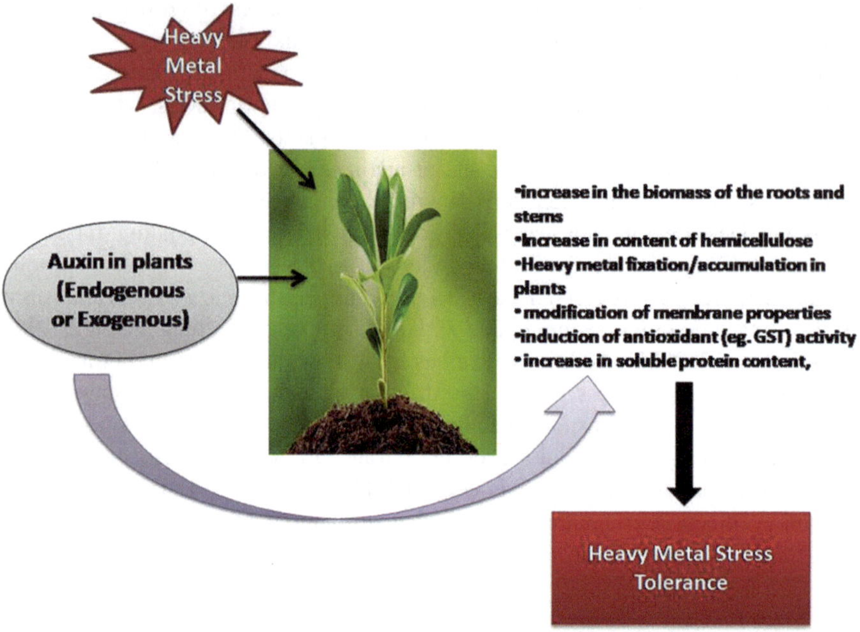

Fig. 3.3 Role of auxin in mitigation of heavy metal stress in plants (auxins play an important role in mitigating the heavy metal stress in plants)

S26 proteasome pathway (Chapman and Estelle 2009). Ultimately, auxin alters the expression of several genes, thus regulating the growth and development of plants (Jain et al. 2006a, b, c).

In earlier studies, it was found that the presence of cadmium (Cd) increased the activity of IAA oxidase, an enzyme responsible for auxin breakdown (Singh and Prasad 2015). Consequently, it was suggested that Cd's activation of IAA oxidase might lead to a decrease in auxin levels, impacting growth. Generally, a reduction in auxin levels was observed in plants exposed to heavy metal stress. For instance, the presence of arsenic (As) in *Brassica juncea* leads to changes in the levels of three auxins, namely, IAA, NAA, and indole-3-butyric acid (IBA).

However, the application of exogenous phytohormones can restore the endogenous levels of auxin within plants, thus helping to mitigate the negative impacts of heavy metal toxicity. It was observed that in sunflower (*Helianthus annuus*) plants subjected to moderate lead (Pb) toxicity, the exogenous application of the phytohormone IAA increased plant biomass (Liphadzi et al. 2006). Similar results were found in *Brassica juncea* plants when exposed to As toxicity (Srivastava et al. 2013). The stress induced by As is mitigated by the auxin transport machinery (Krishnamurthy and Rathinasabapathi 2013).

The toxic heavy metal Cd primarily accumulates in the xylem of roots and is transferred to the tissues of aerial plants, where it inhibits lateral branching and root growth. The expression of auxin reporters is higher in the vascular tissues of poplar roots when plants are exposed to Cd poisoning (Elobeid et al. 2012). Prior to transplantation, the exogenous administration of L-TRP, a precursor of auxin, to plant roots increased plant height, tiller count, panicle count, 1000-grain weight, and *Oryza sativa* production in plants under Cd stress.

During plant development under abiotic stress, reactive oxygen species (ROS) are produced, which can have highly negative effects. However, in environmentally stressed plants, moderate levels of ROS can work in conjunction with auxin to maintain cellular homeostasis and stability. It was investigated that in rice seedlings, ROS (H_2O_2) produced due to Cd toxicity enhanced root growth through the auxin-signaling pathway (Zhao et al. 2012).

Research has unveiled that auxin signaling contributes to activating Cd-induced morphogenic defensive responses in barley (*Hordeum vulgare*) root tips (Tamás et al. 2015). Additionally, auxin reduces the toxicity of Cd in *Arabidopsis* and wheat (*Triticum aestivum*) by enhancing antioxidant defense mechanisms (Agami and Mohamed 2013; Zhu et al. 2013). It is plausible that auxin transport channels are essential for Cd-modulated LR formation in *Arabidopsis* (Hu et al. 2013).

Abiotic stress conditions can induce various phenotypic changes in plants, such as increased lateral root development, decreased primary root elongation, and higher root hair counts.

3.4.1 Phytoremediation by Auxin

Alterations in auxin homeostasis accompanied physiological responses observed in *Arabidopsis* seedlings exposed to substances inducing oxidative stress, suggesting potential crosstalk. In a recent study, plants cultivated with L-TRP supplementation exhibited higher Cd absorption from rice straw compared to untreated plants. Although the L-TRP treatment reduced Cd levels in rice grain, there might have been an elevated Cd accumulation in rice straw due to factors like high transpiration rates, pollutants, and the movement of water-soluble soil components through open stomata.

Auxins have been shown to stimulate algal growth and facilitate Pb biosorption by *A. obliquus* cells. Consistent with these findings, IAA enhanced the absorption of Pb in *Medicago sativa* (López et al. 2005). Hadi et al. (2010) reported that IAA, IBA, and NAA increased Pb and Zn biosorption by *Zea mays*, while Zhu et al. (2013) demonstrated that NAA enhanced Cd accumulation in *Arabidopsis thaliana* roots. Auxin treatment holds promise for enhancing Pb accumulation in plants, thus increasing the efficiency of metal removal.

A recent study revealed that the detrimental effects of Pb on the levels of proteins and chlorophyll in *A. obliquus* can be mitigated by the external application of auxins or cytokinins. These findings align with those observed in vascular plants. For example, supplementing rice treated with As with IAA increased the levels of protein and chlorophyll (Pandey and Gupta 2015). Similarly, the application of IAA to pea seedlings reversed copper-induced changes in thiols and carbonyls (markers of protein oxidation), thereby protecting proteins from oxidation and increasing their abundance (Massoud et al. 2018).

The presence of phytohormones or plant growth regulators (polyamines) in the nutritional medium is essential for the biosorption of heavy metals by *C. vulgaris* cells. In the presence of heavy metals, auxins, cytokinins, gibberellin, and spermidine prevent the accumulation of heavy metals in algal cells. Thus, the inhibition of metal bioaccumulation, which enables cells to maintain precise homeostatic regulation of intracellular heavy metal levels, may explain the protective action of these bioactive compounds against heavy metal stress in *C. vulgaris* cultures. Exploring the potential correlation between *C. vulgaris*'s prevention of metal biosorption from the nutritional medium and the roles of auxins, cytokinins, gibberellin, and spermidine in enhancing plant resistance to toxic metal-induced stress is worthwhile (Piotrowska-Niczyporuk et al. 2012).

López et al. (2005) reported that IAA enhanced Pb uptake by *Medicago sativa* plants, while Fuentes et al. (2000) observed that IBA and NAA increased Pb and Zn biosorption by Zea mays plants. The bioaccumulation of Cd, Cu, and Pb in *C. vulgaris* cells has been documented to negatively impact several crucial metabolic pathways associated with plant physiology. A significant sign of heavy metal poisoning in green algae is the reduction in cell quantity or culture growth. The inhibition of growth and the decrease in the mitotic index in response to heavy metals are likely interconnected (Dalla Vecchia et al. 2005). Conversely, gibberellin, auxins, cytokinins, and Spd exhibit protective properties against heavy metal stress, which aids in restoring algal development. Considering that these phytohormones can promote plant growth and development in the presence of abiotic stress, this finding is not surprising (López et al. 2005; Fuentes et al. 2000; Thomas et al. 2005).

The findings suggest that the application of auxins, cytokinins, gibberellin, and Spd has a beneficial effect on the photosynthetic apparatus of *C. vulgaris* treated with heavy metals. Since photosynthetic pigments shield environmental pollutants from their harmful effects, increasing their abundance improves green microalgae (Piotrowska-Niczyporuk et al. 2012). Table 3.1 shows the role of auxins in plants under heavy metal stress.

Table 3.1 Effect of exogenous/endogenous application of auxin under heavy metal stress

Heavy metal stress	Plant species	Physiological effect on plant	Reference
Zn	*Oryza sativa*	Auxin causes decreased cysteine, glutathione, and PCs content	Begum et al. (2016)
Pb	*Helianthus annuus*	Zn-inhibited root growth was compensated for by the addition of exogenous IAA, which resulted in a considerable increase in the biomass of the roots and stems	Liphadzi et al. (2006)
Zn and Pb	*Helianthus annuus*	IAA stimulates root growth and lessens the detrimental effects of metals on root and shoot dry weight, root length, root volume, and root surface area	
Cu	*Brassica juncea*	Exogenous application of IAA improves plant growth, photosynthesis, and phytoremediation potential	Mir et al. (2022)
Cd	Arabidopsis	When exogenously provided, NAA raises the hemicellulose content and, as a result, the amount of Cd^{2+} fixed in the roots. By raising hemicellulose 1 levels and fixing metal in the root, auxin reduces the amount of Cd^{2+} that is translocated from roots to shoots in this way, hence lessening the toxicity of Cd^{2+} in Arabidopsis	
Pb	–	Auxin alters the characteristics of the membrane, which reduces the harmful effects of exposure to heavy metals. They suggested that a combination of Pb and IAA or NAA reduces the symptoms of heavy metal toxicity while also causing a decrease in the disruption of membrane permeability and organization	Hąc-Wydro et al. (2016)
Cd	–	Supplementing with L-TRP, a precursor of auxin, increased the growth properties of the plants, such as their yield and growth	Farooq et al. (2015)
As	*Oryza sativa*	Application of auxin and selenium simultaneously on the morphological and biochemical characteristics of rice plants under As stress	Pandey and Gupta (2015)
	Pigeon pea	Plant hormone seed priming may enhance crop performance in the presence of heavy metal stress. Concentrations enhanced the time to 50% germination, final germination percentage, average germination time, and germination speed index	Sneideris et al. (2015)
Cd	*Solanum lycopersicum*	Capacity to withstand oxidative stress caused by Cd was improved by the exogenous administration of IAA, which increased ascorbate levels and the activity of the ascorbate-glutathione cycle	Šípošová et al. (2021)
Cd	*Hordeum vulgare* roots	IAA-mediated induction of glutathione sulfotransferase (GST) activity	Bočová et al. (2013)
Cd	*Arabidopsis thaliana* roots	In *A. thaliana* roots, Cd-mediated up-regulation of the gene encoding nitrilase (AtNIT) was found to result in increased IAA concentration. These authors found a correlation between accelerated lateral-root development, higher growth, greater root branching, and elevated levels of IAA	Vitti et al. (2013)
Cd stress	*Vigna radiata*	Auxin plays a role in increasing the amount of soluble protein because it is engaged in transcription and/or translation. It is produced by an increase in the activity of several different enzymes during of 4-CI-IAA treatment	Mazid et al. (2010)

3.5 Molecular Mechanism of Auxin Action

According to Sinha et al. (2011) and Hamel et al. (2006), MAPKs are highly conserved signaling molecules that respond to various environmental pressures by transmitting stress-associated signals and thereby regulating cellular functions. The MAPK cascade consists of three-tier components that mediate phosphorylation processes from an upstream receptor to a downstream target: MAPKKKs, MAPKKs, and MAPKs. Reports suggest that heavy metals such as Cd, Cu, and As activate the MAPK cascade (Jonak et al. 2004; Yeh et al. 2007; Ding et al. 2011; Rao et al. 2011; Smeets et al. 2013). In response to Cu^{2+} and Cd^{2+} treatment in rice leaves and roots, the transcripts of OsMSRMK2 (OsMPK3 homolog), OsMSRMK3 (OsMPK7 homolog), and OsWJUMK1 (OsMPK20-4 homolog) increased (Yeh et al. 2007; Rao et al. 2011).

In plants, heavy metals induce the production of ROS. ROS, in turn, activates the MAPK pathway. Two significant MAPK cascades fully described in *Arabidopsis* are MPK3/6-MKK4/5-MEKK1 (Asai et al. 2002) and MEKK1-MKK2-MPK4/6, which are involved in abiotic stress signaling and are activated subsequent to ROS (Pitzschke et al. 2009; Jalmi and Sinha 2015). After activation, MAPK pathways regulate the formation of ROS through positive feedback. The ROS-activated cascade OXI1-MPK6 also controls ROS production favorably (Asai et al. 2008).

There have been conflicting findings in the past regarding the relationship between auxin and MAPK signaling (Jonak et al. 2002; Mishra et al. 2006). Evidence for the quick and temporary activation of a kinase in *Arabidopsis* roots by low auxin levels was shown by Mockaitis and Howell (2000). Mizoguchi et al. (1994) also reported similar results.

In contrast, Kovtun et al. (1998) demonstrated that the activation of a MAPK cascade by NPK1, a specific plant MAPK kinase kinase (MAPKKK), results in the repression of early auxin-response gene transcription. Additionally, Lee et al. (2009) proposed that MPK12 is a negative regulator of *Arabidopsis* auxin signaling. MKK7 in *Arabidopsis* negatively controls polar auxin transport, according to Dai et al. (2006). On the other hand, Tena and Renaudin (1998) demonstrated that while high auxin concentrations promoted myelin basic protein (MBP) kinase activity in the same lines, low auxin concentrations did not activate MBP kinases in tobacco cell lines. This suggests that auxin signaling involves both positive and negative MAPK pathways (Mishra et al. 2006). Plant growth is significantly regulated by the cell cycle. Plant cell division relies heavily on MAPKs, some of which are momentarily activated during mitosis (Mishra et al. 2006). MMK3 (alfalfa MAPK, Medicago MAP kinase 3), the temporary nature of MMK3 protein kinase activity during mitosis and its correlation with the timing of phragmoplast development provide evidence for MMK3's function in controlling plant cytokinesis (Bögre et al. 1999).

As noted earlier, while MAPKs play crucial roles in signaling responses to heavy metal stress, auxin, and cell cycle cues, the connection between MAPK, auxin distribution, auxin, and cell cycle-related gene expression in Cd-stressed rice roots remains unclear. The expression levels of six MAPK genes, including OsMAPK2, OsMPK12, OsMAPK14, OsMAPK44, *Oryza sativa* putative mitogen-activated protein kinase 3 (OsMSRMK3), and serine/threonine kinase domain and NAF domain (OsMSURPK2), were inhibited. The expression of many auxin-related genes was negatively affected by the alteration in MAPK signaling, such as OsYUCCA, OsPIN, auxin response factors (OsARF), OsIAA, and cyclin-related genes (cyclin-dependent kinases; OsCDK, cyclins; OsCYC, and CDK-like proteins; OsCKL) under Cd stress.

Under heavy metal stress, plants regulate a variety of auxin-related genes, leading to changes in auxin accumulation and distribution within cells and throughout the plant. Auxin biosynthesis genes may be either upregulated or downregulated in response to heavy metal stress. For instance, in *Arabidopsis thaliana* (Arabidopsis) exposed to Cd stress, the expression of the YUCCA6 gene, which encodes proteins like flavin monooxygenase involved in auxin biosynthesis, was increased, resulting in elevated levels of auxin in post-embryonic roots (hypocotyl) (Fattorini et al. 2017). Conversely, application of As leads to the negative regulation of YUCCA6 expression, leading to a decrease in endogenous levels of IAA content.

Furthermore, auxin transport within plants is reduced under Cd/As stress, as indicated by lower transcript levels of the auxin influx carrier and auxin efflux carrier pin-formed 1 (PIN1), as well as like-AUX 1 (LAX3) in the post-embryonic roots of *Arabidopsis*, thereby altering auxin homeostasis (Fattorini et al. 2017). Additionally, under As stress, the auxin transport mutant aux1, deficient in IAA accumulation, exhibited significantly reduced basipetal IAA transfer, resulting in inhibited *Arabidopsis* root growth (Krishnamurthy and Rathinasabapathi 2013). However, it was observed that exogenously applied auxin increased the resistance of aux1 seedlings to heavy metal stress, similar to mutant plants, highlighting the critical role of auxin transport and signaling pathways in regulating root development and resilience to heavy metal stress (Krishnamurthy and Rathinasabapathi 2013).

Under Cd stress, the levels of auxin in plants decrease due to the enhanced expression of gretchenhagen 3 (GH3), which promotes auxin conjugation and auxin degradation. Simultaneously, lignin and peroxidase activity increase, triggering defense responses under heavy metal stress. In rice exposed to Cd and As stress, the expression of auxin biosynthesis genes (anthranilate synthase alpha subunit 2; ASA2 and YUCCA2) and IAA transport genes (auxin permease; AUX1 and PIN5b) is altered, disturbing auxin homeostasis and thereby affecting the root architecture of the plant (Ronzan et al. 2018). This demonstrates how auxin homeostasis contributes to plants' recovery from heavy metal stress (Luo et al. 2019).

3.6 Conclusion

Auxin is an essential phytohormone responsible for regulating various plant processes. Auxin distribution in plants affects a variety of functions performed by auxin in plants. Heavy metal stress is one of the major challenges to plant productivity. However, exogenous application of auxin can restore plant productivity owing to the role of auxin in mitigating the negative effects of HM on plants. Auxin activates various genes as well as various antioxidative enzymes responsible for conferring tolerance to plants under HM toxicity. It was found that the detrimental effects of heavy metals in plants were restored by auxin application in plants, or endogenous levels of auxin increased in plants in response to HM toxicity.

References

Abel S, Theologis A (2010) Odyssey of auxin. Cold Spring Harb Perspect Biol 2(10):a004572

Agami RA, Mohamed GF (2013) Exogenous treatment with indole-3-acetic acid and salicylic acid alleviates cadmium toxicity in wheat seedlings. Ecotoxicol Environ Saf 94:164–171

Asai T, Tena G, Plotnikova J, Willmann MR, Chiu WL, Gómez-Gómez L, Boller T, Ausubel FM, Sheen J (2002) MAP kinase signalling cascade in Arabidopsis innate immunity. Nature 415:977–983

Asai S, Ohta K, Yoshioka H (2008) MAPK signaling regulates nitric oxide and NADPH oxidase-dependent oxidative bursts in Nicotiana benthamiana. Plant Cell 20(5):1390–1406

Bandurski RS, Cohen JD, Slovin JP, Reinecke DM (1995) Plant hormones: physiology, biochemistry and molecular biology, vol 39. Kluwer Academic Publishers, Dordrecht, p 65

Begum MC, Islam M, Sarkar MR, Azad MAS, Huda AN, Kabir AH (2016) Auxin signaling is closely associated with Zn-efficiency in rice (Oryza sativa L.). J Plant Interact 11(1):124–129

Bočová B, Huttová J, Mistrík I, Tamás L (2013) Auxin signalling is involved in cadmium-induced glutathione-S-transferase activity in barley root. Acta Physiol Plant 35(9):2685–2690

Bögre L, Calderini O, Binarova P, Mattauch M, Till S, Kiegerl S et al (1999) A MAP kinase is activated late in plant mitosis and becomes localized to the plane of cell division. Plant Cell 11(1):101–113

Bulak P, Walkiewicz A, Brzezińska M (2014) Plant growth regulators-assisted phytoextraction. Biol Plant 58(1):1–8

Camilleri C, Jouanin L (1991) The TR-DNA region carrying the auxin synthesis genes of the Agrobacterium rhizogenesagropine-type plasmid pRiA4: nucleotide sequence analysis and introduction into tobacco plants. Mol Plant-Microbe Interact 4(2):155–162

Cao X, Yang H, Shang C, Ma S, Liu L, Cheng J (2019) The roles of auxin biosynthesis YUCCA gene family in plants. Int J Mol Sci 20(24):6343

Chapman EJ, Estelle M (2009) Mechanism of auxin-regulated gene expression in plants. Annu Rev Genet 43:265–285

Chaves MM, Oliveira MM (2004) Mechanisms underlying plant resilience to water deficits: prospects for water-saving agriculture. J Exp Bot 55(407):2365–2384

Cheng Y, Dai X, Zhao Y (2006) Auxin biosynthesis by the YUCCA flavin monooxygenases controls the formation of floral organs and vascular tissues in Arabidopsis. Genes Dev 20(13):1790–1799

Cholodny N (1928) Beiträgezurhormonalen Theorie von Tropismen. In: Zeitschrift fur wissenschaftliche Biologie. Abteilung E: Planta. Springer, New York, pp 118–134

Chourey P, Li Q, Bernardi J, Lanubile A, Marocco A, Kumar D, Kladnik A (2012) Impaired auxin biosynthesis in the defective endosperm18 is due to a mutation at the ZmYuc1 gene encoding YUCCA1 protein in maize. In: Plant biology 2012. American Society of Plan Biologists, pp 139–139

Cook SD (2019) An historical review of phenylacetic acid. Plant Cell Physiol 60(2):243–254

Dai Y, Wang H, Li B, Huang J, Liu X, Zhou Y et al (2006) Increased expression of MAP KINASE KINASE7 causes deficiency in polar auxin transport and leads to plant architectural abnormality in Arabidopsis. Plant Cell 18(2):308–320

Dalla Vecchia F, La Rocca N, Moro I, De Faveri S, Andreoli C, Rascio N (2005) Morphogenetic, ultrastructural and physiological damages suffered by submerged leaves of Elodea canadensis exposed to cadmium. Plant Sci 168(2):329–338

Darwin C, Darwin F (1880) Sensitiveness of plants to light: it's transmitted effect. In: The power of movement in plants. Cambridge University Press, Cambridge, UK

Ding Y, Chen Z, Zhu C (2011) Microarray-based analysis of cadmium-responsive microRNAs in rice (Oryza sativa). J Exp Bot 62(10):3563–3573

Du H, Wu N, Fu J, Wang S, Li X, Xiao J, Xiong L (2012) A GH3 family member, OsGH3-2, modulates auxin and abscisic acid levels and differentially affects drought and cold tolerance in rice. J Exp Bot 63(18):6467–6480

Du M, Spalding EP, Gray WM (2020) Rapid auxin-mediated cell expansion. Annu Rev Plant Biol 71:379–402

Eklund DM, Ishizaki K, Flores-Sandoval E, Kikuchi S, Takebayashi Y, Tsukamoto S et al (2015) Auxin produced by the indole-3-pyruvic acid pathway regulates development and gemmae dormancy in the liverwort Marchantia polymorpha. Plant Cell 27(6):1650–1669

Elobeid M, Göbel C, Feussner I, Polle A (2012) Cadmium interferes with auxin physiology and lignification in poplar. J Exp Bot 63(3):1413–1421

Farooq H, Asghar HN, Khan MY, Saleem M, Zahir ZA (2015) Auxin-mediated growth of rice in cadmium-contaminated soil. Turk J Agric For 39(2):272–276

Fattorini L, Ronzan M, Piacentini D, Della Rovere F, De Virgilio C, Sofo A et al (2017) Cadmium and arsenic affect quiescent centre formation and maintenance in Arabidopsis thaliana post-embryonic roots disrupting auxin biosynthesis and transport. Environ Exp Bot 144:37–48

Fuentes HD, Khoo CS, Pe T, Muir S, Khan AG (2000, May) Phytoremediation of a contaminated mine site using plant growth regulators to increase heavy metal uptake. In: Proceedings of the 5th international conference on clean technologies for the mining industry, vol 1. Universidad de Concepción, Santiago, pp 427–435

Fukaki H, Fujisawa H, Tasaka M (1996) Gravitropic response of inflorescence stems in Arabidopsis thaliana. Plant Physiol 110(3):933–943

Gallavotti A, Barazesh S, Malcomber S, Hall D, Jackson D, Schmidt RJ, McSteen P (2008) Sparse inflorescence1 encodes a monocot-specific YUCCA-like gene required for vegetative and reproductive development in maize. Proc Natl Acad Sci 105(39):15196–15201

Guilfoyle TJ, Ulmasov T, Hagen G (1998) The ARF family of transcription factors and their role in plant hormone-responsive transcription. Cell Mol Life Sci 54(7):619–627

Hąc-Wydro K, Sroka A, Jabłońska K (2016) The impact of auxins used in assisted phytoextraction of metals from the contaminated environment on the alterations caused by lead (II) ions in the organization of model lipid membranes. Colloids Surf B: Biointerfaces 143:124–130

Hadi F, Bano A, Fuller MP (2010) The improved phytoextraction of lead (Pb) and the growth of maize (Zea mays L.): the role of plant growth regulators (GA3 and IAA) and EDTA alone and in combinations. Chemosphere 80(4):457–462

Hamel LP, Nicole MC, Sritubtim S, Morency MJ, Ellis M, Ehlting J et al (2006) Ancient signals: comparative genomics of plant MAPK and MAPKK gene families. Trends Plant Sci 11(4):192–198

Hartung W, Witt J (1968) Über den Einfluß der Bodenfeuchtigkeit auf den Wuchsstoffgehalt von Anastatica hierochuntica und Helianthus annuus. Flora oder Allgemeine botanische Zeitung Abt B, Morphologie und Geobotanik 157(4):603–614

Hsiao TC, Acevedo E, Fereres E, Henderson DW (1976) Water stress, growth and osmotic adjustment. Philos Trans R Soc Lond B Biol Sci 273(927):479–500

Hu YF, Zhou G, Na XF, Yang L, Nan WB, Liu X et al (2013) Cadmium interferes with maintenance of auxin homeostasis in Arabidopsis seedlings. J Plant Physiol 170(11):965–975

Huang D, Wang Q, Jing G, Ma M, Li C, Ma F (2021) Overexpression of MdIAA24 improves apple drought resistance by positively regulating strigolactone biosynthesis and mycorrhization. Tree Physiol 41(1):134–146

Hull AK, Vij R, Celenza JL (2000) Arabidopsis cytochrome P450s that catalyze the first step of tryptophan-dependent indole-3-acetic acid biosynthesis. Proc Natl Acad Sci 97(5):2379–2384

Jain M, Kaur N, Garg R, Thakur JK, Tyagi AK, Khurana JP (2006a) Structure and expression analysis of early auxin-responsive Aux/IAA gene family in rice (Oryza sativa). Funct Integr Genomics 6(1):47–59

Jain M, Kaur N, Tyagi AK, Khurana JP (2006b) The auxin-responsive GH3 gene family in rice (Oryza sativa). Funct Integr Genomics 6(1):36–46

Jain M, Tyagi AK, Khurana JP (2006c) Genome-wide analysis, evolutionary expansion, and expression of early auxin-responsive SAUR gene family in rice (Oryza sativa). Genomics 88(3):360–371

Jalmi SK, Sinha AK (2015) ROS mediated MAPK signaling in abiotic and biotic stress-striking similarities and differences. Front Plant Sci 6:769

Jayasinghege C, Ozga JA, Nadeau CD, Kaur H, Reinecke DM (2019) Tir1 auxin receptors implicated in the differential response to 4-chloroindole-3-acetic acid and indole-3-acetic acid in developing pea fruit. J Exp Bot 70(4):1239–1253

Jonak C, Ökrész L, Bögre L, Hirt H (2002) Complexity, cross talk and integration of plant MAP kinase signalling. Curr Opin Plant Biol 5(5):415–424

Jonak C, Nakagami H, Hirt H (2004) Heavy metal stress. Activation of distinct mitogen-activated protein kinase pathways by copper and cadmium. Plant Physiol 136(2):3276–3283

Julkowska MM, Testerink C (2015) Tuning plant signaling and growth to survive salt. Trends Plant Sci 20(9):586–594

Julkowska MM, Koevoets IT, Mol S, Hoefsloot H, Feron R, Tester MA et al (2017) Genetic components of root architecture remodeling in response to salt stress. Plant Cell 29(12):3198–3213

Junghans U, Polle A, Düchting P, Weiler E, Kuhlman B, Gruber F, Teichmann T (2006) Adaptation to high salinity in poplar involves changes in xylem anatomy and auxin physiology. Plant Cell Environ 29(8):1519–1531

Kaldewey H, Ginkel U, Wawczyniak G (1974) Auxin transport and water stress in pea (Pisum sativum L.) 1. Berichte der Deutschen Botanischen Gesellschaft 87(3):563–576

Kovtun Y, Chiu WL, Zeng W, Sheen J (1998) Suppression of auxin signal transduction by a MAPK cascade in higher plants. Nature 395(6703):716–720

Krishnamurthy A, Rathinasabapathi B (2013) Auxin and its transport play a role in plant tolerance to arsenite-induced oxidative stress in Arabidopsis thaliana. Plant Cell Environ 36(10):1838–1849

Lam HK, McAdam SA, McAdam EL, Ross JJ (2015) Evidence that chlorinated auxin is restricted to the Fabaceae but not to the Fabeae. Plant Physiol 168(3):798–803

Lee JS, Wang S, Sritubtim S, Chen JG, Ellis BE (2009) Arabidopsis mitogen-activated protein kinase MPK12 interacts with the MAPK phosphatase IBR5 and regulates auxin signaling. Plant J 57(6):975–985

Lee M, Jung JH, Han DY, Seo PJ, Park WJ, Park CM (2012) Activation of a flavin monooxygenase gene YUCCA7 enhances drought resistance in Arabidopsis. Planta 235(5):923–938

Lehmann T, Janowitz T, Sánchez-Parra B, Alonso MMP, Trompetter I, Piotrowski M, Pollmann S (2017) Arabidopsis NITRILASE 1 contributes to the regulation of root growth and development through modulation of auxin biosynthesis in seedlings. Front Plant Sci 8:36

Leyser O (2009) The control of shoot branching: an example of plant information processing. Plant Cell Environ 32(6):694–703

Liphadzi MS, Kirkham MB, Paulsen GM (2006) Auxin-enhanced root growth for phytoremediation of sewage-sludge amended soil. Environ Technol 27(6):695–704

Liu W, Li RJ, Han TT, Cai W, Fu ZW, Lu YT (2015) Salt stress reduces root meristem size by nitric oxide-mediated modulation of auxin accumulation and signaling in Arabidopsis. Plant Physiol 168(1):343–356

López ML, Peralta-Videa JR, Benitez T, Gardea-Torresdey JL (2005) Enhancement of lead uptake by alfalfa (Medicago sativa) using EDTA and a plant growth promoter. Chemosphere 61(4):595–598

Ludwig-Müller J, Cohen JD (2002) Identification and quantification of three active auxins in different tissues of Tropaeolum majus. Physiol Plant 115(2):320–329

Luo Y, Wei Y, Sun S, Wang J, Wang W, Han D et al (2019) Selenium modulates the level of auxin to alleviate the toxicity of cadmium in tobacco. Int J Mol Sci 20(15):3772

Maeda H, Dudareva N (2012) The shikimate pathway and aromatic amino acid biosynthesis in plants. Annu Rev Plant Biol 63:73–105

Małkowski E, Sitko K, Szopiński M, Gieroń Ż, Pogrzeba M, Kalaji HM, Zieleźnik-Rusinowska P (2020) Hormesis in plants: the role of oxidative stress, auxins and photosynthesis in corn treated with Cd or Pb. Int J Mol Sci 21(6):2099

Mashiguchi K, Tanaka K, Sakai T, Sugawara S, Kawaide H, Natsume M et al (2011) The main auxin biosynthesis pathway in Arabidopsis. Proc Natl Acad Sci 108(45):18512–18517

Massoud MB, Sakouhi L, Karmous I, Zhu Y, El Ferjani E, Sheehan D, Chaoui A (2018) Protective role of exogenous phytohormones on redox status in pea seedlings under copper stress. J Plant Physiol 221:51–61

Matthes MS, Best NB, Robil JM, Malcomber S, Gallavotti A, McSteen P (2019) Auxin EvoDevo: conservation and diversification of genes regulating auxin biosynthesis, transport, and signaling. Mol Plant 12(3):298–320

Mazid M, Ali B, Hayat S, Ahmad A (2010) The effect of 4-chloroindole-3-acetic acid on some growth parameters in mung bean under cadmium stress. Turk J Biol 34(1):9–13

Mikkelsen MD, Hansen CH, Wittstock U, Halkier BA (2000) Cytochrome P450 CYP79B2 from Arabidopsis catalyzes the conversion of tryptophan to indole-3-acetaldoxime, a precursor of indole glucosinolates and indole-3-acetic acid. J Biol Chem 275(43):33712–33717

Mir AR, Alam P, Hayat S (2022) Auxin regulates growth, photosynthetic efficiency and mitigates copper induced toxicity via modulation of nutrient status, sugar metabolism and antioxidant potential in Brassica juncea. Plant Physiol Biochem 185:244–259

Mishra NS, Tuteja R, Tuteja N (2006) Signaling through MAP kinase networks in plants. Arch Biochem Biophys 452(1):55–68

Mizoguchi T, Gotoh Y, Nishida E, Yamaguchi-Shinozaki K, Hayashida N, Iwasaki T et al (1994) Characterization of two cDNAs that encode MAP kinase homologues in Arabidopsis thaliana and analysis of the possible role of auxin in activating such kinase activities in cultured cells. Plant J 5(1):111–122

Mockaitis K, Howell SH (2000) Auxin induces mitogenic activated protein kinase (MAPK) activation in roots of Arabidopsis seedlings. Plant J 24(6):785–796

Moore TC (1989) Ethylene. In: Biochemistry and physiology of plant hormones. Springer, New York, pp 228–254

Morffy N, Strader LC (2020) Old Town Roads: routes of auxin biosynthesis across kingdoms. Curr Opin Plant Biol 55:21–27

Mroue S, Simeunovic A, Robert HS (2018) Auxin production as an integrator of environmental cues for developmental growth regulation. J Exp Bot 69(2):201–212

Naqvi SM (1972) Auxin transport under saline growth conditions. Experientia 28(10):1246–1246

Naqvi SM, Ansari R (1974) Estimation of diffusible auxin under saline growth conditions. Experientia 30(4):350–350

Pandey C, Gupta M (2015) Selenium and auxin mitigates arsenic stress in rice (Oryza sativa L.) by combining the role of stress indicators, modulators and genotoxicity assay. J Hazard Mater 287:384–391

Park JE, Park JY, Kim YS, Staswick PE, Jeon J, Yun J et al (2007) GH3-mediated auxin homeostasis links growth regulation with stress adaptation response in Arabidopsis. J Biol Chem 282(13):10036–10046

Patten CL, Blakney AJ, Coulson TJ (2013) Activity, distribution and function of indole-3-acetic acid biosynthetic pathways in bacteria. Crit Rev Microbiol 39(4):395–415

Peirce GJ (1936) FAFC WENT (1863–1935). Plant Physiol 11(2):i4

Perrot-Rechenmann C (2010) Cellular responses to auxin: division versus expansion. Cold Spring Harb Perspect Biol 2(5):a001446

Phillips KA, Skirpan AL, Liu X, Christensen A, Slewinski TL, Hudson C et al (2011) Vanishing tassel2 encodes a grass-specific tryptophan aminotransferase required for vegetative and reproductive development in maize. Plant Cell 23(2):550–566

Piotrowska-Niczyporuk A, Bajguz A, Zambrzycka E, Godlewska-Żyłkiewicz B (2012) Phytohormones as regulators of heavy metal biosorption and toxicity in green alga Chlorella vulgaris (Chlorophyceae). Plant Physiol Biochem 52:52–65

Pitzschke A, Djamei A, Bitton F, Hirt H (2009) A major role of the MEKK1–MKK1/2–MPK4 pathway in ROS signalling. Mol Plant 2(1):120–137

Pollmann S, Müller A, Piotrowski M, Weiler EW (2002) Occurrence and formation of indole-3-acetamide in Arabidopsis thaliana. Planta 216(1):155–161

Poulet A, Kriechbaumer V (2017) Bioinformatics analysis of phylogeny and transcription of TAA/YUC auxin biosynthetic genes. Int J Mol Sci 18(8):1791

Rao KP, Vani G, Kumar K, Wankhede DP, Misra M, Gupta M, Sinha AK (2011) Arsenic stress activates MAP kinase in rice roots and leaves. Arch Biochem Biophys 506(1):73–82

Ronzan M, Piacentini D, Fattorini L, Della Rovere F, Eiche E, Riemann M et al (2018) Cadmium and arsenic affect root development in Oryza sativa L. negatively interacting with auxin. Environ Exp Bot 151:64–75

Sakata T, Oshino T, Miura S, Tomabechi M, Tsunaga Y, Higashitani N et al (2010) Auxins reverse plant male sterility caused by high temperatures. Proc Natl Acad Sci 107(19):8569–8574

Seregin IV, Ivanov VB (2001) Physiological aspects of cadmium and lead toxic effects on higher plants. Russ J Plant Physiol 48(4):523–544

Shen C, Bai Y, Wang S, Zhang S, Wu Y, Chen M et al (2010) Expression profile of PIN, AUX/LAX and PGP auxin transporter gene families in Sorghum bicolor under phytohormone and abiotic stress. FEBS J 277(14):2954–2969

Shibasaki K, Uemura M, Tsurumi S, Rahman A (2009) Auxin response in Arabidopsis under cold stress: underlying molecular mechanisms. Plant Cell 21(12):3823–3838

Shimizu-Mitao Y, Kakimoto T (2014) Auxin sensitivities of all Arabidopsis Aux/IAAs for degradation in the presence of every TIR1/AFB. Plant Cell Physiol 55(8):1450–1459

Simon S, Petrášek J (2011) Why plants need more than one type of auxin. Plant Sci 180(3):454–460

Singh S, Prasad SM (2015) IAA alleviates Cd toxicity on growth, photosynthesis and oxidative damages in eggplant seedlings. Plant Growth Regul 77:87–98

Sinha AK, Jaggi M, Raghuram B, Tuteja N (2011) Mitogen-activated protein kinase signaling in plants under abiotic stress. Plant Signal Behav 6(2):196–203

Šípošová K, Labancová E, Kučerová D, Kollárová K, Vivodová Z (2021) Effects of exogenous application of indole-3-butyric acid on maize plants cultivated in the presence or absence of cadmium. Plants (Basel) 10(11):2503

Skoog F, Miller C (1957) Chemical regulation of growth and organ formation in plant tissues cultured. In Vitro Symp Soc Exp Biol 11:118–130

Smeets K, Opdenakker K, Remans T, Forzani C, Hirt H, Vangronsveld J, Cuypers A (2013) The role of the kinase OXI1 in cadmium-and copper-induced molecular responses in Arabidopsis thaliana. Plant Cell Environ 36(6):1228–1238

Sneideris LC, Gavassi MA, Campos ML, D'Amico-Damiao V, Carvalho RF (2015) Effects of hormonal priming on seed germination of pigeon pea under cadmium stress. An Acad Bras Cienc 87:1847–1852

Srivastava S, Srivastava AK, Suprasanna P, D'souza SF (2013) Identification and profiling of arsenic stress-induced microRNAs in Brassica juncea. J Exp Bot 64(1):303–315

Stepanova AN, Robertson-Hoyt J, Yun J, Benavente LM, Xie DY, Doležal K et al (2008) TAA1-mediated auxin biosynthesis is essential for hormone crosstalk and plant development. Cell 133(1):177–191
Stepanova AN, Yun J, Robles LM, Novak O, He W, Guo H et al (2011) The Arabidopsis YUCCA1 flavin monooxygenase functions in the indole-3-pyruvic acid branch of auxin biosynthesis. Plant Cell 23(11):3961–3973
Sugawara S, Hishiyama S, Jikumaru Y, Hanada A, Nishimura T, Koshiba T et al (2009) Biochemical analyses of indole-3-acetaldoxime-dependent auxin biosynthesis in Arabidopsis. Proc Natl Acad Sci 106(13):5430–5435
Sun F, Zhang W, Hu H, Li B, Wang Y, Zhao Y et al (2008) Salt modulates gravity signaling pathway to regulate growth direction of primary roots in Arabidopsis. Plant Physiol 146(1):178–188
Sytar O, Kumari P, Yadav S, Brestic M, Rastogi A (2019) Phytohormone priming: regulator for heavy metal stress in plants. J Plant Growth Regul 38(2):739–752
Tamás L, Mistrík I, Alemayehu A, Zelinová V, Bočová B, Huttová J (2015) Salicylic acid alleviates cadmium-induced stress responses through the inhibition of Cd-induced auxin-mediated reactive oxygen species production in barley root tips. J Plant Physiol 173:1–8
Tan CY, Dodd IC, Chen JE, Phang SM, Chin CF, Yow YY, Ratnayeke S (2021) Regulation of algal and cyanobacterial auxin production, physiology, and application in agriculture: an overview. J Appl Phycol 33(5):2995–3023
Tao Y, Ferrer JL, Ljung K, Pojer F, Hong F, Long JA et al (2008) Rapid synthesis of auxin via a new tryptophan-dependent pathway is required for shade avoidance in plants. Cell 133(1):164–176
Tena G, Renaudin JP (1998) Cytosolic acidification but not auxin at physiological concentration is an activator of MAP kinases in tobacco cells. Plant J 16(2):173–182
Thimann KV (1977) Hormones and senescence of leaves. In: Hormonal action in the whole life of plants. University of Massachusetts Press, Amherst, pp 357–380
Thomas JC, Perron M, LaRosa PC, Smigocki AC (2005) Cytokinin and the regulation of a tobacco metallothionein-like gene during copper stress. Physiol Plant 123(3):262–271
Vanneste S, Friml J (2009) Auxin: a trigger for change in plant development. Cell 136(6):1005–1016
Vitti A, Nuzzaci M, Scopa A, Tataranni G, Remans T, Vangronsveld J, Sofo A (2013) Auxin and cytokinin metabolism and root morphological modifications in Arabidopsis thaliana seedlings infected with Cucumber mosaic virus (CMV) or exposed to cadmium. Int J Mol Sci 14(4):6889–6902
Wang B, Chu J, Yu T, Xu Q, Sun X, Yuan J et al (2015) Tryptophan-independent auxin biosynthesis contributes to early embryogenesis in Arabidopsis. Proc Natl Acad Sci 112(15):4821–4826
Went F (1926) On growth-accelerating substances in the coleoptile of Avena sativa. In: Proc Kon Akad Wetensch Amsterdam, vol 30. North Holland Publishing Company, Amsterdam, pp 10–19
Went FW (1928) Wuchsstoff und wachstum. Recueil des travaux botaniquesnéerlandais 25(1/2):1–116
Whippo CW, Hangarter RP (2006) Phototropism: bending towards enlightenment. Plant Cell 18(5):1110–1119
Wittenmayer L, Merbach W (2005) Plant responses to drought and phosphorus deficiency: contribution of phytohormones in root-related processes. J Plant Nutr Soil Sci 168(4):531–540
Won C, Shen X, Mashiguchi K, Zheng Z, Dai X, Cheng Y et al (2011) Conversion of tryptophan to indole-3-acetic acid by TRYPTOPHAN AMINOTRANSFERASES OF ARABIDOPSIS and YUCCAs in Arabidopsis. Proc Natl Acad Sci 108(45):18518–18523
Woodward AW, Bartel B (2005) Auxin: regulation, action, and interaction. Ann Bot 95(5):707–735
Wyatt SE, Rashotte AM, Shipp MJ, Robertson D, Muday GK (2002) Mutations in the gravity persistence signal loci in Arabidopsis disrupt the perception and/or signal transduction of gravitropic stimuli. Plant Physiol 130(3):1426–1435
Yamada M, Greenham K, Prigge MJ, Jensen PJ, Estelle M (2009) The TRANSPORT INHIBITOR RESPONSE2 gene is required for auxin synthesis and diverse aspects of plant development. Plant Physiol 151(1):168–179

Yamamoto Y, Kamiya N, Morinaka Y, Matsuoka M, Sazuka T (2007) Auxin biosynthesis by the YUCCA genes in rice. Plant Physiol 143(3):1362–1371

Yeh CM, Chien PS, Huang HJ (2007) Distinct signalling pathways for induction of MAP kinase activities by cadmium and copper in rice roots. J Exp Bot 58(3):659–671

Yoshikawa T, Ito M, Sumikura T, Nakayama A, Nishimura T, Kitano H et al (2014) The rice FISH BONE gene encodes a tryptophan aminotransferase, which affects pleiotropic auxin-related processes. Plant J 78(6):927–936

Yue J, Hu X, Huang J (2014) Origin of plant auxin biosynthesis. Trends Plant Sci 19(12):764–770

Zhang R, Wang B, Ouyang J, Li J, Wang Y (2008) Arabidopsis indole synthase, a homolog of tryptophan synthase alpha, is an enzyme involved in the trp-independent indole-containing metabolite biosynthesis. J Integr Plant Biol 50(9):1070–1077

Zhao Y (2010) Auxin biosynthesis and its role in plant development. Annu Rev Plant Biol 61:49–64

Zhao Y (2014) Auxin biosynthesis. Arabidopsis book 12:e0173

Zhao Y, Hull AK, Gupta NR, Goss KA, Alonso J, Ecker JR et al (2002) Trp-dependent auxin biosynthesis in Arabidopsis: involvement of cytochrome P450s CYP79B2 and CYP79B3. Genes Dev 16(23):3100–3112

Zhao FY, Han MM, Zhang SY, Wang K, Zhang CR, Liu T, Liu W (2012) Hydrogen peroxide-mediated growth of the root system occurs via auxin signaling modification and variations in the expression of cell-cycle genes in rice seedlings exposed to cadmium stress. J Integr Plant Biol 54(12):991–1006

Zhou J, Cheng K, Huang G, Chen G, Zhou S, Huang Y et al (2020) Effects of exogenous 3-indoleacetic acid and cadmium stress on the physiological and biochemical characteristics of Cinnamomum camphora. Ecotoxicol Environ Saf 191:109998

Zhu XF, Wang ZW, Dong F, Lei GJ, Shi YZ, Li GX, Zheng SJ (2013) Exogenous auxin alleviates cadmium toxicity in Arabidopsis thaliana by stimulating synthesis of hemicellulose 1 and increasing the cadmium fixation capacity of root cell walls. J Hazard Mater 263:398–403

Chapter 4
ABA-Mediated Regulation of Plant Biology Under Heavy Metal Stress: Physiological and Molecular Aspects

Manik Devgan, Arshdeep Kaur, Mohar Singh, Shivam Sharma, Anuj Choudhary, Yashika Dhingra, Rajeev Rathore, Sahil Mehta, Kaveri Umesh Kadam, Bindu Yadav, and Om Prakash Narayan

Abstract Anthropogenic activities have historically allowed heavy metals like Fe, Mn, Cu, Ni, Co, Cd, Zn, Hg, and As to accumulate in the soil. Even while a number of these metals, such as Fe, Mn, Cu, Ni, and Zn, are necessary for a healthy plant's mechanism, excessive exposure to them leads to symptoms including stunted development, chlorosis, browning of the roots, decline, and death. Abscisic acid (ABA), a plant hormone, is engaged in a variety of physical, physiological, and biological processes, including seed dormancy induction, stomatal regulation, and response to

M. Devgan · A. Kaur
Department of Plant Breeding and Genetics, Punjab Agricultural University, Ludhiana, Punjab, India

M. Singh
Division of Agrotechnology, CSIR – Institute of Himalayan Bioresource Technology, Palampur, Himachal Pradesh, India

S. Sharma
Department of Vegetable Sciences, CSK Himachal Pradesh Agricultural University, Palampur, Himachal Pradesh, India

A. Choudhary
Department of Biology and Environmental Sciences, CSK Himachal Pradesh Agricultural University, Palampur, Himachal Pradesh, India

Y. Dhingra
Department of Plant Molecular Biology, South Campus, University of Delhi, Delhi, India

R. Rathore · S. Mehta
Department of Botany, Hansraj College, University of Delhi, Delhi, India

K. U. Kadam
School of Engineering Sciences and Technology, Jamia Hamdard, New Delhi, India

B. Yadav
School of Environmental Sciences, Jawaharlal Nehru University, New Delhi, India

O. P. Narayan (✉)
Division of Chemical Biology and Medicinal Chemistry, University of Texas at Austin, Austin, TX, USA

A. Sharma et al. (eds.), *Hormones and Heavy Metals*, Plant Life and Environment Dynamics, https://doi.org/10.1007/978-981-96-9622-2_4

various biotic and abiotic stresses. In the literature, the exceptional effect of exogenous application of ABA in controlling stress effect in plants is available; however, there is a need to shed light on ABA's role in heavy metal metabolism. In this chapter, we are going to discuss how ABA is important in reducing HM stress in plants. Furthermore, to identify the various underlying biochemical factors involved in the interaction between ABA and HM stress.

Keywords Abiotic stress · Abscisic acid · Heavy metals · Plant physiology · Stress physiology

4.1 Introduction

Multiple heavy metals are being released into the soil nowadays as a result of major industrialization, which has a very severe effect on agricultural yield (Shahid et al. 2015). According to Hassan et al. (2017), damage to soil texture, such as pH fluctuations, the presence of various elements, and the accumulation of heavy metals, can directly or indirectly block several physiological and molecular processes in plants, which can impede plant development. Plants that are exposed to heavy metal toxicity have disruptions in their cellular ionic equilibrium, metabolic balance, cell membrane integrity, protein and enzyme activity, and other related processes that severely impact plant development, physiological processes, and agricultural productivity. Under heavy metal stress, plants naturally develop defense systems that can only mitigate its negative effects (Kumar et al. 2022).

But unlike any other species on earth, plants evolved their safeguards against the hazardous effects of heavy metals. To regulate the uptake, accumulation, trafficking, and detoxification of metals, plants have evolved a complex network of incredibly effective homeostatic processes. Constantly, new elements of this network have been discovered, such as chaperons, vacuolar compartmentalization, and cytosolic chelates (Yang and Chu 2011). Abscisic acid (ABA), a significant phytohormone, is one of the key players in regulating the metal imbalance in plants and minimizing its detrimental consequences (Vishwakarma et al. 2017). Through this chapter, we will attempt to defoliate the complex activities that are physiologically and molecularly controlled by ABA in regulating heavy metal stress in plants.

4.2 Heavy Metal's Effects on Plants

These harmful substances create morphological defects and metabolic problems in plants that lower yield (Amari et al. 2017). The redox equilibrium of the cells is disrupted by the production of reactive oxygen species (ROS), including superoxide anion radicals (O_2^-), H_2O_2, and hydroxyl radicals (OH–). An increase in the deadly chemical methylglyoxal has also been connected to heavy metal stress. An important contributor to the toxicity of heavy metals in plants is recognized to be this redox state imbalance (Ibrahim et al. 2015). Particularly in cellular membranes, these highly reactive species induce lipid peroxidation, which results in membrane leakage (Fig. 4.1). In addition, they cause oxidative stress by cleaving DNA strands and damaging biomolecules (Ahmad et al. 2012) (Table 4.1).

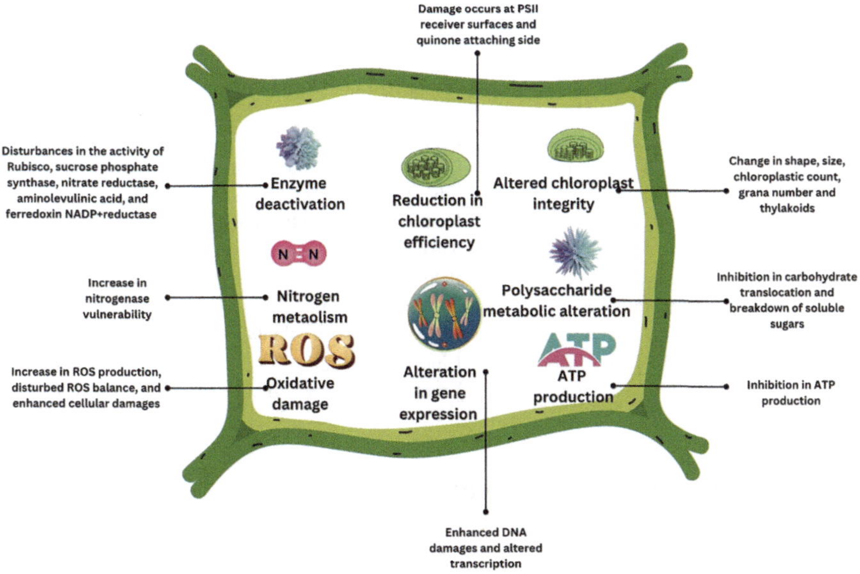

Fig. 4.1 The impact of heavy metals at the cellular level

Table 4.1 The ill effects of various heavy metals in different plants

Heavy metal	Species reported	Region affected	Effect	References
Cd (Cadmium)	*Glycine max* *Silene cucubalus*	Shoot and Roots	Inhibition of root Fe(III) reductase, Browning of root tips, N fixation	Guo et al. (2008)
Hg (Mercury)	*Vigna radiata* *Zea mays*	Primary Roots	Leaf stomata closure Inhibited Mitochondria activity	Azevedo and Rodriguez (2012)
Fe (Iron)	Tobacco Canola Soybean Rice	Roots and leaves	Excess free radical production leading to DNA damage	Mukti (2014)
As (Arsenic)	Rice Lettuce Spinach Carrot	Leaves	Interrupting intra-chromosomal homologous recombination Inhibiting respiration	Kumar et al. (2015)

(continued)

Table 4.1 (continued)

Heavy metal	Species reported	Region affected	Effect	References
Pb (Lead)	*Vicia faba*	Seeds and Roots	Reacting with the phosphate group of ADP and ATPs Seedling germination stops	Kumar et al. (2017)
Mn (Manganese)	Rice *Arabidopsis* Caribbean stylo	Shoots and Roots	Chlorosis and necrosis Photosynthesis disruption	Li et al. (2019)
Ni (Nickel)	*Nicotiana tabacum* *Brassica juncea*	Roots and Seeds	Biomass accumulation reduced Seed germination reduced	Hassan et al. (2019)
Zn (Zinc)	*Phaseolus vulgaris* *Brassica juncea*	Young leaves	Chlorosis Increased accumulation of other toxic metals Displacement of Mg^{2+} ions in RuBisCO	Ghori et al. (2019), Nagajyoti et al. (2010)
Co (Cobalt)	*Avena sativa* *Oryza sativa*	Leaves	Pale colored leaves Discolored viens Iron deficiency in plants	Hu et al. (2021)
Cu (Copper)	*Arabidopsis*	Roots	Lipid peroxidation Production of free radicals	José Rodrigues Cruz et al. (2022)

4.3 Phytohormones and Heavy Metal Stress

Recent research has also demonstrated the usefulness of biochar, compost, and nanoparticles in improving plant growth and mitigating the negative impacts of heavy metal toxicity (Ur Rehman et al. 2021). Still, a workable, affordable, practical, and eco-friendly technique is lacking that could address the lethality of heavy metals and their impact on plants. Phytohormones are another ecologically safe tactic that has gained popularity recently for boosting plant resistance to heavy metal stress (Sytar et al. 2019).

Interestingly, exogenous administration of phytohormones has recently been shown to mitigate the deleterious effects of heavy metal stress. Signaling chemicals called phytohormones help plants grow and develop while lessening their toxicity to heavy metals. Auxins, gibberellins, cytokinins, jasmonates, polyamines, brassinosteroids, salicylic acid, and ethylene are some of these signaling chemicals. All plants need precise locations and timings of signaling molecules for optimal growth and development (Smith et al. 2017). In addition to numerous defense systems, external exposure to heavy metals causes several biochemical responses in plants that alter signaling pathways and cause the creation of stress-mediating phytohormones. These simulated phytohormones prevented heavy metals from entering and building up in plants while allowing plants to tolerate them. This allowed heavy metals to detoxify and, in turn, prevented heavy metals from contaminating the food chain (Saini et al. 2021).

Exogenous treatment options and alterations to the endogenous status of phytohormones via the control of their signaling/biosynthesis-related genes are effective strategies for protecting plants against the harmful effects of heavy metals. Exogenous phytohormone treatment in plants has been found to have several advantageous impacts. For example, auxins, cytokinins, and gibberellins can positively affect the regulation of the ascorbate-glutathione cycle, transpiration rate, cell division, and nitrogen metabolism and absorption activities, all of which promote plant development (Sytar et al. 2019). Salicylic acid, ethylene, and brassinosteroids have all been demonstrated to enhance photosynthesis, minimize ROS and lipid peroxidation levels, and enhance the antioxidant system when administered exogenously to plants that are impacted by heavy metals. Upon exogenous administration, crosstalk between phytohormones is activated (Bücker-Neto et al. 2017). However, a thorough understanding of their potential routes is required to get the most out of phytohormone-mediated subcellular mechanisms to attenuate heavy metal toxicity (Ur Rahman et al. 2023). Abscisic acid represents a major phytohormone that poses significant endogenous as well as exogenous effects in regulating heavy metal stress in plants. The potential and detailed processes by which ABA occurs in plants under this abiotic stress are discussed in the next paragraphs.

4.4 ABA: "The ACE"

Biochemically characterized, ABA is a minuscule sesquiterpene molecule (15-carbon terpenoid) that was first identified as a cotton constituent that inhibited its growth. All cells containing chloroplasts or amyloplasts synthesize ABA, which is present in the most important organs and tissues (Taiz et al. 2015). Abscisic acid is produced by a multitude of genes, beginning in the chloroplast and ending in the cytoplasm. ABA production in Arabidopsis thaliana L. is controlled by AtNCED2, 3, 5, 6, and 9—a small family of NCED genes. These genes are found in plastids (Kumar et al. 2022). These genes are affected by an array of factors, thereby influencing the indigenous synthesis of ABA in plants. Abiotic stressors that plants frequently experience include high salt concentrations (salinity), low temperatures (chilling or freezing), normal temperatures (warmth), and water scarcity (drought or dehydration). According to Fujita et al. (2011), plants use ABA to simulate the impacts of stressful events and have the ability to continually modify their ABA levels in response to shifting physiological and environmental conditions. The extent to which plant ABA biosynthesis is impacted by heavy metal stress varies depending on the kind, species, and intensity of the stress. Abscisic acid synthesis is inversely correlated with heavy metal stress, increasing when levels are lower and decreasing when levels are higher (Fig. 4.1). Different heavy metals with various molecular weights have varying effects on ABA production and are simultaneously modulated by ABA.

It has been well established that abscisic acid signaling has a role in the resistance of rice species to cadmium (Cd, divalent form), a heavy metal with the highest toxicity for plants (Hsu and Kao 2008), particularly under high-temperature conditions (30–35 °C). Additionally, *Boehmeria nivea* L. roots showed a greater rise in ABA content under Cd stress than older leaves, with newer leaves showing the lowest levels. After prolonged exposure to high Cd concentrations, endogenous ABA production was reduced (Chen et al. 2021). Another heavy metal, copper (Cu^{2+}), has demonstrated substantial coordination with ABA in plants. In one study, abscisic acid up-regulated a gene called GPX-6 (which codes for isoforms of the antioxidant enzyme glutathione peroxidase found in the cytosol and mitochondria) as a result of copper toxicity. This gene, which is connected to redox, serves two functions in plants: (i) relaying signals in guard cells and (ii) controlling hydrogen peroxide (Vishwakarma et al. 2017). Similarly, it has been discovered that ABA correlates with arsenic toxicity in several species. An ABA-controlled gene called WRKY6 has been discovered to modify As (V) (arsenic) uptake in plant roots in Arabidopsis species. According to Sun et al. (2019), this gene is a positive regulator whose overexpression or inhibition changes the concentration of arsenic in plants. As a result of ABA-mediated Ni tolerance, Parwez et al. (2023) reported decreased Ni accumulation and a commensurate decrease in reactive oxygen species (ROS), such as superoxide and hydrogen peroxide radicals. In addition, proline, NO (nitric oxide), and enhanced ROS-detoxification enzyme activity were among the antioxidant activities that resulted from the use of ABA under Ni stress conditions.

Plants with increased endogenous ABA concentrations include *Helianthus annuus L.*, *Lactuca sativa*, *Populus canescens*, *Triticum aestivum*, *Oryza sativa*, *Cucumis sativus*, and *Solanum tuberosum*. This suggests that ABA is important in mitigating the adverse consequences of heavy metal stress (Tang et al. 2020). Furthermore, elevated ABA levels in plants can initiate signaling cascades downstream of phytohormones such as auxin (IAA) and salicylic acid (SA), potentially mitigating the toxicity of heavy metals. In research by Atici et al., it was also shown that abscisic acid moderated the toxicity effects of various heavy metals, including zinc (Zn), lead (Pb), etc. (2005).

Endogenous ABA levels have reportedly increased concurrently with the inhibitory effects of heavy metals in plant tissues, indicating that this phytohormone may be mediating some of the phytotoxicity caused by heavy metals such as lead and zinc. Abscisic acid must be used deliberately and judiciously in plants to manage heavy metal stressors, which necessitates a thorough understanding of the signaling pathways triggered by the former. The following paragraphs will examine two potential strategies that could be utilized to successfully develop heavy metal tolerance in plants through abscisic acid.

4.5 Physiological Approaches to Improve Heavy Metal Tolerance Under Strict Regulation of ABA

The regulation of heavy metal stress in plants by abscisic acid involves several physiological attributes. Scientists from all around the world have conducted extensive research on various methods that demonstrate the useful role of the ABA molecule and the signaling cascade it triggers when a plant is exposed to heavy metal stress, some of them are mentioned as follows:

4.5.1 Exogenous Application of ABA at Low Concentrations

Exogenous ABA treatment can increase plants' resistance to stress caused by toxic metals. Tang et al. (2020) observed that foliar application of abscisic acid at $<20\ \mu M$ to *Lactuca sativa* mitigated cadmium toxicity in the plant in addition to fostering its development, chlorophyll content, photosynthetic efficiency, catalase activity, and proline and protein levels. Similarly, 5–10 μM application of ABA to *Malus hupehensis* and *Vigna radiata* also increased their tolerance to cadmium toxicity (Deng et al. 2021; Leng et al. 2021) (Table 4.2).

4.5.2 Phytoremediation via Abscisic Acid-Catabolizing Bacteria

Phytostabilization, phytoextraction, and phytovolatilization are examples of effective phytoremediation mechanisms exhibited by several bacterial species. In addition to possessing robust enzymatic and non-enzymatic defense mechanisms that counteract the oxidative damage caused by an excess of reactive oxygen species (ROS), these bacterial species also have physiological processes that aid in the absorption, translocation, and accumulation of toxic heavy metals (HMs) into low-activity cell organelles (Bortoloti and Baron 2022). Recent studies in Arabidopsis have demonstrated the effect of ABA-generating bacteria in increasing endogenous abscisic acid concentration and therefore decreasing the concentration of heavy metals such as cadmium in plants (Lu et al. 2019). Additionally, an earlier study found that the inoculation of ABA-producing bacteria could significantly decrease the expression of Fe-regulated Transporter 1 (IRT1, a metal transporter with a broad substrate range) in plants stressed by Cd, which mainly results in the inhibition of cadmium uptake by the plant roots (Xu et al. 2018).

Table 4.2 The effects of ABA in reducing heavy metal stress in plants through exogenous application

Heavy metal	Concentration	Species	Improvement	References
Cd	10 ppm	*Oryza sativa*	Reduced transpiration rate	Hsu and Kao (2003)
Cd	5 μmol/L	*Brassica campestris ssp. chinensis*	More shoot growth	Qian et al. (2008)
Pb	5–10 mg/L	*Atractylodes macrocephala*	Enhanced plant growth, antioxidant enzyme activity, soluble sugars	Wang et al. (2013)
Hg	10 μmol/L	*Triticum aestivum*	More shoot growth	Kang et al. (2015)
Cd	0.1–0.5 μmol/L	*Arabidopsis*	Increased iron uptake	Fan et al. (2014)
Zn	10 μM	*Vitis vinifera*	Reducing zinc accumulation in roots	Song et al. (2019)
Zn	10 μmol/L	*Triticum aestivum*	Improved root and shoot growth	Vasyuk et al. (2019)
Cu	100 μmol/soil sample	*Artemisia annua*	Increase antioxidant activity, area and density of glandular trichomes, security of plant cells	Zehra et al. (2020)
Cd, Zn	20–60 μmol/L	*Sedum alfredii*	More leaf chlorophyll content, antioxidant activity	Cheng et al. (2022)
Se	5–10 μmol/L	*Perilla frutescens*	Increased shoot biomass, increased content of photosynthetic pigments, peroxidase activity, superoxide dismutase activity	Liao (2022)
Cd, Cr, Hg, Sn	40 μmol/L	*Fragaria* x *annasa*	More leaf chlorophyll content, antioxidant activity	Kocaman (2023)

The ability of different plant growth-promoting bacteria (PGPB) to catabolize ABA over a prolonged period is worth noting and is a promising possibility (Belimov Andrey et al. 2014). Most rhizospheric bacteria have the potential to alter the bioavailability of heavy metals (HMs) in soil by the excretion of various substances such as organic acids, siderophores, and biosurfactants. Furthermore, research found that the same bacteria that promote plant development also function as a stimulant for the roots of plants to absorb heavy metals (Lu et al. 2019). In several plant species, including vetiver, brassica, ryegrass, Solanum nigrum, and Sedum alfredo, Rhodococcus qingshengii has been shown to have an active role in promoting the absorption of heavy metals by plant roots, including cadmium, zinc, and nickel (Wang et al. 2022).

4.5.3 Decreasing the Accumulation of Heavy Metals in Shoots Through Synergy Between Arbuscular Mycorrhizal and ABA

A newly discovered synergy between arbuscular mycorrhizal fungi and ABA has been shown to reduce the accumulation of heavy metals in plant shoots. Instead, these metals are sequestered within the arbuscular mycorrhizal structures, thereby limiting their translocation to the host plant (Lou et al. 2021). Riaz et al. (2020) reported high content of heavy metals, such as zinc/cadmium, to be accumulated in fungal hyphae of arbuscular mycorrhiza. Therefore, host plants that have been inoculated with AM fungus and exogenous ABA at the same time have an increased probability of mitigating the effects of heavy metal stress.

4.6 Molecular Approaches to Improve Heavy Metal Tolerance Under Strict Regulation of ABA

Numerous plants exposed to heavy metals (HMs) have higher endogenous ABA concentrations, including sunflower, lettuce, poplar, wheat, rice, cucumber, and potato. Certain findings imply that specific genes play a critical role in elevating endogenous abscisic acid expression and accordingly minimizing the toxicities brought on by heavy metals (Shi et al. 2019). A crucial enzyme in the synthesis of ABA, 9-cis-epoxy carotenoid dioxygenase (NCED), modulates a plant's reaction to multiple abiotic stresses (Nambara and Marion-Poll 2005). In Arabidopsis thaliana, the gene encoding NCED is part of a small multigene family that includes five members (NCED2, 3, 5, 6, and 9); NCED3 is primarily in charge of ABA buildup in response to drought stress in *Arabidopsis* (Gavassi et al. 2021). Furthermore, up-regulation of the stomata ClNCED3, ClNCED1, and ClNCED5 resulted in their closing concurrent with a rise in ABA concentration, demonstrating the involvement of ABA in modulating plant hydration in the presence of Al. In addition to these, MhNCED3, when expressed ectopically in *Arabidopsis*, increased its antioxidant capacity and decreased the amount of apoptosis brought on by cadmium stress in *Malushu pehensis* (Zhang et al. 2019). Numerous studies have been carried out in various species to examine the function of ABA in controlling genes and their expression in the presence of heavy metal stress. According to Huang et al. (2012), arsenic-induced oxidative homeostasis in As-stressed rice seedlings was controlled transcriptomically by up-regulating the genes OsNCED2, 3 (9-cis-epoxy carotenoid dioxygenase 2, 3), and OsABA4 (abscisic acid-deficient 4).

The vital role of ABA in the reduction of Cd toxicity was highlighted by the transcriptomics-based study in *Oryza sativa*, which revealed an increase in the gene (OsNCED4) responsible for ABA biosynthesis (Tan et al. 2017). An investigation into Cd stress in many mutants of Arabidopsis sp. (ABA-deficient mutants, aba-1, 3, 4 and ABA-insensitive mutants, abi2–1, abi3–1) was carried out using a mutant-based approach (Sharma and Kumar 2002), and the results indicated that these mutants were more susceptible to Cd toxicity than the wild-type. It was found in *Sedum alfredi* that ABA protects plants from Cd stress, foliar application of ABA combined with Cd enhanced the endogenous status of ABA by upregulating various enzymes like NCED, AAO, and ZEP that synthesize ABA and by downregulating the ABA inhibitors like ABA 8-hydroxylase (Tao et al. 2021). As HMA4 (HM ATPase 4) is accountable for the translocation of Cd from the root to the shoot through apoplast and its deposition in shoot tissues, it was shown that increasing the expression of this enzyme alleviated the deleterious effects of oxidative damage induced by Cd (Tao et al. 2021). The regulation of stomatal conductance, water status, and stomatal functioning are among the advantageous effects of foliar application of ABA during heavy metal stress on plants, as demonstrated by studies conducted by Bücker-Neto et al. (2017), Cheng et al. (2022), and Zehra et al. (2020).

Under Cd and Zn stress, ABA treatment in grapevine and lettuce plants dramatically reduced Zn and Cd bioaccumulation and simultaneously increased the overall biomass of root and shoot (Tao et al. 2021). The zinc/iron transporter protein (ZIP) family, which is critical for cross-cellular HM translocation and deposition during Zn cytotoxicity, was also expressed more when exposed to ABA. Many HM decontaminated genes, including YSL (yellow stripe-like), NRAMP3 (natural resistance-associated macrophage), and PCR2 (plant-cadmium resistance 2) in grapevine plants, were upregulated by the application of ABA and Zn (Song et al. 2019). By modifying genes linked to detoxification and zinc absorption, ABA plays a significant role in lowering zinc toxicity. In the lab, applying ABA to the leaves raised the expression of critical genes involved in lead (Pb) absorption, transit, and elimination (Shi et al. 2019). Natural resistance-associated macrophage protein (NRAMP1.4), ATP-binding cassette transporter G40 (ABCG40), phytochelatin synthase (PCS1.1), ferric reductase deficient protein (FRD3.1), and ATP-binding cassette transporter C 1.1 (ABCC1.1) have all been associated with these genes in *Populus canescens* (Fig. 4.2). Additionally, the ABA treatment decreased the overproduction of reactive oxygen species (ROS) by increasing the activity of antioxidant enzymes like ascorbate peroxidase (APX), POD (peroxidases), CAT (catalase), and SOD (superoxide dismutase). It also increased photosynthetic rate and root biomass, which decreased Pb toxicity (Shi et al. 2019). Many plants exposed to heavy metals (HMs), including *Lactuca sativa, Helianthus annuus L., Populus canescens,* and *Oryza sativa*, have demonstrated a significant increase in ABA concentration under heavy metal stress, suggesting that ABA is important in reducing the toxic effects of HMs (Kim et al. 2014; Lou et al. 2021; Shi et al. 2019; Tang et al. 2020).

It was recently demonstrated that treating *Solanum lycopersicum* with ABA and 6-benzyl amino purine (BAP) simultaneously reduced the adverse effects of cobalt (Co) stress (Kamran et al. 2021). The findings demonstrated that administering ABA to plants increased their growth and chlorophyll content while lowering CO deposition. Following the administration of ABA and BAP (6-benzyl amino purine) together under co-stress conditions, levels of proline, H_2O_2, malate dehydrogenase, and antioxidant enzymes such as SOD (superoxide dismutase), POD (peroxidases), CAT (catalase), and APX (ascorbate peroxidase) increased.

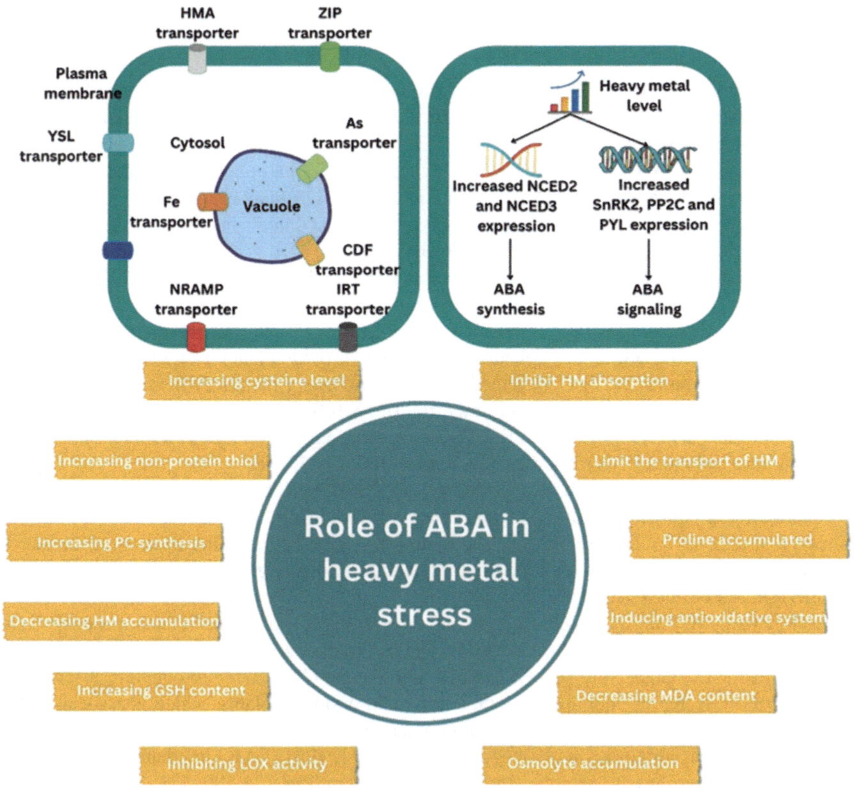

Fig. 4.2 Illustration showing cellular transport, impact of HM on ABA, and role of ABA in heavy metal tolerance

4.7 Conclusion

As a consequence of mankind's continued exploitation of the earth, significant unnatural environmental challenges have arisen, significantly upsetting our ecological equilibrium. This is a significant issue for all terrestrial ecosystems. Plants under heavy metal stress drastically reduce their yield and become unfit for human consumption. Plants still struggle for existence in the face of this abiotic stress despite a variety of adaptations to deal with the metal-induced toxicity, such as conditioning or in situ fixation, excavation to remediate HMs, adsorption, ion exchange, filtration, electrochemical reactions, chemical precipitation, solvent extraction, coagulation/flocculation, bioremediation, soil cleaning, and conditioning. But in the end, we need tolerant genotypes that can withstand adverse conditions. The discussion on the physiological and molecular insight in this chapter will aid in providing an angle regarding the importance of ABA in eradicating the threat of HM stress. Additionally, the pathways triggered and downstream components need to be explored more toward the goal of sustainability.

References

Ahmad P, Ozturk M, Gucel S (2012) Oxidative damage and antioxidants induced by heavy metal stress in two cultivars of mustard (*Brassica juncea* L.) plants. Fresenius Environ Bull 21(10):2953–2961

Amari T, Ghnaya T, Abdelly C (2017) Nickel, cadmium and lead phytotoxicity and potential of halophytic plants in heavy metal extraction. S Afr J Bot 111:99–110

Atici Ö, Ağar G, Battal P (2005) Changes in phytohormone contents in chickpea seeds germinating under lead or zinc stress. Biol Plant 49:215–222

Azevedo R, Rodriguez E (2012) Phytotoxicity of mercury in plants: a review. J Bot 2012(1):848614

Belimov Andrey A, Dodd Ian C, Safronova Vera I, Dumova Valentina A, Shaposhnikov Alexander I, Ladatko Alexander G, Davies William J (2014) Abscisic acid metabolizing rhizobacteria decrease ABA concentrations in planta and alter plant growth. Plant Physiol Biochem 74:84–91

Bortoloti GA, Baron D (2022) Phytoremediation of toxic heavy metals by Brassica plants: a biochemical and physiological approach. Environ Adv 8:100204

Bücker-Neto L, Paiva ALS, Machado RD, Arenhart RA, Margis PM (2017) Interactions between plant hormones and HMs responses. Genet Mol Biol 40:373–386

Chen K, Chen P, Qiu X, Chen J, Gao G, Wang X et al (2021) Regulating role of abscisic acid on cadmium enrichment in ramie (*Boehmeria nivea* L.). Sci Rep 11:1–10

Cheng L, Pu L, Li A, Zhu X, Zhao P, Xu X, Lei N, Chen J (2022) Implication of exogenous abscisic acid (ABA) application on phytoremediation: plants grown in co-contaminated soil. Environ Sci Pollut Res 29:8684–8693

Deng B, Zhang W, Yang H (2021) Abscisic acid decreases cell death in Malushupehensis Rehd. under Cd stress by reducing root Cd^{2+} influx and leaf transpiration. J Plant Growth Regul 40:1–8

Fan SK, Fang XZ, Guan MY, Ye YQ, Lin XY, Du ST, Jin CW (2014) Exogenous abscisic acid application decreases cadmium accumulation in Arabidopsis plants, which is associated with the inhibition of IRT1-mediated cadmium uptake. Front Plant Sci 5:721

Fujita Y, Fujita M, Shinozaki K, Yamaguchi-Shinozaki K (2011) ABA-mediated transcriptional regulation in response to osmotic stress in plants. J Plant Res 124:509–525

Gavassi MA, Silva GS, da Silva CMS, Thompson AJ, Kyle M, Paulo M, Rayner O, MarianaFeitosa C, Domingues DS, Habermann G (2021) NCED expression is related to increased ABA biosynthesis and stomatal closure under aluminum stress. Environ Exp Bot 185:104404

Ghori NH, Ghori T, Hayat MQ et al (2019) Heavy metal stress and responses in plants. Int J Environ Sci Technol 16:1807–1828

Guo J, Dai X, Xu W, Ma M (2008) Over-expressing GSHI and AsPCSI simultaneously increases the tolerance and accumulation of cadmium and arsenic in *Arabidopsis thaliana*. Chemosphere 72:1020–1026

Hassan TU, Bano A, Naz I (2017) Alleviation of heavy metals toxicity by the application of plant growth promoting rhizobacteria and effects on wheat grown in saline sodic field. Int J Phytoremediation 19:522–529

Hassan MU, Chattha MU, Khan I, Chattha MB, Aamer M, Nawaz M, Ali A, Khan MAU, Khan TA (2019) Nickel toxicity in plants: reasons, toxic effects, tolerance mechanisms, and remediation possibilities-a review. Environ Sci Pollut Res Int 26:12673–12688

Hsu YT, Kao CH (2003) Role of abscisic acid in cadmium tolerance of rice (*Oryza sativa* L.) seedlings. Plant Cell Environ 26:867–874

Hsu YT, Kao CH (2008) Distinct roles of abscisic acid in rice seedlings during cadmium stress at high temperature. Bot Stud 49:335–342

Hu X, Wei X, Ling J, Chen J (2021) Cobalt: an essential micronutrient for plant growth? Front Plant Sci 12:768523

Huang TL, Nguyen QTT, Fu SF, Lin CY, Chen YC, Huang HJ (2012) Transcriptomic changes and signaling pathways induced by arsenic stress in rice roots. Plant Mol Biol 80(6):587–608

Ibrahim MA, Khan PR, Hegazy SS, Hashim EA, Azamal H, Ansari MK et al (2015) Improving the phytoextraction capacity of plants to scavenge heavy-metal infested sites. Environ Rev 23:1–22

José Rodrigues Cruz F, Leone da Cruz Ferreira R, Silva Conceição S, Ugulino Lima E, Ferreira de Oliveira Neto C, Rodrigues Galvão J, de Jesus Matos Viegas I (2022) Copper toxicity in plants: nutritional, physiological, and biochemical aspects. London, UK. IntechOpen

Kamran M, Danish M, Saleem MH, Malik Z, Parveen A, Abbasi GH, Jamil M, Ali S, AfzalS RM (2021) Application of abscisic acid and 6-benzylaminopurine modulated morpho-physiological and antioxidative defense responses of tomato (*Solanum lycopersicum* L.) by minimizing cobalt uptake. Chemosphere 263:128169

Kang G, Li G, Wang L, Wei L, Yang Y, Wang P, Guo T (2015) Hg-responsive proteins identified in wheat seedlings using iTRAQ analysis and the role of ABA in Hg stress. J Proteome Res 14(1):249–267

Kim YH, Khan AL, Kim DH, Lee SY, Kim KM, Waqas M, Jung HY, Shin JH, Kim JG, Lee IJ (2014) Silicon mitigates heavy metal stress by regulating P-type heavy metal ATPases, *Oryza sativa* low silicon genes, and endogenous phytohormones. BMC Plant Biol 14(1):1–13

Kocaman A (2023) Effects of foliar application of abscisic acid on antioxidant content, phytohormones in strawberry shoots, and translocation of various heavy metals. Sci Hortic 314:e111943

Kumar S, Dubey RS, Tripathi RD, Chakrabarty D, Trivedi PK (2015) Omics and biotechnology of arsenic stress and detoxification in plants: current updates and prospective. Environ Int 74:221–230

Kumar B, Smita K, Flores LC (2017) Plant mediated detoxification of mercury and lead. Arab J Chem 10:S2335–S2342

Kumar S, Shah SH, Vimala Y, Jatav HS, Ahmad P, Chen Y, Siddique KHM (2022) Abscisic acid: metabolism, transport, crosstalk with other plant growth regulators, and its role in heavy metal stress mitigation. Front Plant Sci 13:972856

Leng Y, Li Y, Ma YH, He LF, Li SW (2021) Abscisic acid modulates differential physiological and biochemical responses of roots, stems, and leaves in mung bean seedlings to cadmium stress. Environ Sci Pollut Res 28:6030–6043

Li J, Jia Y, Dong R, Huang R, Liu P, Li X, Wang Z, Liu G, Chen Z (2019) Advances in the mechanisms of plant tolerance to manganese toxicity. Int J Mol Sci 20:5096

Liao R (2022) Effects of abscisic acid on growth and selenium uptake in medicinal plant *Perilla frutescens*. PLoS One 17(10):e0275813

Lou X, Zhang X, Zhang Y, Tang M (2021) The synergy of arbuscular mycorrhizal fungi and exogenous abscisic acid benefits Robinia pseudoacacia L. growth through altering the distribution of Zn and endogenous abscisic acid. J Fungi (Basel) 7:671

Lu Q, Weng Y, You Y, Xu Q, Li H, Li Y, Du S (2019) Inoculation with abscisic acid (ABA)-catabolizing bacteria can improve phytoextraction of heavy metal in contaminated soil. Environ Pollut 257:113497

Mukti G (2014) Heavy metal stress in plants: a review. Int J Adv Res 2:1043–1055

Nagajyoti PC, Lee KD, Sreekanth TVM (2010) Heavy metals, occurrence and toxicity for plants: a review. Environ Chem Lett 8:199–216

Nambara E, Marion-Poll A (2005) Abscisic acid biosynthesis and catabolism. Annu Rev Plant Biol 56:165–185

Parwez R, Aftab TM, Khan MA, Naeem N (2023) Exogenous abscisic acid fine-tunes heavy metal accumulation and plant's antioxidant defense mechanism to optimize crop performance and secondary metabolite production in *Trigonella foenumgraecum* L. under nickel stress. Plant Sci 332:111703

Qian HS, Chen YH, Wang GP, Shen ZG (2008) Cadmium accumulation and effect of exogenous abscisic acid on cadmium accumulation in cadmium-treated *Brassica campestris* ssp. *chinensis* L. J Nanjing Agric Uni 31(4):61–65

Riaz M, Kamran M, Fang YZ, Wang QQ, Cao HY, Yang GL, Deng LL, Wang YJ, Zhou YY, Znastopoulos I et al (2020) Arbuscular mycorrhizal fungi-induced mitigation of heavy metal phytotoxicity in metal contaminated soils: a critical review. J Hazard Mater 402:123919
Saini S, Kaur N, Pati PK (2021) Phytohormones: key players in the modulation of heavy metal stress tolerance in plants. Ecotoxicol Environ Saf 223:112578
Shahid M, Khalid S, Abbas G, Shahid N, Nadeem M, Sabir M et al (2015) Heavy metal stress and crop productivity. In: Hakeem KR (ed) Crop production and global environmental issues. Springer International Publishing, Cham, pp 1–25
Sharma SS, Kumar V (2002) Responses of wild type and abscisic acid mutants of Arabidopsis thaliana to cadmium. J Plant Physiol 59(12):1323–1327
Shi WG, Liu W, Yu W, Zhang Y, Ding S, Li H, Mrak T, Kraigher H, Luo ZB (2019) Abscisic acid enhances lead translocation from the roots to the leaves and alleviates its toxicity in Populus× canescens. J Hazard Mater 362:275–285
Smith SM, Li C, Li J (2017) Hormone function in plants. In: Li J, Li C, Smith SM (eds) Hormone metabolism and signaling in plants. Academic, London, pp 1–38
Song C, Yan Y, Rosado A, Zhang Z, Castellarin SD (2019) ABA alleviates uptake and accumulation of zinc in grapevine (*Vitis vinifera* l.) by inducing expression of ZIP and detoxification-related genes. Front Plant Sci 10:872
Sun D, Feng HY, Li XY, Ai H, Sun S, Chen Y et al (2019) Expression of new Pterisvittata phosphate transporter PvPht1;4 reduces arsenic translocation from the roots to shoots in tobacco plants. Environ Sci Technol 54:1045–1053
Sytar O, Kumari P, Yadav S, Brestic M, Rastogi A (2019) Phytohormone priming: regulator for heavy metal stress in plants. J Plant Growth Regul 38:739–752
Taiz L, Zeiger E, Møller IM, Murphy A (2015) Plant physiology and development, 6th edn. Sinauer Associates Incorporated, Sunderland
Tan M, Cheng D, Yang Y, Zhang G, Qin M, Chen J, Chen Y, Jiang M (2017) Co-expression network analysis of the transcriptomes of rice roots exposed to various cadmium stresses reveals universal cadmium-responsive genes. BMC Plant Biol 17(1):1–18
Tang Y, Wang L, Xie Y, Yu X, Li H, Lin L, Liao MA, Wang Z, Sun G, Wang X (2020) Effects of exogenous abscisic acid on the growth and cadmium accumulation of lettuce under cadmium-stress conditions. Int J Environ Anal Chem 100(6):720–731
Tao Q, Jupa R, Dong Q, Yang X, Liu Y, Li B, Yuan S, Yin J, Xu Q, Li T (2021) Abscisic acid-mediated modifications in water transport continuum are involved in cadmium hyperaccumulation in *Sedum alfredii*. Chemosphere 268:129339
Ur Rahman S, Li Y, Hussain S, Hussain B, Khan W, Riaz L, Ashraf MN, Khaliq MA, Du Z, Cheng H (2023) Role of phytohormones in heavy metal tolerance in plants: a review. Eco Indic 146:109844
Ur Rehman MZ, Waqar M, Bashir S, Rizwan M, Ali S, El-Baroudy AAEF, Khalid H, Ayub MA, Usman M, Jahan S (2021) Effect of biochar and compost on cadmium bioavailability and its uptake by wheat–rice cropping system irrigated with untreated sewage water: a field study. Arab J Geosci 14:1–12
Vasyuk VA, Voytenko LV, Shcherbatiuk MM, Kosakivska IV (2019) Effect of exogenous abscisic acid on seed germination and growth of winter wheat seedlings under zinc stress. J Stress Physiol Biochem 15(2):68–78
Vishwakarma K, Upadhyay N, Kumar N, Yadav G, Singh J, Mishra RK, Kumar V, Verma R, Upadhyay RG, Pandey M, Sharma S (2017) Abscisic acid signaling and abiotic stress tolerance in plants: a review on current knowledge and prospects. Front Plant Sci 8:161
Wang J, Chen J, Pan K (2013) Effect of exogenous abscisic acid on the level of antioxidants in Atractylodes macrocephala Koidz under lead stress. Environ Sci Pollut Res 20:1441–1449
Wang Y, Li Z, Wu J, Liu H, Sun X, Liu L, Du S (2022) Abscisic acid-catabolizing bacteria: a useful tool for enhancing phytoremediation. Sci Total Environ 812:151474

Xu QR, Pan W, Zhang RR, Lu Q, Xue WL, Wu CN, Song BX, Du ST (2018) Inoculation with *Bacillus subtilis* and *Azospirillum brasilense* produces abscisic acid that reduces Irt1-mediated cadmium uptake of roots. J Agric Food Chem 66:5229–5236

Yang Z, Chu C (2011) Towards understanding plant response to heavy metal stress. InTech. https://doi.org/10.5772/24204

Zehra A, Choudhary S, Wani KI, Naeem M, Khan MMA, Aftab T (2020) Exogenous abscisic acid mediates ROS homeostasis and maintains glandular trichome to enhance artemisinin biosynthesis in *Artemisia annua* under copper toxicity. Plant Physiol Biochem 156:125–134

Zhang W, Wang Z, Song J, Yue S, Yang H (2019) Cd^{2+} uptake inhibited by MhNCED3 from *Malushu pehensis* alleviates Cd-induced cell death. Environ Exp Bot 166:103802

Chapter 5
Brassinosteroids-Mediated Regulation of Plant Biology Under Heavy Metal Stress: Physiological and Molecular Aspects

Sadaf Shehzad, Lalit Kumar, Romila Rawat Bisht, Sheen Tickoo, Sahil Mehta, Kaveri Umesh Kadam, Bindu Yadav, and Om Prakash Narayan

Abstract Brassinosteroids (BRs) are a universal group of plant hormones known to regulate the growth and development of plants and shield them from both biotic and abiotic stresses. The heavy metal pollution caused by ever-increasing industrialization is one of the major causes of stress in plants. The uptake of heavy metals by some plants beyond critical levels leads to oxidative stress. It is in this scenario where BRs come into the picture, safeguarding the plant by inhibiting lipid degradation and promoting the synthesis of phytochelates. Significant studies on the Arabidopsis plant have led to the divulgence of information on the regulatory mechanism of BR biosynthesis, metabolism, and signaling. In this review, we have provided updates on molecular and physiological aspects of BR signaling under stress induced by heavy metals.

Keywords Plant · Hormones · Brassinosteroids · Heavy metal stress · Signaling

Authors Sadaf Shehzad and Lalit Kumar have equally contributed to this chapter.

S. Shehzad
University of Texas Rio Grande Valley, Edinburg, TX, USA

L. Kumar
Department of Botany, Dharam Samaj College, Aligarh, Uttar Pradesh, India

R. R. Bisht · S. Tickoo · S. Mehta
Department of Botany, Hansraj College, University of Delhi, Delhi, India

K. U. Kadam
School of Engineering Sciences and Technology, Jamia Hamdard, New Delhi, India

B. Yadav
School of Environmental Sciences, Jawaharlal Nehru University, New Delhi, India

O. P. Narayan (✉)
Division of Chemical Biology and Medicinal Chemistry, University of Texas at Austin, Austin, TX, USA

A. Sharma et al. (eds.), *Hormones and Heavy Metals*, Plant Life and Environment Dynamics, https://doi.org/10.1007/978-981-96-9622-2_5

5.1 Introduction

To adapt to the ever-changing environment, plants synthesize primary and secondary metabolites. While primary metabolites play a major role in regulating growth, development, and reproduction, secondary metabolites regulate biotic-abiotic stress, defense mechanism, and adaptation to environmental changes (Samuni-Blank et al. 2012; Stamp 2003). Primary metabolites in basic growth and development include plant hormones like gibberellins, brassinosteroids, abscisic acid, and strigolactone.

Brassinosteroids (BRs) were discovered from *Brassica napus* pollen based on their ability to promote growth (Mitchell et al. 1970). They belong to a group of plant steroidal hormones and are known to regulate processes as diverse as xylem differentiation, cell elongation, and fruit ripening. They have been detected in a large number of species, varied among different taxonomic groups (Fig. 5.1) and thus are considered to be ubiquitous in nature (Bajguz and Tretyn 2003). The advancement in technology in the last decades has led to a significant increase in our understanding of BR biology, viz., its distribution, biosynthesis, signal transduction, and cross-talk between BR and other hormones (Belkhadir and Chory 2006; Bishop et al. 2006; Fujioka and Yokota 2003; Li and Jin 2007).

5.2 Distribution of BRs

BRs are widely distributed in most plant tissues (Bajguz and Tretyn 2003). Their presence has been reported both in lower and higher plants, viz., angiosperms; and within all plant organs, including roots, stems, leaves, flowers, anthers, pollen, seeds, and grain (Bajguz 2019; Bajguz and Tretyn 2003; Yokota et al. 2017; Zullo and Bajguz 2019) (Fig. 5.1). However, the level of bioactive BRs castasterone (CS) and brassinolide (BL), varies among different parts of the plant (Bajguz and Tretyn 2003; Bancosi et al. 2002; Shimada et al. 2003; Symons and Reid 2004). Reproductive organs such as pollens are known to have the highest level [5–1000 ng/g fresh weight (FW)] of bioactive BRs seeds (0.3–1600 ng/g FW), and fruits (0.2–3.5 ng/g FW) (0.2–3.5 ng/g FW) (Bajguz and Tretyn 2003; Montoya et al. 2005; Nomura et al. 2007; Symons et al. 2006). However, vegetative tissues (shoot) have generally much lower levels of bioactive BRs (0.12–2 ng/g FW) while the roots account for the lowest levels (<0.05 ng/g FW) (Bancosi et al. 2002; Shimada et al. 2003; Symons and Reid 2004). Also, the shoot tip (apex) and the young internodes account for the region with the greatest accumulation within the shoot region, thereby contributing to its role in the promotion of cell division and elongation (Shimada et al. 2003; Symons and Reid 2004). However, this widespread distribution raises the question of whether the BRs are synthesized in a localized and then transported to distant sites of action.

Fig. 5.1 Distribution of BRs among lower and higher plants, and are localized in almost all the plant parts; however, the levels of bioactive BRs vary based on localization

5.3 Biosynthetic Pathway

BRs share a striking resemblance in chemical structure with corticoids, estrogens, androgens, and ecdysteroids (Bajguz 2019; Bajguz and Tretyn 2003; Yokota et al. 2017; Zullo and Bajguz 2019). They are categorized into C27, C28, and C29 based on carbon atom count, with 5α-cholestane, 5α-ergostane, and 5α-stigmastane forming the foundational structure for C27, C28, and C29-BRs, respectively. Differences in these hormones' structures primarily arise from variations in oxygenated functions in the A- and B-rings and the number and positioning of functional groups in the molecule's side chain (Bajguz 2007; Fujioka and Yokota 2003; Piotrowska and Bajguz 2011; Wendeborn et al. 2017; Zullo and Bajguz 2019). Currently, three pathways are recognized for BR biosynthesis, resulting in C27-, C28-, or C29-type BRs as final products. These pathways, initiated either via the mevalonate (MVA) or non-MVA pathway, share common initial steps but diverge to differentiate the BR biosynthesis pathways (cycloartenol- and cycloartenol-dependent).

Isopentenyl pyrophosphate (IPP) biosynthesis occurs in lower plants and most higher plants through two pathways: non-MVA and MVA, respectively. IPP undergoes conversion to squalene via geranyl pyrophosphate and farnesyl pyrophosphate. Squalene epoxidase then oxidizes squalene to squalene-2,3-oxide, which, through the action of cycloartenol synthase, converts to cycloartenol. Cycloartenol serves as a substrate for multistep reactions in BR biosynthesis, leading to the synthesis of C27-, C28-, and C29-BRs.

5.3.1 Biosynthesis of C27-BRs

The conversion of cycloartenol to C4-sterol methyloxidase3 (SMO3) by the sterol side chain reductase 2 (SSR2) marks the first step in the C_{27}-BR biosynthesis pathway. The C4-sterol methyloxidase3 (SMO3) then synthesizes 31-norcycloartanol from cycloartenol which on further biochemical changes synthesizes 31-nor-24(25)-dihydro lanosterol, 4α-methylcholestadienol, 4α-methyl-24(25)-dihydrozymosterol, 4α-methylcholestenol, 5α-cholest-7-en-3β-ol, 7-dehydroCR and finally CR, respectively (Fig. 5.2). The C4-sterol methyloxidase 4 (SMO4) catalyzes the synthesis of 5α-cholest-7-en-3β-ol, while the synthesis of 7-dehydrocholesterol is catalyzed by sterol C5(6) desaturase. Then, the 7-dehydrocholesterol reductase 2 converts 7-dehydrocholesterol to cholesterol (Sonawane et al. 2016).

Also, C_{27}-BRs biosynthesis might occur by the late C6 oxidation pathway. The first step is marked by the 5α-reductase (encoded by the *DET2* gene) catalyzing the conversion of cholestanol (a C_{27}-BR biosynthesis precursor) to cholesterol. In the consecutive steps of the C_{27}-BR biosynthesis pathway, 6-deoxo-28-norcathasterone,

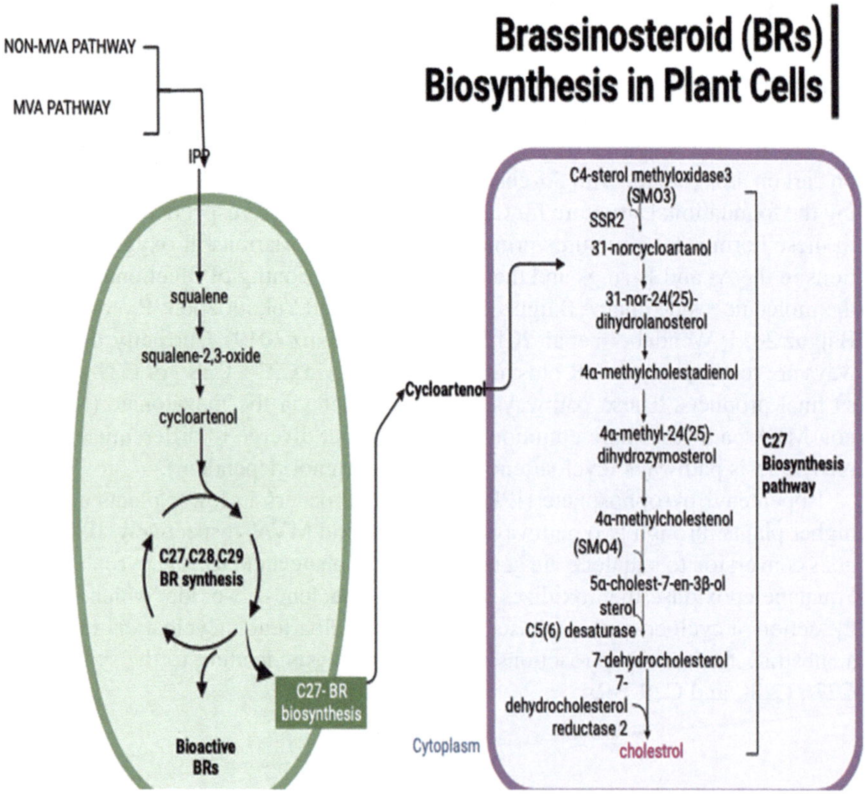

Fig. 5.2 Diagrammatic illustration of C27 Biosynthetic pathway of BRs

6-deoxo-28-norteasterone, 6-deoxo-28-nor-3-dehydroteasterone, 6-deoxo-28-nortyphasterol, and 6-deoxo-28-norcastasterone are synthesized.

Also, the oxidation of cholestanol into 6-oxocholestanol initiates the early C6 oxidation pathway, followed by the synthesis of 28-norcathasterone, 28-norteasterone, 28-nor-3-dehydroteasterone, 28-nortyphasterol, 28-norcastasterone, and 28-norbrassinolide. These compounds also provide evidence regarding the crosstalk between late and early C6 oxidation pathways.

5.3.2 Biosynthesis of C28-BRs

The first characteristic metabolite in the C28-BR biosynthesis is episterol. C-4 sterol methyl oxidase 2 (SMO2) catalyzes the conversion of 24-methylenelophenol to episterol which by the sterol C-5(6) desaturase is converted to 5-dehydroepisterol, which later gets converted to 24-methyleneCR by 7-dehydrocholesterol reductase (Choe 2006; Ohnishi 2018) (Fig. 5.3). Similar to the C27-BR biosynthesis, the later steps of C_{28}-BR biosynthesis are also performed through the late and early C-22 oxidation pathway catalyzed by different enzymes.

5.3.3 Biosynthesis of C29-BRs

In this pathway, sterol methyltransferase 2 (SMT2) converts 24-methylenelophenol into 24-ethylidenelophenol which in subsequent steps is converted into avenasterol by the sterol methyl oxidase 2. Subsequently, a series of reactions produces isofucosterol and β-sitosterol catalyzed by the DWF7, DWF5, and SSR1 (DWF1) enzymes. Precursor of the C_{29}-BRs, β-sitosterol is hydroxylated into 6-deoxo-28-homoTY and oxygenated into 28-homoTY by the CYP724B2 and CYP90B3 C-22 hydroxylase, respectively (Ohnishi et al. 2006).

The cross-linking between C_{27}- and C_{29}-BRs with C_{28}-BRs occurs mainly between the end products of the pathways. The production of CS, an active BR, in plants is carried out by five different pathways via 28-norCS, DS, 6-deoxoCS by 6α-hydroxyCS, 28-homoDS, and 28-homoCS (Fujita et al. 2006; Joo et al. 2015, 2012; Kim et al. 2018).

5.3.4 Inhibitors of BR Biosynthesis

A total of 17 inhibitors, viz., KM-01, brassinozole (Brz), propiconazole, Brz220, Brz2001, yucaizol, YCZ-18, fenarimol, triadimefon, spironolactone, imazalil, 4-MA, DSMEM21, VG106, finastride, brassinopride, and AFA76 are known to

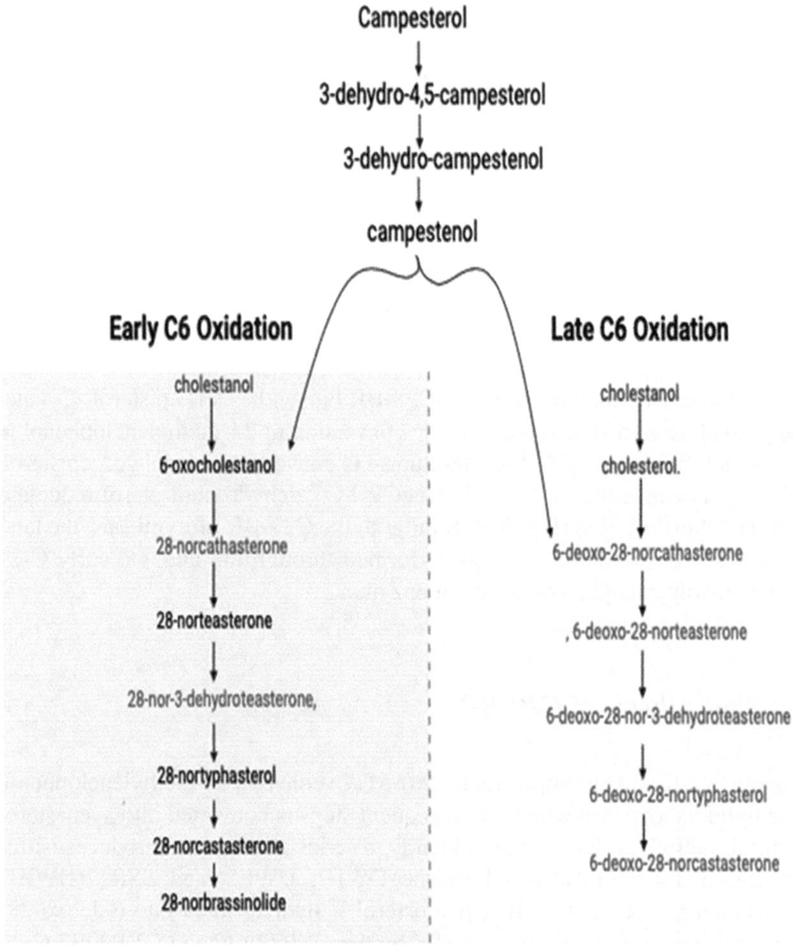

Fig. 5.3 Diagrammatic representation of the synthesis of cholesterol in early and late oxidation pathways and crosstalk among them

date, however, the knowledge of only nine inhibitors regarding their site of action is known (Rozhon et al. 2019).

5.4 Insights into BR Signaling and Heavy Metal Stress

Plant steroidal hormones, known as BRs, play pivotal roles in modulating developmental processes and managing stress. BRI1 and BAK1 (BRI1-associated receptor kinase 1), two cell surface receptor kinases, recognize BRs, initiating a phosphorylation cascade. This cascade involves the phosphorylation of the BSU1 protein and

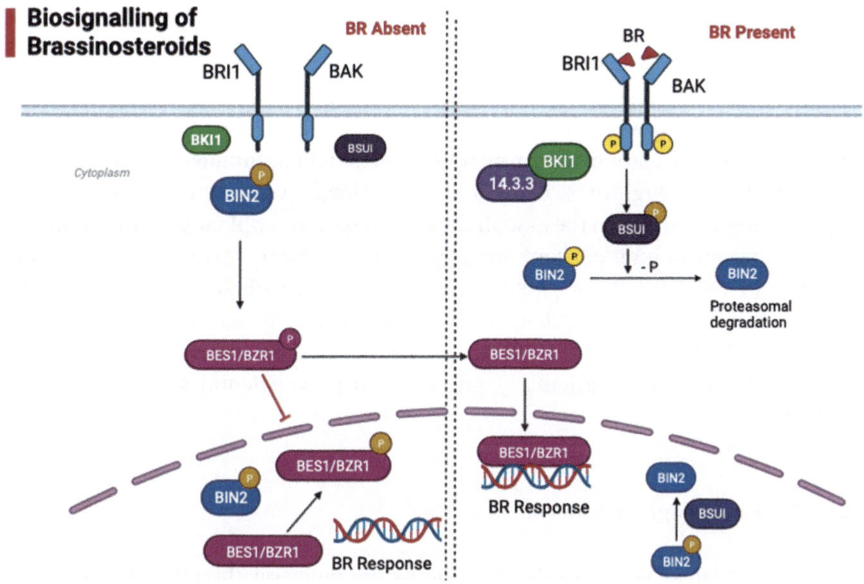

Fig. 5.4 Stress-induced physiological mechanisms in BRs-mediated action in plants

the proteasomal degradation of BIN2 proteins (Chakraborty et al. 2015). Degradation of BIN2 is followed by the entry of BES1/BZR1 in the nucleus and regulation of the target genes' expression (Fig. 5.4). BRs also play a significant role by cross-talking with other phytohormones to combat abiotic and heavy metal stress.

BRI1, an LCC receptor kinase, possesses an extracellular domain where BRs bind and undergo trans-phosphorylation in association with a co-receptor, SERK (somatic embryogenesis receptor kinase). Together, they stimulate a downstream signaling network of kinases and phosphatases. Upon BR binding to BRI1, it triggers the association of BRI1 with BAK1 and signals the disassociation of BRI1 kinase inhibitor 1 (BKI1). In plants like Oryza sativa and Arabidopsis thaliana, BKI1 serves as a membrane-bound negative regulator of BR signals, binding to BRI1 to prevent interference with the BRI1-BAK1 interaction. In *Oryza sativa* and *A. thaliana* plants, BKI1 is a membrane-bound negative regulator of BR signals, which binds to BRI1 thereby preventing the BRI1- BAK1 interference. BRI1 is a multi-function Ser-Thr-Tyr kinase; however, studies report that the phosphorylation of Tyr residue is vital for specific aspects of BR signaling in plants (Oh et al. 2012). BRI1, a homodimer, is known to be localized to the plasma membrane and is ligand-independent. BRI1 is responsible for the stabilization and initiation of BR binding and signaling (Albrecht et al. 2008). BAK1 (also termed SERK3) is known to undergo auto- and cross-phosphorylation (Wang et al. 2008). It also interacts with other LRR RLK, FLS2 known to be involved in plant defense (Bar et al. 2010; Karlova et al. 2006).

5.4.1 Effect of BRs on Heavy Metal Stress

Increased human interventions and the Industrial Revolution have amplified environmental pollution with metals paving their way into our food chains. Plants are known to have the astounding property to uptake and accumulate metals from the environment. Although most of the redox reactions involved in cellular functions require some metals, even the essential metals required for plant growth and metabolism if present in high concentrations cause toxic effects. These effects lead to a lowering of the contents of photosynthetic pigments, carbohydrates, and proline; the reduction in net photosynthetic rate; growth inhibition; and increased content of malondialdehyde. These effects are caused by the interaction with functional groups of molecules in cells, particularly proteins and polynucleotides (Sridhara Chary et al. 2008).

5.4.1.1 Physiological Aspects

Heavy metal stress is one of the most important concerns threatening plants' survival in agricultural research. BRs are the primary hormones mitigating the effects of metal contamination on plants by scavenging reactive oxygen species (ROS) and activating the antioxidant defense enzymes, viz., peroxidase (POD), superoxide dismutase (SOD), catalase (CAT), glutathione reductase (GR), ascorbate peroxidase (APOX), glutathione-S-transferase (GST), and guaiacol peroxidase (GPOX).

BRs such as 24-epibrassinolide (EBL) and 28-homobrassinolide (HBL) contribute to enhancing phytoremediation processes, aiding plants in detoxifying toxic metals (Barbafieri and Tassi 2011). They achieve this by modifying cell membrane permeability, inducing the expression of defensive enzymes, and stimulating the production of stress proteins, thereby increasing ATPase expression to reduce heavy metal uptake (Madhan et al. 2014). Some studies suggest that externally applied BRs induce short-lived H_2O_2 formation, activating MAPK and leading to the production of NADPH oxidase, consequently upregulating stress proteins and defense enzymes (Jiang et al. 2012; Yin et al. 2016).

Research has investigated metal accumulation, including cadmium (Cd), lead (Pb), copper (Cu), and zinc (Zn), in various cultivated plants such as barley, tomato, and radish following BR application (Hasan et al. 2011; Hayat et al. 2010; Rajewska et al. 2016; Ramakrishna and Rao 2013, 2015). Studies indicate that the application of 24-epibrassinolide (EBL), a type of BR, reduced lead content by 50% in beetroot and decreased copper uptake and accumulation in Indian mustard (Brassica juncea) seeds, while also enhancing shoot generation and biomass production (Sharma et al. 2007). *Chlorella vulgaris* exhibited lower bioaccumulation of metals, thereby preventing the loss of sugar, chlorophyll, and protein, and improving the synthesis of phytochelatins (PC) (Bajguz 2000, 2002). Also, mung beans exposed to aluminum (Al) stress, when treated with brassinolide, promoted its growth (Abdullahi et al.

2002), along with considerably increasing the fresh mass of shoots and roots, and chlorophyll content (Ali et al. 2008).

28-homobrassinolide (HBL) improves seed germination and the lengths of shoots and roots in Indian mustard under nickel (Ni) stress (Yusuf et al. 2011). In rape cotyledons, BR was observed to eliminate the toxic effects of Cd on photochemical pathways by relieving the reaction centers of their damage and O_2-evolving complexes, and by ensuring efficient electron transport (Janeczko et al. 2005). Therefore, in plants such as Indian mustard, radish, wheat, and maize, HBL can reduce metal stress (Bhardwaj et al. 2007; Hayat et al. 2007; Rajewska et al. 2016).

In wheat and Indian mustard under abiotic stresses, HBL is known to regulate the activities of various enzymes involved in photosynthesis and plant defense mechanisms (Hayat et al. 2007). Under the exposure of metal in pre-soaked seeds, HBL improves seedling growth and chlorophyll content; however, in radish or rice seedlings, BR lowers the increased uptake of Cr^{2+}, thus reducing toxicity (Sharma et al. 2011, 2016).

5.4.1.2 Molecular Aspects

Metal toxicity is a result of oxidative stress, generated by reactive oxygen species (ROS) (Krishna 2003). Plants, on exposure to metal stress, induce the Ca^{2+} ion signaling cascades involving calcium-dependent protein kinases, calmodulins, calmodulin-like proteins, and calcineurin B-like proteins (Ludwig 2003). In *Arabidopsis*, the Cd^{2+} stress triggers ROS accumulation, which in turn activates two mitogen-activated protein kinases (MAPKs), MPK3 and MPK6 (Liu et al. 2010) (Fig. 5.5). However, to date, there is no evidence of BRs being involved in the above-stated activation.

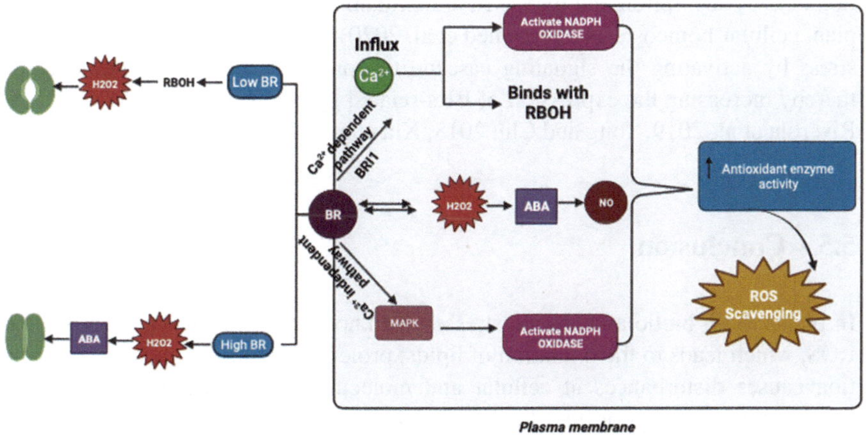

Fig. 5.5 Stress-induced molecular mechanisms in BRs

On exposure to metal toxicity, overproduction of ROS leads to a cascade of events causing oxidative damage to membranes, followed by alteration in the levels of antioxidants and metalloenzymes or antioxidant enzymes (Pandey et al. 2005). However, to study the effects of BR under oxidative stress, studies were carried out using mustard plants, and it was observed that BRs increased the level of antioxidants which protected its photosynthetic machinery, the plant growth, and increased its tolerance for salt and Ni stress. Similarly, under Cd stress, HBL-supplemented mustard plants exhibited upregulation of CAT, APOX, and SOD, and the increased content of proline.

However, the photosynthetic efficiency and stomatal opening are regulated by the concentrations of BRs, like low doses of EBR (0.1–0.15 μM) suppress the photosynthetic efficiency, while 0.1 μM EBR stimulates stomatal opening, while 1.0 μM EBR facilitates the closing of stomata.

The heavy metal toxicity in plants is regulated by BRs implementing H_2O_2 production, ROS scavenging, upregulation of MAPK, and by increasing the concentration of Na^+ and K^+ ions, along with osmolytes, and antioxidants (Rajewska et al. 2016). Studies have shown that in the plasma membrane of plant cells, O_2^- is converted to H_2O_2 by superoxide dismutase and NADPH oxidase, thereby further increasing H^+-ATPase activity by upregulation of the *CsHA* gene (Jakubowska et al. 2015).

With the uptake of toxic metals, the plant cell membranes undergo peroxidation of lipids and oxidation of proteins by the generation of ROS, negatively affecting the plant cell membranes. Phytohormones like BRs are known to counterfeit this effect by boosting the level of antioxidants and defensive enzymes, thereby mitigating the toxicity and restoring normal osmoregulation (Shahzad et al. 2018). This has been proven in studies carried out in *A. thaliana* by the application of a BR-specific inhibitor, brassinazole (Brz), and the bioactive form, 24-brassinolide (EBL) (Wu et al. 2019). BRs are also known to overcome heavy metal toxicity through restoration of CO_2 absorption and increased antioxidants, which in turn help maintain the plant cellular homeostasis (Ahammed et al. 2020). The BRs also mitigate the metal stress by activating the signaling cascade to upregulate the transcription factors thereby increasing the expression of BRs-related genes (Nolan et al. 2020; Planas-Riverola et al. 2019; Tong and Chu 2018; Xia et al. 2018).

5.5 Conclusion

In plants, both biotic and abiotic stresses are known to generate large amounts of ROS, which leads to the oxidation of lipids, proteins, and nucleic acid. This oxidation causes disturbances at cellular and molecular levels, leading to apoptosis. However, the absorption and bioaccumulation of metals in plants determine the toxicity. This toxicity, in turn, depends upon the uptake route, bioavailability, storage, immobilization, degradation, and excretion of metal by the tolerance mechanism. BR signal pathway on stress induction initiates MAPKs, resulting in a cascade of protein phosphorylation, resulting in the activation of NADPH oxidase releasing ROS. However, there are still lacunae in understanding anti-stress mechanistic pathways. More studies are needed for a better understanding of molecular and physiological aspects under metal stress. Given the importance of BRs in crop improvement, there is a need to identify and study the key regulatory elements of the BR signaling pathway modulating growth and development.

Also, based on the property of BRs to modify some agronomic traits of the plants, the application of chemical additives to the soil, such as chelators, organic acids, etc., the studies may prove beneficial in deciphering the bioavailability of metals for plant uptake and in designing the strategies to complement phytoremediation and crop improvement under stress.

References

Abdullahi BA, Gu X, Gan Q, Yang Y (2002) Brassinolide amelioration of aluminum toxicity in Mungbean seedling growth. J Plant Nutr 26:1725–1734

Ahammed GJ, Li X, Liu A, Chen S (2020) Brassinosteroids in plant tolerance to abiotic stress. J Plant Growth Regul 39:1451–1464

Albrecht C, Russinova E, Kemmerling B, Kwaaitaal M, De Vries SC (2008) Arabidopsis SOMATIC EMBRYOGENESIS RECEPTOR KINASE proteins serve brassinosteroid-dependent and -independent signaling pathways. Plant Physiol 148:611–619

Ali B et al (2008) A role for brassinosteroids in the amelioration of aluminium stress through the antioxidant system in mung bean (Vigna radiata L. Wilczek). Environ Exp Bot 62:153–159

Bajguz A (2000) Blockade of heavy metals accumulation in Chlorella vulgaris cells by 24-epibrassinolide. Plant Physiol Biochem 38:797–801

Bajguz A (2002) Brassinosteroids and lead as stimulators of phytochelatins synthesis in Chlorella vulgaris. J Plant Physiol 159:321–324

Bajguz A (2007) Metabolism of brassinosteroids in plants. Plant Physiol Biochem 45:95–107

Bajguz A (2019) Brassinosteroids in microalgae: application for growth improvement and protection against abiotic stresses. In: Hayat S, Yusuf M, Bhardwaj R, Bajguz A (eds) Brassinosteroids: plant growth and development. Springer, Singapore, pp 45–58

Bajguz A, Tretyn A (2003) The chemical characteristics and distribution of brassinosteroids in plants. Phytochemistry 62:1027–1046

Bancosi S et al (2002) Regulation of transcript levels of the Arabidopsis cytochrome P450 genes involved in brassinosteroid biosynthesis. Plant Physiol 130:504–513

Bar M, Sharfman M, Ron M, Avni A (2010) BAK1 is required for the attenuation of ethylene-inducing xylanase (Eix)-induced defense responses by the decoy receptor LeEix1: BAK1 in Eix signaling. Plant J 63:791–800

Barbafieri M, Tassi E (2011) Brassinosteroids for phytoremediation application. In: Hayat S, Ahmad A (eds) Brassinosteroids: a class of plant hormone. Springer Netherlands, Dordrecht, pp 403–437

Belkhadir Y, Chory J (2006) Brassinosteroid signaling: a paradigm for steroid hormone signaling from the cell surface. Science 314:1410–1411

Bhardwaj R, Arora N, Sharma P, Arora HK (2007) Effects of 28-homobrassinolide on seedling growth, lipid peroxidation and antioxidative enzyme activities under nickel stress in seedlings of Zea mays L. Asian J Plant Sci 6:765–772

Bishop G et al (2006) Dwarfism and cytochrome P450-mediated C-6 oxidation of plant steroid hormones. Biochem Soc Trans 34:1199–1201

Chakraborty N et al (2015) G-protein α-subunit (GPA1) regulates stress, nitrate and phosphate response, flavonoid biosynthesis, fruit/seed development and substantially shares GCR1 regulation in A. thaliana. Plant Mol Biol 89:559–576

Choe S (2006) Brassinosteroid biosynthesis and inactivation. Physiol Plant 126:539–548

Fujioka S, Yokota T (2003) Biosynthesis and metabolism of brassinosteroids. Annu Rev Plant Biol 54:137–164

Fujita S et al (2006) Arabidopsis CYP90B1 catalyses the early C-22 hydroxylation of C_{27}, C_{28} and C_{29} sterols. Plant J 45:765–774

Hasan SA, Hayat S, Ahmad A (2011) Brassinosteroids protect photosynthetic machinery against the cadmium induced oxidative stress in two tomato cultivars. Chemosphere 84:1446–1451

Hayat S, Ali B, Aiman Hasan S, Ahmad A (2007) Brassinosteroid enhanced the level of antioxidants under cadmium stress in Brassica juncea. Environ Exp Bot 60:33–41

Hayat S, Hasan SA, Hayat Q, Ahmad A (2010) Brassinosteroids protect Lycopersicon esculentum from cadmium toxicity applied as shotgun approach. Protoplasma 239:3–14

Jakubowska D, Janicka-Russak M, Kabała K, Migocka M, Reda M (2015) Modification of plasma membrane NADPH oxidase activity in cucumber seedling roots in response to cadmium stress. Plant Sci 234:50–59

Janeczko A, Koscielniak J, Pilipowicz M, Szarek-Lukaszewska G, Skoczowski A (2005) Protection of winter rape photosystem 2 by 24-epibrassinolide under cadmium stress. Photosynthetica 43:293–298

Jiang Y et al (2012) Cellular glutathione redox homeostasis plays an important role in the brassinosteroid-induced increase in CO_2 assimilation in *Cucumis sativus*. New Phytol 194:932–943

Joo S-H et al (2012) Biosynthesis of a cholesterol-derived brassinosteroid, 28-norcastasterone, in Arabidopsis thaliana. J Exp Bot 63:1823–1833

Joo S-H, Jang M-S, Kim MK, Lee J-E, Kim S-K (2015) Biosynthetic relationship between C28-brassinosteroids and C29-brassinosteroids in rice (Oryza sativa) seedlings. Phytochemistry 111:84–90

Karlova R et al (2006) The *Arabidopsis* SOMATIC EMBRYOGENESIS RECEPTOR-LIKE KINASE1 protein complex includes BRASSINOSTEROID-INSENSITIVE1. Plant Cell 18:626–638

Kim S, Moon J, Roh J, Kim S-K (2018) Castasterone can be biosynthesized from 28-homodolichosterone in Arabidopsis thaliana. J Plant Biol 61:330–335

Krishna P (2003) Brassinosteroid-mediated stress responses. J Plant Growth Regul 22:289–297

Li J, Jin H (2007) Regulation of brassinosteroid signaling. Trends Plant Sci 12:37–41

Liu X-M et al (2010) Cadmium activates Arabidopsis MPK3 and MPK6 via accumulation of reactive oxygen species. Phytochemistry 71:614–618

Ludwig AA (2003) CDPK-mediated signalling pathways: specificity and cross-talk. J Exp Bot 55:181–188

Madhan M, Mahesh K, Rao SS (2014) Effect of 24-epibrassinolide on aluminium stress induced inhibition of seed germination and seedling growth of Cajanus cajan (L.) Millsp. Int J Multidiscipl Current Res 2:286–290

Mitchell JW, Mandava N, Worley JF, Plimmer JR, Smith MV (1970) Brassins—a new family of plant hormones from rape pollen. Nature 225:1065–1066

Montoya T et al (2005) Patterns of *Dwarf* expression and brassinosteroid accumulation in tomato reveal the importance of brassinosteroid synthesis during fruit development. Plant J 42:262–269

Nolan TM, Vukašinović N, Liu D, Russinova E, Yin Y (2020) Brassinosteroids: multidimensional regulators of plant growth, development, and stress responses. Plant Cell 32:295–318

Nomura T et al (2007) Roles of brassinosteroids and related mRNAs in pea seed growth and germination. Plant Physiol 143:1680–1688

Oh E, Zhu J-Y, Wang Z-Y (2012) Interaction between BZR1 and PIF4 integrates brassinosteroid and environmental responses. Nat Cell Biol 14:802–809

Ohnishi T (2018) Recent advances in brassinosteroid biosynthetic pathway: insight into novel brassinosteroid shortcut pathway. J Pestic Sci 43:159–167

Ohnishi T et al (2006) C-23 hydroxylation by *Arabidopsis* CYP90C1 and CYP90D1 reveals a novel shortcut in brassinosteroid biosynthesis. Plant Cell 18:3275–3288

Pandey V, Dixit V, Shyam R (2005) Antioxidative responses in relation to growth of mustard (Brassica juncea cv. Pusa Jaikisan) plants exposed to hexavalent chromium. Chemosphere 61:40–47

Piotrowska A, Bajguz A (2011) Conjugates of abscisic acid, brassinosteroids, ethylene, gibberellins, and jasmonates. Phytochemistry 72:2097–2112

Planas-Riverola A et al (2019) Brassinosteroid signaling in plant development and adaptation to stress. Development 146:dev151894

Rajewska I, Talarek M, Bajguz A (2016) Brassinosteroids and response of plants to heavy metals action. Front Plant Sci 7:629

Ramakrishna B, Rao SSR (2013) Preliminary studies on the involvement of glutathione metabolism and redox status against zinc toxicity in radish seedlings by 28-Homobrassinolide. Environ Exp Bot 96:52–58

Ramakrishna B, Rao SSR (2015) Foliar application of brassinosteroids alleviates adverse effects of zinc toxicity in radish (Raphanus sativus L.) plants. Protoplasma 252:665–677

Rozhon W, Akter S, Fernandez A, Poppenberger B (2019) Inhibitors of brassinosteroid biosynthesis and signal transduction. Molecules 24:4372

Samuni-Blank M et al (2012) Intraspecific directed deterrence by the mustard oil bomb in a desert plant. Curr Biol 22:1218–1220

Shahzad B et al (2018) Role of 24-epibrassinolide (EBL) in mediating heavy metal and pesticide induced oxidative stress in plants: a review. Ecotoxicol Environ Saf 147:935–944

Sharma P, Bhardwaj R, Arora N, Arora HK (2007) Effect of 28-homobrassinolide on growth, zinc metal uptake and antioxidative enzyme activities in Brassica juncea L. seedlings. Braz J Plant Physiol 19:203–210

Sharma I, Pati PK, Bhardwaj R (2011) Effect of 28-homobrassinolide on antioxidant defence system in Raphanus sativus L. under chromium toxicity. Ecotoxicology 20:862–874

Sharma P, Kumar A, Bhardwaj R (2016) Plant steroidal hormone epibrassinolide regulate—heavy metal stress tolerance in Oryza sativa L. by modulating antioxidant defense expression. Environ Exp Bot 122:1–9

Shimada Y et al (2003) Organ-specific expression of brassinosteroid-biosynthetic genes and distribution of endogenous brassinosteroids in Arabidopsis. Plant Physiol 131:287–297

Sonawane PD et al (2016) Plant cholesterol biosynthetic pathway overlaps with phytosterol metabolism. Nat Plants 3:16205

Sridhara Chary N, Kamala CT, Samuel Suman Raj D (2008) Assessing risk of heavy metals from consuming food grown on sewage irrigated soils and food chain transfer. Ecotoxicol Environ Saf 69:513–524

Stamp N (2003) Out of the quagmire of plant defense hypotheses. Q Rev Biol 78:23–55

Symons GM, Reid JB (2004) Brassinosteroids do not undergo long-distance transport in pea. Implications for the regulation of endogenous brassinosteroid level. Plant Physiol 135:2196–2206

Symons GM et al (2006) Grapes on steroids. Brassinosteroids are involved in grape berry ripening. Plant Physiol 140:150–158

Tong H, Chu C (2018) Functional specificities of brassinosteroid and potential utilization for crop improvement. Trends Plant Sci 23:1016–1028

Wang X et al (2008) Sequential transphosphorylation of the BRI1/BAK1 receptor kinase complex impacts early events in brassinosteroid signaling. Dev Cell 15:220–235

Wendeborn S et al (2017) Biological activity of brassinosteroids—direct comparison of known and new analogs *in planta*. Helv Chim Acta 100:hlca.201600305

Wu C et al (2019) The potential role of brassinosteroids (BRs) in alleviating antimony (Sb) stress in Arabidopsis thaliana. Plant Physiol Biochem 141:51–59

Xia X et al (2018) Brassinosteroid-mediated apoplastic H_2O_2 -glutaredoxin 12/14 cascade regulates antioxidant capacity in response to chilling in tomato. Plant Cell Environ 41:1052–1064

Yin Y-L et al (2016) Interplay between mitogen-activated protein kinase and nitric oxide in brassinosteroid-induced pesticide metabolism in Solanum lycopersicum. J Hazard Mater 316:221–231

Yokota T et al (2017) Occurrence of brassinosteroids in non-flowering land plants, liverwort, moss, lycophyte and fern. Phytochemistry 136:46–55

Yusuf M, Fariduddin Q, Ahmad A (2011) 28-Homobrassinolide mitigates boron induced toxicity through enhanced antioxidant system in Vigna radiata plants. Chemosphere 85:1574–1584

Zullo MAT, Bajguz A (2019) The Brassinosteroids family—structural diversity of natural compounds and their precursors. In: Hayat S, Yusuf M, Bhardwaj R, Bajguz A (eds) Brassinosteroids: plant growth and development. Springer, Singapore, pp 1–44

Chapter 6
Gibberellins-Mediated Regulation of Plant Biology Under Heavy Metal Stress: Physiological and Molecular Aspects

Amisha Rani, Kiran Soni, Radhika Sharma, Mohar Singh, Anuj Choudhary, Shivam Sharma, Mohit Uppal, Monika Koul, Sahil Mehta, Kaveri Umesh Kadam, and Om Prakash Narayan

Abstract The contamination of heavy metals (HMs) has become a global concern regarding crop quality and productivity and is more intense with the population explosion. Heavy metals may come from natural and anthropogenic processes and end up in various environmental compartments (soil, water, air, and their interface). Numerous studies have shown that natural sources of the environment's heavy metals are frequently of low significance when compared to anthropogenic activity. Several studies have been reporting the negative impact of the morphological, physiological, biochemical, and molecular responses expressed in terms of crop produc-

---

A. Rani · K. Soni
Department of Forest Products, Dr YS Parmar University of Horticulture & Forestry, Nauni, Solan, Himachal Pradesh, India

R. Sharma
Department of Soil Science, Punjab Agricultural University, Ludhiana, Punjab, India

M. Singh
Division of Agrotechnology, CSIR – Institute of Himalayan Bioresource Technology, Palampur, Himachal Pradesh, India

A. Choudhary
Department of Biology and Environmental Sciences, CSK Himachal Pradesh Agricultural University, Palampur, Himachal Pradesh, India

S. Sharma
Department of Vegetable Sciences, CSK Himachal Pradesh Agricultural University, Palampur, Himachal Pradesh, India

M. Uppal · M. Koul · S. Mehta
Department of Botany, Hansraj College, University of Delhi, Delhi, India

K. U. Kadam
School of Engineering Sciences and Technology, Jamia Hamdard, New Delhi, India

O. P. Narayan (✉)
Division of Chemical Biology and Medicinal Chemistry, University of Texas at Austin, Austin, TX, USA

A. Sharma et al. (eds.), *Hormones and Heavy Metals*, Plant Life and Environment Dynamics, https://doi.org/10.1007/978-981-96-9622-2_6

tivity. These metals enter the plant cell and disrupt cell membrane integrity, leading to the overproduction of reactive oxygen species (ROS), altered enzymatic activity, pigment degradation, damage to photosystems, and impaired water and nutrient supply. Inside the cell, several signaling loops get activated and alter the cellular responses according to their genetic potential. Among them, hormones have an important role in regulating signaling pathways. Gibberellin is a well-known classic hormone and has been known for a long time for its remarkable contribution to plant improvement. Mainly, its role was more restricted to growth and development. Therefore, it is very interesting to know its less recognized potential in HM stress.

Keywords Heavy metals · Gibberellin · Mitigation · Tolerance · Morpho-physiological alterations.

6.1 Introduction

Earth is made up of igneous rocks, and only 5% is composed of sedimentary rocks. In general, heavy metals like Co, Ni, Cd, and Cu are abundant in basaltic rocks, whereas Cd, Mn, Zn, Pb, and Cu are abundant in shales (Singh et al. 2016; Sytar et al. 2019). Some soils pick up such elements from parent material that is a prominent natural source of heavy metals. Even as parent material typically has large levels of these metals, this inheritance may be harmful to plants and other species. The parent substance from which they were formed is the main source of soil heavy metals. By natural phenomena such as meteoric, biological, terrestrial, and volcanic processes, as well as erosion, leaching, and surface winds, heavy metals contained in rocks can infiltrate the soil environment. Although human impacts comprise domestic effluents, mining, industry, and agriculture, natural sources comprise sedimentary processes, volcanic activity, soil formation, and rock weathering (Ghori et al. 2019). However, using pollution indicators is useful for distinguishing between anthropogenic and geogenic soil pollution (Shinozaki et al. 2015). Despite the use of sophisticated research techniques, it should be recognized that source allocation may be challenging in many situations (Rhaman et al. 2020).

6.2 Sources of Heavy Metal

6.2.1 Natural Source of Heavy Metals

In terms of heavy metals' natural sources, igneous as well as sedimentary rocks are thought to be the most prevalent. It has been discovered that the ratios for different components change from one rock to another as well as within a single rock type. The kind of rock and the state of the local environment can be used to determine the concentration of heavy metals. Along with river sediments, soil formation is another important factor in the buildup of heavy metals (Jia et al. 2020; Zhang and Wang 2020).

6.2.2 Sources of Heavy Metals Caused by Anthropogenic Activity

One or more heavy metals accumulate in the soil as a result of anthropogenic actions that disrupt the naturally occurring, slowly recurring biogeochemical cycle of heavy metals. Heavy metals are thought to come from anthropogenic sources such as industry, agriculture, mining, and wastewater (Vareda et al. 2019). These factors greatly contribute to the increase in the concentration of heavy metals and ecosystem pollution, such as smelting, which releases Cu, Zn, and As pesticides, which help to generate As; burning fossil fuels, which creates Hg; and vehicle exhaust, which helps to release Pb (Alengebawy et al. 2021; Rajendran et al. 2022). Additionally, routine human activities like farming, manufacturing, and industrial processes harm the biosphere's ability to maintain its balance.

6.2.3 Sources of Heavy Metals in Agriculture

Agricultural pollutants, also referred to as biotic and abiotic consequences of farming methods, are frequently harmful to agroecosystems. The agroecosystem in the area is typically contaminated and degraded by these contaminants. The most prevalent agricultural sources of metals are sewage sludge, herbicides, and fertilizers (Afonne et al. 2020; Hanfi et al. 2020). Whether they accumulate in soil or plants, poisonous heavy metals differ in their behavior and mechanisms of accumulation. Heavy metals interact with each other and are dispersed throughout soil and sediment components by a variety of processes, including complexation, precipitation, ion exchange, and adsorption. Heavy metals take a long time to degrade; thus they remain in the soil for a long time (Dhaliwal et al. 2020). Among the physicochemical properties, pH and organic matter are important parameters that play a key role in the accumulation and availability of heavy metals in the soil environment.

Phosphorus is frequently utilized in fertilizer production, but it also contributes significantly to the buildup of heavy metals in the soil. It has been demonstrated that water-insoluble phosphorus fertilizers create phosphate rocks, which are crucial for the soil's ability to immobilize metals through the precipitation of metal phosphates.

Heavy metal deposition in agricultural soils caused by prolonged, excessive fertilizer application diminishes soil fertility, which in turn reduces plant growth and output (Ashraf et al. 2019; Afonne and Ifediba 2020). After heavy metals have contaminated the soil, it is very difficult to restore the soil ecosystem. Due to the prolonged use of fertilizers, agricultural soil has a larger potential for Cu, Zn, and Cd buildup. The main inorganic fertilizer types that lead to the production of heavy metals in soil and their subsequent uptake by plants are phosphate fertilizers, liming substances, and bio-fertilizers. As a result, they move through the food chain, as well as the food web (Fei et al. 2022).

6.3 Bioavailability of HMs

Heavy metals contaminate the natural environment through anthropogenic and natural processes and can follow a long route in the ecosystem (Ashraf et al. 2015). They cause a serious impact on the environmental entities of the ecosystem and can directly or indirectly target the food security of the global population (Altaf et al. 2022). HM pollution has emerged as a threat due to industrial processes, driven by increased utilization of minerals, mineral resources, and their abstraction. Several cases of atmospheric contamination have shown a high contribution as central HMs sources, such as organic pollutants, biodegradation, and atmospheric dust particles (Pandey et al. 2023). The availability of these metals for their absorption is mainly governed by several factors. The retention of metals in the soil is maintained at an equilibrium and controlled by different chemicals because of soil properties. In the soils, HMs are observed as complexation with humus and in precipitated forms. Only a fraction of the metal is available for plant uptake (Khan et al. 2021a).

6.4 Entry and Translocation of HM

Heavy metals are most preferably found in insoluble forms in soils, making them non-available for plants. The bioavailability of plants can be enhanced by discharging several root exudates to modify the rhizosphere pH and thus increase the solubility of heavy metals (Dalvi and Bhalerao 2013). It follows a sequence of events starting from root uptake, loading to xylem, followed by root to shoot, cellular compartmentalization, and further sequestration. Once the metal is adsorbed on the surface of the root, it transfers across the membrane inside the root cells. There are two possibilities for the heavy metal uptake into the roots: active transport against the gradient (symplastic pathway) and passive movement (apoplastic pathway). The symplastic pathway is a more common approach for the uptake of HMs in an energy-dependent manner by ion carriers (Shi et al. 2019). Inside the root cells, HMs are complexed into several chelators, including different organic acids, to form sulfate, phosphate precipitate, and carbonate. These are immobilized in the intracellular spaces such as vacuoles and extracellular spaces such as apoplastic cell walls (Pasricha et al. 2021). Then, it may sequester into the root cell vacuole or be transported into the stele tissue to further enter the xylem stream. They are further transferred to the shoot and distributed in the leaves, where they are sequestered into cell walls or vacuoles to reduce their accumulation in the cytosol.

The specialized transporter and carrier proteins are essential for metal uptake localized on the cell membrane. These can transfer specific metals to efflux and influx the metal from the root to the above-ground part (DalCorso et al. 2019). Based on sequence homology, several metal transporters have been identified belonging to different families, such as the metal transporter proteins (MTPs) family, heavy metal transporting ATPases (HMAs) family, naturally resistant associated macrophage proteins (NRAMPs) family, and Zn-regulated, Fe-regulated transporter-like protein (ZIP) family. The ZiP family proteins are involved in the accumulation of cations, such as Zn, Mn, and Fe, and their transportation from the root to above-ground parts (Guerinot 2000). The roots of *Arabidopsis halleri* and *Thlaspi caerulescens* (Zn hyperaccumulator) have a higher level of Zn uptake as compared to non-hyper-accumulated ones and are highly correlated with the elevated level of ZIP members in the hyper-accumulator species (Assuncao et al. 2001). The overexpression of the ZNT1 gene in Arabidopsis enhances the accumulation and tolerance of Cd and Zn, suggesting its involvement in long-distance transport from the root to the shoot via xylem loading (Lin et al. 2016). A vacuolar P1B-ATPase—HMA3—is involved in the vacuolar sequestration of Zn, Cd, Pb, and Co (Hanikenne and Baurain 2014). Guo et al. (2017) experimented on *Festulolium loliaceum* to correlate the activity of Cd stress to the FlHMA3 expression (P1B2-ATPase). They suggested the role of these transporter proteins in the Cd sequestration inside the root cell vacuoles, thus reducing the concentration to reduce their toxicity.

6.5 Plant Responses to Heavy Metal

6.5.1 Morpho-Physiological Alteration

6.5.1.1 Chloroplast Structure

Metal phytotoxicity causes leaf chlorosis and growth suppression and has been thoroughly researched for its impact on photosynthesis. Heavy metals may have negative effects on chloroplast ultrastructure, including the possibility of complete chloroplast destruction from metal overdose (Rai et al. 2016). It includes altered chloroplast shapes, reduced chloroplast size and number, decreased grana and thylakoid content, disorganized thylakoid structures, increased size and number of plastoglobuli, and damage to the chloroplast membrane. For some plant species, like barley, peas, and wheat, these reactions, including associated disruption of photosynthetic processes as well as chloroplast replication, are typical (Souri et al. 2019). Plants that are exposed to soils contaminated with heavy metals have fewer grana and a disordered thylakoid system. Additionally, chloroplasts frequently include multiple, sizable plastoglobuli, a sign of plant stress. Lastly, under extreme

Table 6.1 Insights into morpho-physiological and biochemical alteration in plants upon heavy metal exposure

Heavy metal	Plants	Morphological changes	Physiological changes	Biochemical changes	References
Ni	*Populus nigra*	Caused visible injury in developing leaves	Ni stress significantly decreased photosynthesis	Metabolism impairment, so isoprene emission in developing leaves, and cis-β-ocimene emission in mature leaves	Velikova et al. (2011)
Ni	*Capsicum annuum*	Reduced root growth and development and decreased dry and fresh biomass of both root and shoot	The degree of stomatal opening and intercellular CO_2 and transpiration rate decreased considerably	Induction of oxidative stress/ damage	Altaf et al. (2022)
Pb	*Oryza sativa*	Reduction in plant growth	Reduction of nutrient uptake	Deactivation of cell membrane permeability	Ashraf et al. (2015)
Pb	*Oryza sativa*	Pb causes oxidative stress which further reduces rice productivity and biomass and also inhibits plant growth	Pb causes oxidative stress	Pb reduced protein and chlorophyll accumulation and decreased proline content	Khan et al. (2021b)
As As(III) and As(V)	*Medicago sativa and Hordium vulgare*	Chlorosis, necrosis, reduced biomass, epinasty, and stunted growth	Causes disturbances in mineral uptake and homeostasis	Inactivate essential biomolecules by blocking essential functional groups or displacing essential metal ions	Wang et al. (2017)

(continued)

Table 6.1 (continued)

Heavy metal	Plants	Morphological changes	Physiological changes	Biochemical changes	References
As(III) and As(V)	*Oryza sativa*	Causes growth reduction in shoot and root	As(V) causes interference in ATP synthesis and protein phosphorylation, while As(III) disrupts the electron transport chain in photosynthesis and respiration	As(III) binds to thiol groups in proteins and enzymes, thus altering their chemical functionality and it also binds to glutathione resulting in oxidative stress	Pandey et al. (2023)
V	*Cicer arietinum*	Stunted the growth of plants and roots extensively and showed a significant reduction in the growth of lateral roots	Vanadium showed positive change with the enzymes performing antioxidant activities at low concentration	Protein contents reduced significantly ($P \leq 0.05$) due to vanadium stress in all chickpea genotypes	Imtiaz et al. (2015)
Cd	*Triticum durum* (Durum Wheat)	Cd declined the shoot length, dry weight of the shoot	Cd toxicity suppresses growth development by inhibiting cell division and some enzymes related to growth	Cd decreased the content of Ca and N, P, K in the shoot of durum wheat	Yourtchi and Bayat (2013)
Cd	*Glycine max*	Visible symptoms in terms of chlorosis, growth inhibition, browning of root tips, and finally death	Nitrogen fixation and primary ammonia assimilation decreased and chlorophyll biosynthesis is inhibited	Reduction in the activity of enzymes involved in CO_2 fixation	Raziuddin et al. (2011), Balestrasse et al. (2003), Guo et al. (2008)
Cd	*Phaseolus vulgaris*	Cd causes a reduction of 83.9%, 66.3%, and 52.3% in root length, shoot length, and the ratio of R/S seedlings, respectively	Seed germination percentage and fresh weight also decreased by 9.9% and 42.3%, respectively	Cd causes oxidative damage in bean plants	Bahmani et al. (2012)

(continued)

Table 6.1 (continued)

Heavy metal	Plants	Morphological changes	Physiological changes	Biochemical changes	References
Zn and Cd	*Nicotiana tabacum*	Decrease in growth, and development of both root and shoot and results in senescence	Decrease in plant metabolism	Induction of oxidative stress/ damage	Malik et al. (2011), Tkalec et al. (2014)
Cd and Zn	*Pisum sativum*	Chlorosis and retardation in the growth of both root and shoot	Obstructs the flow of water	Alternation in catalytic efficiency of enzymes	Romero-Puertas et al. (2004), Malik et al. (2011)
Cr	*Allium sativum*	Stunted plant growth	The first physiological process affected by Cr stress is seed germination because it obstructs the transport of sugars to the embryo axes	Cr stress increases the activity of the protease enzyme and decreases the activity of the amylase enzyme and both of these actions reduce seed germination	Peralta et al. (2001), Nematshahi et al. (2012), Zeid (2001)
Co	*Hordeum vulgare, Brassica napus, Lycopersicon esculentum, Brassica oleracea* var. *botrytis*	Adverse effect on shoot growth and biomass	Decreased water potential and transpiration rate and affected the translocation of P, S, Mn, Zn, and Cu from root to shoot	Reduces the activity of catalase enzyme and thus enhances the oxidative stress	Li et al. (2009), Chatterjee and Chatterjee (2000)
Cd, Cr, Cu, Fe, Pb, Zn	*Taraxacum officinale*	Reduction of leaf thickness and foliar thickness proportional to HM concentration	Less organized palisade parenchyma Lacking cristae in the mitochondria of parenchyma cells	Increasing concentration of heavy metals increases pH and causes alteration in mitochondrial structure	Maleci et al. (2014)
Hg	*Fagopyrum tataricum*	Growth and biomass are both affected negatively at high Hg concentration	Photosynthesis is adversely affected due to the decline in the chlorophyll pigment	As the concentration of Hg increases, proline and other protein content decrease	Pirzadah et al. (2018)

(continued)

Table 6.1 (continued)

Heavy metal	Plants	Morphological changes	Physiological changes	Biochemical changes	References
Hg	*Lycopersicon esculentum*	Chlorosis and decreases in plant height and fruit weight	Reduction in germination percentage and flowering	Displaces the essential molecule Mg^{2+} ion in chlorophyll	Shekar et al. (2011)
Fe	*Oryza sativa*	Decrease in overall plant growth, biomass, and yield	Disruption of the root cell membrane changes xylem and phloem flow and decreases photosynthesis	Modification of various genes and transporters, reduction of enzymatic activity, and production of free radical species	Ahmed et al. (2023)

stress, broken chloroplast membranes can be detected. Table 6.1 entails the morpho-physiological and biochemical alteration in plants upon heavy metal exposure.

6.5.1.2 Seed Germination

From seed germination until senescence, the impacts of HM poisoning can be seen in practically all plant tissues. However, these effects are strongest throughout germinating seeds and root growth (Seneviratne et al. 2019). Since seed germination is susceptible to the process conditions of the rhizobacteria, HM toxicity has typically been implicated in reports of declines in seed germination and vigor and consequent seedling growth in *Raphanussativus, Lens culinaris, Oryza sativa, Hordeum vulgare, Vigna radiata, Medicago sativa, and Zea mays* are just a few of the crops that have shown that Pb-toxicity can have negative impacts on seed morphology (Sethy and Ghosh 2013). Exogenous administration of other HMs also has impacts on seedling architecture, physiology, and germination that are comparable to those of Pb (Fig. 6.1). For example, a combination of Cu and Cd treatment to *Solanum melongena* significantly decreased the number of germinating seeds, seedling growth, and the number of lateral roots (Kumar and Chopra 2016).

6.5.1.3 Root Architecture

In addition to the seeds, alterations in the root system of the plants that grow in the HM-contaminated soils have also been noted. For example, in the presence of Cu, Pb, Cr, Zn, and Cd stress, several plants, including *Arabidopsis thaliana, Triticum aestivum, Sesbania rostrata, Sesbania cannabina, Pinus sylvestris, and Lupinus luteus*, have shown decreased root elongation and increased lateral root formation

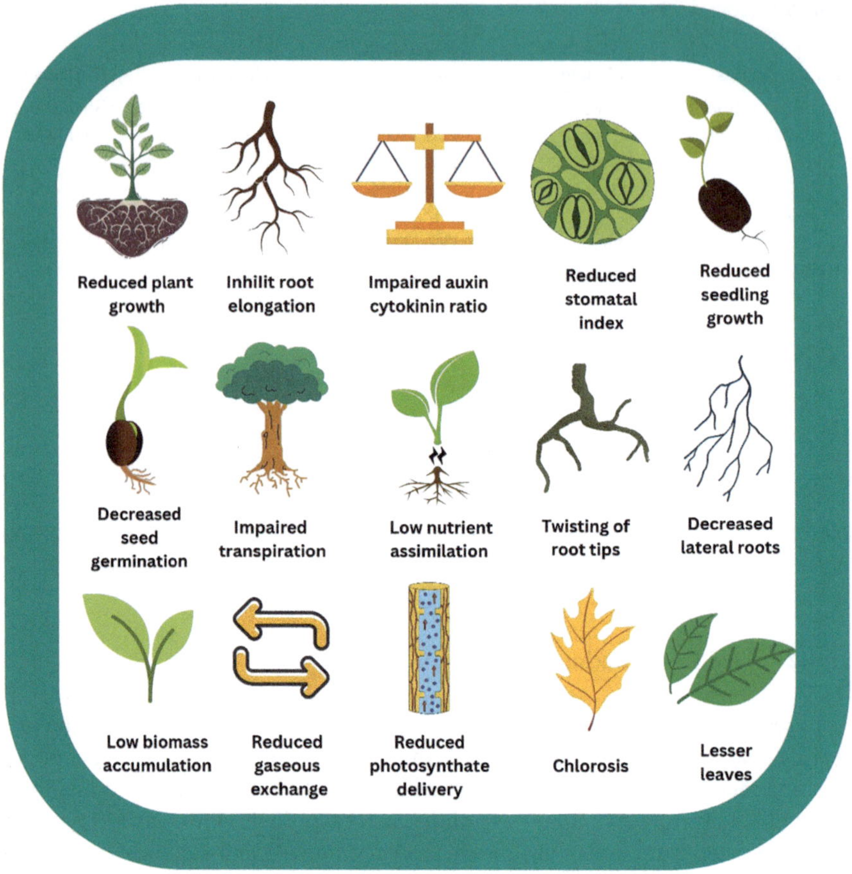

Fig. 6.1 Overview of alterations due to heavy metal exposure in the plant

(Riyazuddin et al. 2021). Localized swelling was observed behind some root tips in *Vigna aunguiculata* due to the commencement of lateral roots, the twisting of some root tips, the decrease in fresh biomass, and the decreased root growth caused by Pb toxicity. According to one theory, HMs, in particular Zn, build up in the epidermis and pericycle and cause the development of the root system (Hu et al. 2012).

The first sign of HM toxicity is the development of lateral roots, which then affects water absorption and conduction, which, in turn, reduces the delivery of photosynthetic activity to roots. During HM toxicity, roots absorb less photosynthetic material, which inhibits the growth of primary and secondary roots, resulting in a shorter effective root length and a higher root/shoot area ratio. Later studies revealed that indole acetic acid (IAA) and phenols generated in the presence of HMs are partially responsible for these HM toxicity-induced alterations in root architecture. When HM toxicity was present, it was discovered that genetic variants in IAA production were activated, increasing the amount of auxin and the auxin/cytokinin ratio, important determinants of the morphological changes in roots (Ghuge et al. 2023). Additionally, new data emphasizes the part nitric oxide (NO) plays in the reduction of root meristem growth during stress conditions in *Arabidopsis*.

6.5.1.4 Stunted Growth

Under HM toxicity, slowed plant growth has also been noted in addition to root inhibition. As seen in a variety of crop and non-crop plants, including mung bean, oats, *Curcumas sativus, Lactuca sativa, Panicum miliaceum,* wheat, *Jatropha curcas*, maize, rice, and poplar (Riyazuddin et al. 2021). HM transport from contaminated soils via roots to aerial parts and their accumulation in plant cells directly interferes with the cellular metabolism. Even though the precise molecular method of HM-induced inhibition of growth is currently unknown, it has been reported that some HMs, including As and Hg, cause cells to lose turgor pressure, a decrease in mitotic index, and inhibition of cell elongation as a result of their selective binding to the -SH group of proteins involved in the control of cell division and replacement of the phosphate group of ATPs. These effects are thought to be responsible for the reduction in plant growth under HM toxicity (Noor et al. 2022).

6.5.1.5 Transpiration and Photosynthesis

By disrupting plant-water relationships, which in turn influence numerous physiological processes including transpiration and photosynthesis, HM toxicity has an impact on the area, number, color, and thickness of leaves (Anjum et al. 2016). Generally speaking, under HM toxicity, a decrease in several leaves, area, and biomass has been noted in several plants, including *Albiziale beck, Arabidopsis thaliana, Spinacea oleracea, Brassica oleracea, Oryza sativa, Acacia holosericea,* and *Leucaena leucocephala*. For sugarcane, a lower Cr concentration (40 ppm) led to chlorosis of the leaves while a greater Cr concentration (80 ppm) led to necrosis (Radha et al. 2000). Additionally, HM-toxicity reports have been made regarding the browning of the leaf tip of *Phaseolus vulgaris* owing to Hg toxicity and the small, fragile purple leaves of *Jatropha curcas* in the influence of Pb (Karthik et al. 2016). In *Beta vulgaris* and *Sorghum vulgaris*, a reduction in the stomatal index was seen during Zn and Cd-Cu stress. In the early phases of HM toxicity, it has been observed that several plants, like *Helianthus annuus* and *Vigna radiata*, have more stomata when Pb and As are present (Khan et al. 2010). *Vigna radiata* also develops stopped, fused, and aberrant stomata at this time. Similar to *Zea mays*, Zn and Cd poisoning led to the growth of abnormal and non-functional stomata (da Cunha and do Nascimento 2009). Additionally, HM poisoning causes the growth of reduced mesophyll tissue as well as the loss of parenchymatous tissues and xylem arteries (Fig. 6.2).

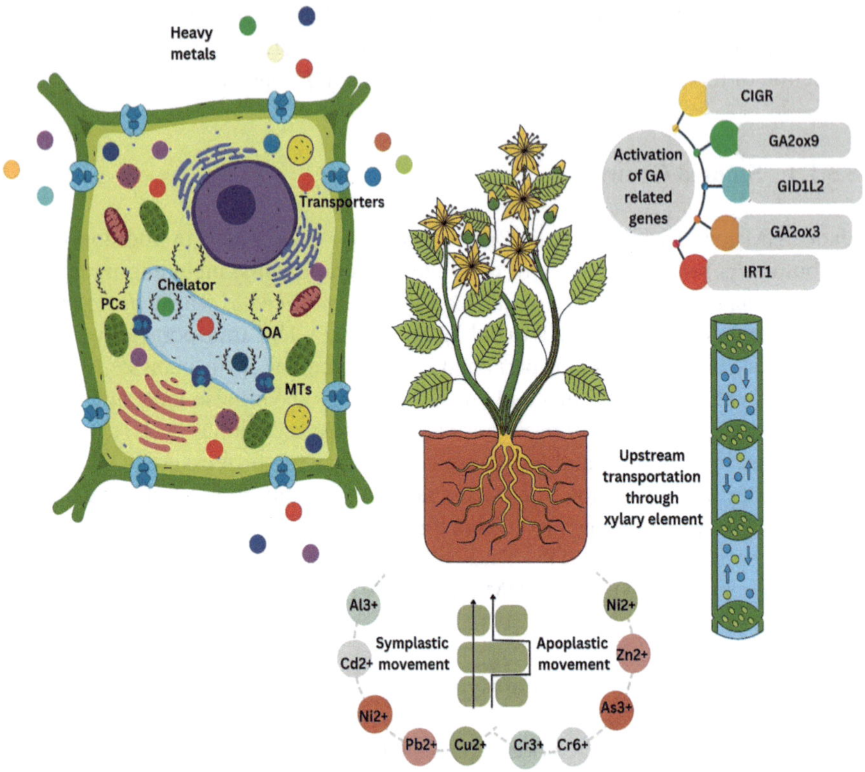

Fig. 6.2 Pathway and movement of heavy metals into the plants and involvement of various GA-related genes (*gibberellin-2-oxidase (GA2ox), gibberellin insensitive dwarf1 L2 (GID1L2), chitin-inducible gibberellin-responsive* (*CIGR*), and *iron-regulated transporter* (*IRT*))

6.5.1.6 Gas Exchange

Higher stomatal resistance, a decrease in the rate of transpiration, and altered water relations are all observed in plants that are more frequently exposed to environments that are contaminated with heavy metals (Sharma and Agrawal 2005). It leads to limiting photosynthesis resulting in decreased photosynthesis and mesophyll conductivity to CO_2, in addition to biochemical inhibition (Anjum et al. 2016).

6.5.2 Biochemical Alteration

6.5.2.1 Carbohydrates

The total carbohydrate values in seedlings rise when exposed to low concentrations of copper treatments, while the opposite is true when exposed to high concentrations. Plants' amylase activity decreased as the concentration of heavy metals rose (Khan et al. 2015). As a result, soluble saccharides and polysaccharides accumulate. The amylase activity of Maize seedlings treated with Cu and Mn was decreased, which could be related to increased peroxidant status (Maksimović et al. 2007). By decreasing the antioxidant glutathione pool and disrupting the iron-mediated defense mechanisms, heavy metal stress reduces enzyme activity. Cd poisoning significantly hindered the transfer of soluble carbohydrates to the developing embryonic axis as well as the breakdown of soluble sugars (Smiri et al. 2009). Cd significantly limits the mobilization of reserves, including soluble sugars and amino acids in pea seeds.

6.5.2.2 Proteins

The primary proteases of the plant proteolytic system are found inside organelles, and the ubiquitin-proteasome pathway is found in both the cytoplasm and the nucleus (Hasan et al. 2017). Although the impact of heavy metals on the proteolytic system cannot be generalized, reduced protease activity and impaired proteasome functionality appear to be typical aspects of metal toxicity in plants. Heavy metal stress due to Zn results in a decrease in protein concentration and a matching increase in the activity of hydrolytic enzymes like protease; this fact strongly implies catabolic activities (Rau et al. 2007). Exposure to heavy metals likely speeds up the breakdown of important metabolites, including protein, RNA, and chlorophyll. Since nitrate reductase is thought to be rate-limiting in the overall assimilation of nitrate, a decline in activity could be caused by a rise in protein levels, which would then affect total protein and plant development. It's also possible that heavy metals caused lipid peroxidation in plants and caused proteins to fragment as a result of ROS's harmful effects, which resulted in lower protein content (Zhang et al. 2020).

6.5.2.3 Nitrogen Assimilation

The SH-enzyme nitrate reductase is vulnerable to oxidative conditions. There is a lot of interest in the relationship between heavy metals, active oxygen species, and nitrate-reductase activities. Heavy metal stress has been observed to inhibit nitrate reductase function in higher plants (Hussain et al. 2020). The nitrate reductase enzyme is substantially more vulnerable to heavy metal contamination than other nitrate assimilation factors like protein and organic nitrogen concentration (Mokhele et al. 2012). Several authors have already observed that heavy metals decrease nitrate reductase function in beans and other plant species.

6.6 Heavy Metal Tolerance Mediated by Gibberellin

Due to environmental contamination, heavy metals are gaining momentum in their toxicity day by day, which is a significant threat to many plant species consumable by human beings. This toxicity is the repercussion of nationwide mining industries, foundries, smelters, coal-burning power plants, and agriculture practices, which escalate the concentration of heavy metals, rather than their natural baseline. Heavy metals like As, Cd, Cr, Pb, and Hg are systemic which enter in food chain and affect human health, and prolonged exposure to heavy metals may result in multiple organ failure (Bücker-Neto et al. 2017). So, there is an essentiality of taking measures to mitigate heavy metal toxicity. Gibberellic acid (GA_3) is a prime phytohormone that can absorb heavy metals from the environment (phytoextraction). It regulates several physiological mechanisms in plants, viz., plant growth and composition, flowering, leaf expansion stimulation, elongation, cell division, plant height, tissue differentiation, osmoregulation stimulation in internodes, dry matter, photosynthesis, transpiration rate, biomass composition, germination, and also increases sink space improvement. Several studies have shown the phytoremediation potential of GA_3 in various crops (Table 6.2) (Shah et al. 2021).

6.6.1 Against Cd Toxicity

The cumulative effect of GA_3 and ethylenediaminetetraacetic acid (EDTA) was investigated in the *Parthenium hysterophorus* plant to eradicate Cd toxicity from the oil by phytoremediation. Elevation in plant growth and dry biomass was recognized by the application of GA_3 and EDTA under various combinations (Hadi et al. 2014). GA_3 elevated the adverse effects of Cd toxicity in two varieties of Brassica, including Indian mustard (*Brassica juncea* (L.) Czern.) commonly known as "Raya" and rapeseed (*Brassica rapa* L.) as "Toria" mitigates the adverse effect of Cd toxicity and enhances the total chlorophyll, protein content, carbohydrates, and CAT of studied crops (Iqbal et al. 2022). Maize, being the third most important crop in the world after rice and wheat, has gained attention. To evaluate the effect of cadmium chloride stress, gibberellin was applied to two different maize varieties—Neelam and Desimakai. Along with morphological characters, chlorophyll content, carotenoids, net photosynthetic rate, transpiration rate, gas conduct types of stomatal conductance, total soluble proteins, and peroxidase, etc., were also evaluated (Shah et al. 2021; Naqvi et al. 2020). GA_3 is an adequate phytohormone that has the potential to limit the negative effects of heavy metals. On barley, the unfavorable effect of Cd and molybdenum contained amination concerning total protein and amino acid

Table 6.2 Gibberellic acid mediated mitigation of heavy metal stress in some commercial crops

Effective conc.	Plant	Heavy metal	Improvements	References
200 mg/L	*Vicia faba*	Cd and Pb	Increased mitotic activity, carbohydrate metabolism, protein and RNA contents	Mansour and Kamel (2005)
10^{-5} M	*Parthenium hysterophorus*	Cd	Improved plant growth and dry biomass	Hadi et al. (2014)
10^{-4} M	*Vigna radiata*	Ni	Increased the growth and yield, reduced Ni uptake	Ali et al. (2015)
0.5 Mm	*Hordeum vulgare*	Cd and Mo	Elevated total protein and amino acid	Amri et al. (2016)
–	*Zea mays*	Cd	Improved transpiration rate, zeatin riboside (ZR) content, photosynthetic rate, and balanced phytohormones	He et al. (2017)
0.25 mg	*Zea mays*	$CdCl_2$	Upregulated transpiration rate, gas conductance, stomatal conductance, total soluble proteins, and peroxidase	Naqvi et al. (2020)
100 mg/L	*Solanum lycopersicum*	B	Improved antioxidant enzymes (catalase, superoxide dismutase, and peroxidase) and antioxidant defense system	Kaya et al. (2020)
40 mg/L	*Zea mays*	Cd	Increased antioxidative enzymatic genes and net photosynthetic rate	Sun et al. (2020)
5 mm	*Zea mays*	Pb	Increased osmolytes, and photosynthetic pigments	Adejumo (2020)
50 µmol/L	*Sedum alfredii*	Cd and Pb	Enhanced chlorophyll content, potassium contents	Liang et al. (2021)
50 mg/L	*Coriandrum sativum*	B	Enhanced all enzymatic and non-enzymatic antioxidative-related genes	Saleem et al. (2021)
50 µM	*Vigna radiata*	Cd	Improved plant metabolism and photosynthetic pigments	Hakla et al. (2021)
10^{-6} M	*Cicer arietinum*	Cd	Enhanced proline, glycine betaine, ascorbate glutathione, and glyoxalase	Ahmed et al. (2021)
40 mg/L	*Spinacia oleracea*	Cu	Enhanced proline concentration, antioxidant enzyme activities, photosynthetic parameters, glycine betaine, ascorbate glutathione, proline, and glyoxylate	Gong et al. (2021)
100 mg/L	*Pisum sativum*	Cu	Elevated growth, physiology, enzymatic activities CAT, GSH, SOD, APX, and POD	Javed et al. (2021)
0.25 mg	*Zea mays*	$CdCl_2$	Enhanced chlorophyll content, carotenoids, and net photosynthetic rate	Shah et al. (2021)

(continued)

Table 6.2 (continued)

Effective conc.	Plant	Heavy metal	Improvements	References
300 mg/L	*Cyphomandra betacea*	Cd	Enhanced net photosynthetic rate, stomatal conductance, intercellular CO_2 concentration, transpiration rate, and peroxidase, superoxide dismutase, and catalase activities	Yang et al. (2021)
100 mg/L	*Brassica juncea* (L.) *Czern.* and *Brassica rapa*	Cd	Elevated total chlorophyll, protein content, carbohydrates, and CAT	Iqbal et al. (2022)
10^{-6} M	*Vicia faba*	Cr	Improved tolerance mechanisms, pigment contents, photosynthetic rate, antioxidant system, glyoxylate cycle, and osmolyte metabolism	Alam et al. (2022)
10 μM	*Medicago sativa*	Cd	Enhanced proline contents, SOD, and CAT activities	Najafi and Taghizadeh (2022)
–	*Solanum lycopersicum*	Cd, Cr, Cu, Ni, and Zn	Improved growth and altered heavy metals accumulation	Vitus Ogugua et al. (2022)
100 mg/L	*Corchorus capsularis*	Cu	Improved morpho-physiological, antioxidative defense attributes	Alatawi et al. (2023)
150 ppm	*Triticum aestivum*	Ni	Upregulated total chlorophyll, carotenoids, and carbohydrates	Gulnaz et al. (2023)
–	*Glycine max*	Ni	Improved response of photosynthetic pigments, antioxidant enzyme (CAT, SOD, APX, and GSH), glyoxalase I & glyoxalase II activities, and phytochelatins (PCs) genes	Bhat et al. (2023)
10 μM	*Vigna radiata*	Cd	Enhanced amylase activity, total chlorophyll content, carbohydrate metabolism, growth, ROS-detoxifying enzymes (SOD, APX, and GR), antioxidant enzyme, and morpho-physiological characteristics	Kaur et al. (2023)
150 ppm	*Triticum aestivum*	Ni	Mitigation of the venomous effect of Ni along with enhanced biomass and reduced concentration of Ni in plants	Gulnaz et al. (2023)

was attenuated by the application of GA_3 (Amri et al. 2016). Cd contamination affects crop growth, and to diminish Cd toxicity, the application of gibberellic acid and jasmonic acid individually and in combination to regulate Cd stress in chickpeas was studied. GA_3 enhanced various parameters, including proline, glycine betaine, ascorbate glutathione, glyoxalase, and reduced EL, MDA, and ROS-mediated genes (Ahmed et al. 2021). Cd toxicity effects were seen on mungbean, differentiating it with or without applying GA_3. GA_3 mitigates Cd toxicity by harming shoot and root growth, boosting overall plant metabolism, and improving photosynthetic pigments. GA_3 being righteous to alleviate phytotoxicity also resulted in improving shoot and root morphometry and transcript abundance of VrPCS1, VrIRT1, VrIRT2, and VrCD29 (Hakla et al. 2021).

Exogenous foliar application of GA_3 along with uniconazole alleviates Cd contamination in *Zea mays* L. crop. It showed supremacy in improving biomass, plant height, stem diameter, leaf area, and root parameters along with increasing the SPAD value, antioxidative enzymatic genes, and net photosynthetic rate in plants under various concentrations (Sun et al. 2020). Phytoextraction of Cd and Pb toxicity with exogenous application of GA_3 alone and in combination with oxalic acid in *Sedum alfredii* leads to improved shoot biomass, enhanced chlorophyll content, potassium contents, and decreased malondialdehyde content (Liang et al. 2021). *Chlorella vulgaris* showed the influence of GA_3 application concerning adaptation and stress tolerance against Cd and Pb stress. It resulted in modulating growth, metal biosorption, cell number, proteins, monosaccharides, chlorophyll a and b, and total carotenoids (Falkowska et al. 2011). Cd toxicity in seedlings of *Cyphomandra betacea* was mitigated by applying GA_3 under different concentrations which resulted in increased stomatal conductance, photosynthetic rate, intercellular CO_2 concentration, transpiration rate, and peroxidase, superoxide dismutase, and catalase activities (Yang et al. 2021).

6.6.2 Against Cu Toxicity

The essentiality of copper in plant growth is known, but an excess amount leads to negative effects. Copper toxicity in peas affects the growth, physiology, and enzymatic activities of the crops, which was alleviated by applying GA_3, which escalated the activity of antioxidant enzymes, i.e., superoxide dismutase, peroxidase, and catalase along with growth (Javed et al. 2021). Copper toxicity in spinach seedlings was restricted by GA_3 application, which in response enhanced proline concentration, antioxidant enzyme activities, photosynthetic parameters, and total biomass of the crop (Gong et al. 2021). In copper-contaminated soil, the response of the pea crop was observed, where, under various concentrations, gibberellin was applied to the crop at different intervals. Foliar application of GA_3 concentration leads to alleviated CAT, GSH, SOD, APX, and POD and aggravated MAD and H_2O_2 to mitigate copper-induced phytotoxicity (Javed et al. 2021). Measures were taken for developing copper stress tolerance in spinach seedlings and concerning phytoextraction in copper-contaminated soil. Exogenous GA_3 foliar application was applied in various concentrations to mitigate copper stress, where GA_3 refined the expression of glycine betaine, ascorbate glutathione, proline, and glyoxylate and suppressed EL, MDA, and ROS-mediated genes (Gong et al. 2021). The combined effect of phosphorus and GA_3 at varying concentrations was analyzed on the growth of plants in copper-contaminated soil. It leads to modulating morpho-physiological, defense attributes, photosynthetic pigments, and gas exchange characteristics of jute crops (Alatawi et al. 2023).

6.6.3 Against Ni and Se Toxicity

GA_3 reduced the negative effect of nickel toxicity on many crops. Total chlorophyll, carotenoids, and carbohydrates, along with growth parameters, were enhanced by GA_3 application on contaminated and non-contaminated land (Gulnaz et al. 2023). Nickel toxicity in the soybean crop affected its growth and several other factors, causing lethal effects on human health. It was mitigated by GA_3, which resulted in enhanced seed germination, plant growth, biomass indices, and water contents (Alam et al. 2022). Wheat exposed to nickel toxicity, which poses lethal effects to humans, showed mitigation when GA_3 was applied. This treatment reduced the toxic effects of Ni, enhanced biomass, and lowered Ni concentration in the plants (Gulnaz et al. 2023). Proline and GA_3 played an important role in maize growth under Pb stress. Earlier mentioned two had increased osmolytes (P, phenolics; PH and Glycine betaine; GB), photosynthetic pigments (chlorophyll and carotenoids), along with compost in combination in various treatments (Adejumo 2020). The role played by *Cyphomandra betaca* to uptake Selenium (Se) under Se stress by applying GA_3 in different concentrations was studied. Along with enhancing Se content in *C. betacea*, GA_3 simultaneously alleviated the photosynthetic pigment content, photosynthetic parameters, and antioxidant enzyme activity of *C. betacea* (Xu et al. 2022).

In tomato, boron toxicity leads to a reduction of its dry biomass, leaf water potential, and leaf relative water content, which was elevated by foliar application of gibberellic acid along with other factors (Kaya et al. 2020). Reduction in total biomass, growth, and net yield due to Cr toxicity was identified. Exogenous application of GA3 individually and in combination with kinetin upregulated the net growth and total biomass of the crop (Alam et al. 2022). The fatal effect of nickel toxicity in soybean was mitigated by gibberellic acid treatment. It improved crop tolerance against nickel toxicity along with regulating photosynthetic pigments, antioxidant enzymes (CAT, SOD, APX, and GSH), glyoxalase I and glyoxalase II activities, phytochelatins (PCs) genes and suppressed activity of MDA level, ROS, electrolyte, and methylglyoxal contents (Bhat et al. 2023).

6.6.4 Against B and Cr Toxicity

Boron toxicity in coriander was alleviated by applying GA_3 along with NPK against varying levels of boron toxicity in soil. It resulted in enhancing all enzymatic and non-enzymatic antioxidative-related genes and decreased ROS (EL and H_2O_2) and MDA-mediated genes compared with boron toxicity alone (Saleem et al. 2021). In tomato, boron toxicity resulted in a reduction of chlorophyll content, photosynthetic quantum yield (Fv/Fm), ascorbate (AsA), and glutathione (GSH). Foliar application of Gibberellic acid alone or in combination with NaHS improved the activity of antioxidant enzymes (catalase, superoxide dismutase, and peroxidase) and antioxidant defense system (Kaya et al. 2020). Cr toxicity in faba beans affected the overall growth of the crop, simultaneously decreased the photosynthetic pigments, and inhibited photosynthesis, gas exchange parameters, antioxidant enzymes, and the glyoxylate cycle. Gibberellic acid alone and in combination with kinetin was supplemented, which improved tolerance mechanisms, pigment contents, and photosynthetic rate, antioxidant system, glyoxylate cycle, and osmolyte metabolism (Alam et al. 2022).

6.7 Conclusion

This review highlights the role of GA in the light of heavy metal stress, and we come to know several important aspects regarding the perception of GA. However, the gaps in the understanding of GA-regulated pathways were filled with numerous unanswered questions. Furthermore, a better understanding is slightly developed regarding the physiological responses of plants. Several attempts have been made to illustrate the significance of GA in monitoring plant survival, growth, and productivity. Positively, progress is underway, and many more components still need to be highlighted, such as early perception of heavy metal (HM) stress and more precise gibberellin (GA)- mediated strategies. Interestingly, future research can be directed toward the heavy metal-specific response and the cumulative response to various heavy metals in a more remarkable visualization.

References

Adejumo SA (2020) Exogenous proline and hormone in combination with compost improves growth and tolerance of maize under heavy metal stress. Plants Environ 2:40–53

Afonne OJ, Ifediba EC (2020) Heavy metals risks in plant foods–need to step up precautionary measures. Curr Opin Toxicol 22:1–6

Ahmed T, Noman M, Manzoor N, Shahid M, Abdullah M, Ali L, Li B (2021) Nanoparticle-based amelioration of drought stress and cadmium toxicity in rice via triggering the stress responsive genetic mechanismsand nutrient acquisition. Ecotoxicol Environ Saf 209:111829

Ahmed SF, Ullah H, Aung MZ, Tisarum R, Cha-Um S, Datta A (2023) Iron toxicity tolerance of rice genotypes about growth, yield and physiochemical characters. Rice Sci 30(4):321–334

Alam P, Azzam MA, Balawi T, Raja V, Bhat JA, Ahmad P (2022) Mitigation of negative effects of chromium (VI) toxicity in faba bean (Vicia faba) plants through the supplementation of kinetin (KN) and gibberellic acid (GA3). Plants (Basel) 11:3302

Alatawi A, Wang X, Saleem MH, Mohsin M, Rehman M, Usman K, Fahad S, Mfarrej MFB, Hefft DI, Ali S (2023) Individual and synergic effects of phosphorus and gibberellic acid on organic acids exudation pattern, ultra-structure of chloroplast and stress response gene expression in Cu-stressed jute (Corchorus Capsularis L.). J Plant Growth Regul 42:1186–1211

Alengebawy A, Abdelkhalek ST, Qureshi SR, Wang MQ (2021) Heavy metals and pesticides toxicity in agricultural soil and plants: ecological risks and human health implications. Toxics 9(3):42

Ali MA, Asghar HN, Khan MY, Muhammad Saleem MS, Muhammad Naveed MN, Niazi NK (2015) Alleviation of nickel-induced stress in mungbean through application of gibberellic acid

Altaf MA, Hao Y, He C, Mumtaz MA, Shu H, Fu H, Wang Z (2022) Physiological and biochemical responses of pepper (Capsicum annuum L.) seedlings to nickel toxicity. Front Plant Sci 13:950392

Amri B, Khamassi K, Ali MB, Teixeira da Silva JA, Bettaieb Ben Kaab L (2016) Effects of gibberellic acid on the process of organic reserve mobilization in barley grains germinated in the presence of cadmium and molybdenum. S Afr J Bot 106:35–40

Anjum SA, Ashraf U, Khan I, Tanveer M, Saleem MF, Wang L (2016) Aluminum and chromium toxicity in maize: implications for agronomic attributes, net photosynthesis, physiobiochemical oscillations, and metal accumulation in different plant parts. Water Air Soil Pollut 227:1–14

Ashraf U, Kanu AS, Mo Z, Hussain S, Anjum SA, Khan I, Tang X (2015) Lead toxicity in rice: effects, mechanisms, and mitigation strategies—a mini review. Environ Sci Pollut Res 22:18318–18332

Ashraf S, Ali Q, Zahir ZA, Ashraf S, Asghar HN (2019) Phytoremediation: environmentally sustainable way for the reclamation of heavy metal polluted soils. Ecotoxicol Environ Saf 174:714–727

Assuncao AG, Da Costa MP, De Folter S, Vooijs R, Schat H, Aarts MGM (2001) Elevated expression of metal transporter genes in three accessions of the metal hyperaccumulator Thlaspi caerulescens. Plant Cell Environ 24:217–226

Bahmani R, Bihamta MR, Habibi D, Forozesh P, Ahmadvand S (2012) Effect of cadmium chloride on growth parameters of different bean genotypes (Phaseolus vulgaris L.). ARPN J Agric Biol Sci 7:35–40

Balestrasse KB, Benavides MP, Gallego SM, Tomaro ML (2003) Effect on cadmium stress on nitrogen metabolism in nodules and roots of soybean plants. Funct Plant Biol 30:57–64

Bhat JA, Basit F, Alyemeni MN, Mansoor S, Kaya C, Ahmad P (2023) Gibberellic acid mitigates nickel stress in soybean by cell wall fixation and regulating oxidative stress metabolism and glyoxalase system. Plant Physiol Biochem 198:107678

Bücker-Neto L, Paiva ALS, Machado RD, Arenhart RA, Margis-Pinheiro M (2017) Interactions between plant hormones and heavy metals responses. Genet Mol Biol 40:373–386

Chatterjee J, Chatterjee C (2000) Phytotoxicity of cobalt, chromium and copper in cauliflower. Environ Pollut 109:69–74

da Cunha, do Nascimento CWA (2009) Silicon effects on metal tolerance and structural changes in maize (Zea mays L.) grown on a cadmium and zinc enriched soil. Water Air Soil Pollut 197:323–330

DalCorso G, Fasani E, Manara A, Visioli G, Furini A (2019) Heavy metal pollutions: state of the art and innovation in phytoremediation. Int J Mol Sci 20:1–17

Dalvi AA, Bhalerao SA (2013) Response of plants towards heavy metal toxicity: an overview of avoidance, tolerance and uptake mechanism. Ann Plant Sci 2(9):362–368

Dhaliwal SS, Singh J, Taneja PK, Mandal A (2020) Remediation techniques for removal of heavy metals from the soil contaminated through different sources: a review. Environ Sci Pollut Res 27(2):1319–1333

Falkowska M, Pietryczuk A, Piotrowska A, Bajguz A, Grygoruk A, Czerpak R (2011) Introduction the effect of gibberellic acid (GA 3) on growth, metal biosorption and metabolism of the green algae Chlorella vulgaris (Chlorophyceae) Beijerinck exposed to cadmium and lead stress

Fei X, Lou Z, Xiao R, Ren Z, Lv X (2022) Source analysis and source-oriented risk assessment of heavy metal pollution in agricultural soils of different cultivated land qualities. J Clean Prod 341:130942

Ghori NH, Ghori T, Hayat MQ, Imadi SR, Gul A, Altay V, Ozturk M (2019) Heavy metal stress and responses in plants. Int J Environ Sci Technol 16(3):1807–1828

Ghuge SA, Nikalje GC, Kadam US, Suprasanna P, Hong JC (2023) Comprehensive mechanisms of heavy metal toxicity in plants, detoxification, and remediation. J Hazard Mater 450:131039

Gong Q, Li ZH, Wang L, Zhou J-Y, Kang Q, Niu D-D (2021) Gibberellic acid application on biomass, oxidative stress response, and photosynthesis in spinach (Spinacia oleracea L.) seedlings under copper stress. Environ Sci Pollut Res 28:53594–53604

Guerinot ML (2000) The ZIP family of metal transporters. Biochim Biophys Acta 1465:190–198

Gulnaz T, Hayat Bhatti K, Hussain Siddiqi E, Iqbal I, Hussain K, Zafar M, Ajaib M, Zeb Khan S, Hussain T (2023) Interactive effect of gibberellic acid and nickel stress at germination stages in two varieties of wheat interactive effect of gibberellic acid and nickel stress at germination stages in two varieties of wheat (Triticum aestivum L.). GU J Phytosci 3(1):53–59

Guo J, Dai X, Xu W, Ma M (2008) Over expressing GSHI and AsPCSI simultaneously increases the tolerance and accumulation of cadmium and arsenic in Arabidopsis thaliana. Chemosphere 72:1020–1026

Guo Q, Meng L, Humphreys MW, Scullion J, Mur LA (2017) Expression of FlHMA3, a P1B2-ATPase from Festulolium loliaceum, correlates with response to cadmium stress. Plant Physiol Biochem 112:270–277

Hadi F, Ali N, Ahmad A (2014) Enhanced phytoremediation of cadmium-contaminated soil by parthenium hysterophorus plant: effect of gibberellic acid (GA3) and synthetic chelator, alone and in combinations. Biorem J 18:46–55

Hakla HR, Sharma S, Urfan M, Yadav NS, Rajput P, Kotwal D, Latef AAHA, Pal S (2021) Gibberellins target shoot-root growth, morpho-physiological and molecular pathways to induce cadmium tolerance in Vigna radiata L. Agronomy 11:896

Hanfi MY, Mostafa MY, Zhukovsky MV (2020) Heavy metal contamination in urban surface sediments: sources, distribution, contamination control, and remediation. Environ Monit Assess 192(1):1–21

Hanikenne M, Baurain D (2014) Origin and evolution of metal P-type ATPases in plantae (Archaeplastida). Front Plant Sci 4:544

Hasan MK, Cheng Y, Kanwar MK, Chu XY, Ahammed GJ, Qi ZY (2017) Responses of plant proteins to heavy metal stress—a review. Front Plant Sci 8:1492

He Y, Yang Z, Li M, Jiang M, Zhan F, Zu Y, Li T, Zhao Z (2017) Effects of a dark septate endophyte (DSE) on growth, cadmium content, and physiology in maize under cadmium stress. Environ Sci Pollut Res 24:18494–18504

Hu YT, Ming F, Chen WW, Yan JY, Xu ZY, Li GX, Xu CY, Yang JL, Zheng SJ (2012) TcOPT3, a member of oligopeptide transporters from the hyperaccumulator Thlaspi caerulescens, is a novel Fe/Zn/cd/cu transporter. PLoS One 7(6):e38535

Huang Y, Chen Q, Deng M, Japenga J, Li T, Yang X, He Z (2018) Heavy metal pollution and health risk assessment of agricultural soils in a typical peri-urban area in Southeast China. J Environ Manag 207:159–168

Hussain S, Khaliq A, Noor MA, Tanveer M, Hussain HA, Hussain S, Shah T, Mehmood T (2020) Metal toxicity and nitrogen metabolism in plants: an overview. In: Carbon and nitrogen cycling in soil. Springer, Singapore, pp 221–248

Imtiaz M, Tu S, Xie Z, Han D, Ashraf M, Rizwan MS (2015) Growth, V uptake, and antioxidant enzymes responses of chickpea (Cicer arietinum L.) genotypes under vanadium stress. Plant Soil 390:17–27

Iqbal MS, Bashir F, Akbar M, Ahmad KS, Ali MA, Hussain SA, Arshad N, Masood H, Munir S, Ahmad T, Islam M (2022) Cadmium toxicity alleviation through exogenous application of gibberellic acid (GA3) in mustard (Brassica juncea (L.) Czern.) and rapeseed (Brassica rapa L.). Sains Malays 51:3607–3619

Javed T, Ali MM, Shabbir R, Anwar R, Afzal I, Mauro RP (2021) Alleviation of copper-induced stress in pea (Pisum sativum l.) through foliar application of gibberellic acid. Biology (Basel) 10:1–14

Jia X, Fu T, Hu B, Shi Z, Zhou L, Zhu Y (2020) Identification of the potential risk areas for soil heavy metal pollution based on the source-sink theory. J Hazard Mater 393:122424

Jiang HH, Cai LM, Wen HH, Hu GC, Chen LG, Luo J (2020) An integrated approach to quantifying ecological and human health risks from different sources of soil heavy metals. Sci Total Environ 701:134466

Karthik C, Oves M, Thangabalu R, Sharma R, Santhosh SB, Arulselvi PI (2016) Cellulosimicrobium funkei-like enhances the growth of Phaseolus vulgaris by modulating oxidative damage under Chromium (VI) toxicity. J Adv Res 7(6):839–850

Kaur H, Nazir F, Hussain SJ, Kaur R, Rajurkar AB, Kumari S, Khan MIR (2023) Gibberellic acid alleviates cadmium-induced seed germination inhibition through modulation of carbohydrate metabolism andantioxidant capacity in mung bean seedlings. Sustainability 15(4):3790

Kaya C, Sarıoğlu A, Ashraf M, Alyemeni MN, Ahmad P (2020) Gibberellic acid-induced generation of hydrogen sulfide alleviates boron toxicity in tomato (Solanum lycopersicum L.) plants. Plant Physiol Biochem 153:53–63

Khan NA, Syeed S, Masood A, Nazar R, Iqbal N (2010) Application of salicylic acid increases contents of nutrients and antioxidative metabolism in mungbean and alleviates adverse effects of salinity stress. Int J Plant Biol 1(1):e1

Khan A, Khan S, Khan MA, Qamar Z, Waqas M (2015) The uptake and bioaccumulation of heavy metals by food plants, their effects on plants nutrients, and associated health risk: a review. Environ Sci Pollut Res 22:13772–13799

Khan M, Al Azzawi TNI, Imran M, Hussain A, Mun BG, Pande A, Yun BW (2021a) Effects of lead (Pb)-induced oxidative stress on morphological and physio-biochemical properties of rice. Biocell 45(5):1413

Khan M, Rolly NK, Al Azzawi TNI, Imran M, Mun BG, Lee IJ, Yun BW (2021b) Lead (Pb)-induced oxidative stress alters the morphological and physio-biochemical properties of rice (Oryza sativa L.). Agronomy 11(3):409

Kumar V, Chopra AK (2016) Agronomical performance of high yielding cultivar of eggplant (Solanum melongena L.) grown in sewage sludge amended soil. Res Agric 1(1):1–24

Li HF, Gray C, Mico C, Zhao FJ, McGrath SP (2009) Phytotoxicity and bioavailability of cobalt to plants in a range of soils. Chemosphere 75:979–986

Li C, Zhou K, Qin W, Tian C, Qi M, Yan X, Han W (2019) A review on heavy metals contamination in soil: effects, sources, and remediation techniques. Soil Sediment Contam Int J 28(4):380–394

Liang Y, Xiao X, Guo Z, Peng C, Zeng P, Wang X (2021) Co-application of indole-3-acetic acid/gibberellin and oxalic acid for phytoextraction of cadmium and lead with Sedum alfredii Hance from contaminated soil. Chemosphere 285:131420

Lin YF, Hassan Z, Talukdar S, Schat H, Aarts MG (2016) Expression of the ZNT1 zinc transporter from the metal hyperaccumulator Noccaea caerulescens confers enhanced zinc and cadmium tolerance and accumulation to Arabidopsis thaliana. PLoS One 11(3):e0149750

Maksimović I, Kastori R, Krstić L, Luković J (2007) Steady presence of cadmium and nickel affects root anatomy, accumulation and distribution of essential ions in maize seedlings. Biol Plant 51:589–592

Maleci L, Buffa G, Wahsha M, Bini C (2014) Morphological changes induced by heavy metals in dandelion (Taraxacumofficinale Web.) growing on mine soils. J Soils Sediments 14:731–743

Malik NJ, Chamon AS, Mondal MD, Elahi SF, Faiz SMA (2011) Effect of different levels of zinc on growth and yield of red amaranthus and rice. J Bangladesh Assoc Young Res 1(1):79–91

Mansour MM, Kamel EAR (2005) Interactive effect of heavy metals and gibberellic acid on mitotic activity and some metabolic changes of Vicia faba L plants. Cytologia, 70(3):275–282

Mokhele B, Zhan X, Yang G, Zhang X (2012) Nitrogen assimilation in crop plants and its affecting factors. Can J Plant Sci 92(3):399–405

Najafi F, Taghizadeh Z (2022) The effects of gibberellic acid on certain physiological parameters in alfalfa (Medicago sativa L.) under cadmium stress. J Plant Environ 17(66):163–175

Naqvi ZB, Shamsi L, Nazir S, Idrees A, Abid R, Yaqoob S, Mumtaz A, Firdous H (2020) Effects of gibberellic acid under cadmium chloride stress on maize (Zea mays L.). Int J Biol Res 3(1):2618–1444

Nematshahi N, Lahouti M, Ganjeali A (2012) Accumulation of chromium and its effects on growth of Allium cepa cv. Hybrid. Eur J Exp Biol 2(4):969–974

Noor I, Sohail H, Sun J, Nawaz MA, Li G, Hasanuzzaman M, Liu J (2022) Heavy metal and metalloid toxicity in horticultural plants: tolerance mechanism and remediation strategies. Chemosphere 303:135196

Pandey A, Wu LB, Murugaiyan V, Schaaf G, Ali J, Frei M (2023) Differential effects of arsenite and arsenate on rice (Oryzasativa) plants differing in glutathione S-transferase gene expression. Environ Sci Pollut Res: 30:1–14

Pasricha S, Mathur V, Garg A, Lenka S, Verma K, Agarwal S (2021) Molecular mechanisms underlying heavy metal uptake, translocation and tolerance in hyperaccumulators-an analysis: heavy metal tolerance in hyperaccumulators. Environ Chall 4:100197

Peralta JR, GardeaTorresdey JL, Tiemann KJ, Gomez E, Arteaga S, Rascon E (2001) Uptake and effects of five heavy metals on seed germination and plant growth in alfalfa (Medicago sativa L.). Bull Environ Contam Toxicol 66:727–734

Pirzadah TB, Malik B, Tahir I, Irfan QM, Rehman RU (2018) Characterization of mercury-induced stress biomarkers in Fagopyrumtataricum plants. Int J Phytoremediation 20(3):225–236

Radha J, Srivastava S, Madan VK (2000) Influence of chromium on growth and cell division of sugarcane. Indian J Plant Physiol 5(3):228–231

Rai R, Agrawal M, Agrawal SB (2016) Impact of heavy metals on physiological processes of plants: with special reference to photosynthetic system. In: Plant responses to xenobiotics. Springer, Singapore, pp 127–140

Rajendran S, Priya TA, Khoo KS, Hoang TK, Ng HS, Munawaroh HS, Karaman C, Orooji Y, Show PL (2022) A critical review on various remediation approaches for heavy metal contaminants removal from contaminated soils. Chemosphere 287:132369

Rau S, Miersch J, Neumann D, Weber E, Krauss GJ (2007) Biochemical responses of the aquatic moss Fontinalis antipyretica to Cd, Cu, Pb and Zn determined by chlorophyll fluorescence and protein levels. Environ Exp Bot 59(3):299–306

Raziuddin F, Akmal M, Shah SS, Mohammad F, Zhou W (2011) Effects of cadmium and salinity on growth and photosynthetic parameters of brassica species. Pak J Bot 43(1):333–340

Rhaman MS, Imran S, Rauf F, Khatun M, Baskin CC, Murata, Hasanuzzaman M (2020) Seed priming with phytohormones: an effective approach for the mitigation of abiotic stress. Plants (Basel) 10(1):37

Riyazuddin R, Nisha N, Ejaz B, Khan MIR, Kumar M, Ramteke PW, Gupta R (2021) A comprehensive review on the heavy metal toxicity and sequestration in plants. Biomolecules 12(1):43

Romero-Puertas MC, Rodriquez-Serrano M, Corpas FJ, Gomez M, Del Rio LA, Sandalio LM (2004) Cadmium-induced subcellular accumulation of O_2 –and H_2O_2 in pea leaves. Plant Cell Environ 27(9):1122–1134

Saleem MH, Wang X, Ali S, Zafar S, Nawaz M, Adnan M, Fahad S, Shah A, Alyemeni MN, Hefft DI, Ali S (2021) Interactive effects of gibberellic acid and NPK on morpho-physio-biochemical traits and organic acid exudation pattern in coriander (*Coriandrum sativum* L.) grown in soil artificially spiked with boron. Plant Physiol Biochem 167:884–900

Seneviratne M, Rajakaruna N, Rizwan M, Madawala HMSP, Ok YS, Vithanage M (2019) Heavy metal-induced oxidative stress on seed germination and seedling development: a critical review. Environ Geochem Health 41:1813–1831

Sethy SK, Ghosh S (2013) Effect of heavy metals on germination of seeds. J Nat Sci Biol Med 4(2):272

Shah O, Khan S, Haq F, Malak S, Qadir A, Farhan Khan M, Naqvi Z, Ali S, Khan MM, Shah U, Awan AW, Usman M, Shahzad Z (2021) Effects of gibberellic acid under cadmium chloride stress on maize (Zea mays L.). Ann Rom Soc Cell Biol 25(7):1036–1054

Sharma RK, Agrawal M (2005) Biological effects of heavy metals: an overview. J Environ Biol 26(2):301–313

Shekar CC, Sammaiah D, Shasthree T, Reddy KJ (2011) Effect of mercury on tomato growth and yield attributes. Int J Pharm Bio Sci 2(2):B-358–BB364

Shi P, Xiao J, Wang Y, Chen L (2014) Assessment of ecological and human health risks of heavy metal contamination in agriculture soils disturbed by pipeline construction. Int J Environ Res Public Health 11(3):2504–2520

Shi W, Zhang Y, Chen S, Polle A, Rennenberg H, Luo ZB (2019) Physiological and molecular mechanisms of heavy metal accumulation in nonmycorrhizal versus mycorrhizal plants. Plant Cell Environ 42(4):1087–1103

Shinozaki K, Uemura M, Bailey-Serres J, Bray EA, Weretilnyk E (2015) Responses to abiotic stress. In: Biochemistry and molecular biology of plants, pp 1051–1100. Willey Hoboken, NJ, USA

Singh S, Parihar P, Singh R, Singh VP, Prasad SM (2016) Heavy metal tolerance in plants: role of transcriptomics, proteomics, metabolomics, and ionomics. Front Plant Sci 6:1143

Smiri M, Chaoui A, El Ferjani E (2009) Respiratory metabolism in the embryonic axis of germinating pea seed exposed to cadmium. J Plant Physiol 166(3):259–269

Souri Z, Cardoso AA, da Silva CJ, de Oliveira LM, Dari B, Sihi D, Karimi N (2019) Heavy metals and photosynthesis: recent developments. In: Photosynthesis, productivity and environmental stress, pp 107–134. Scientific Research Publishing 1521 Melwood Drive, Glendale CA 91207, USA

Sun K, Yue Y, Zhang H, Yang N, Wen D, Li X, Chen N, Yang Y, Wang K (2020) Potential of gibberellic acid (Ga3) and uniconazole for enhancing the cd absorption efficiency of maize (Zea mays L.). Pol J Environ Stud 30:851–861

Sytar O, Kumari P, Yadav S, Brestic M, Rastogi A (2019) Phytohormone priming: regulator for heavy metal stress in plants. J Plant Growth Regul 38(2):739–752

Tkalec M, Styefanic PP, Cvjetko P, Sikic S, Paulica M, Balew B (2014) The effect of cadmium-zinc interactions on biochemical responses in tobacco seedlings and adult plants. PLoS One 9(1):e87582

Vareda JP, Valente AJ, Durães L (2019) Assessment of heavy metal pollution from anthropogenic activities and remediation strategies: a review. J Environ Manag 246:101–118

Velikova V, Tsonev T, Loreto F, Centritto M (2011) Changes in photosynthesis, mesophyll conductance to CO_2, and isoprenoid emissions in Populus nigra plants exposed to excess nickel. Environ Pollut 159(5):1058–1066

Ogugua UV, Kanu SA, Ntushelo K (2022) Gibberellic acid improves growth and reduces heavy metal accumulation: a case study in tomato (Solanum lycopersicum L.) cultivar (Heinz 1370) seedlings exposed to acid mine water. Heliyon 8(12)

Wang X, Ma R, Cui D, Cao Q, Shan Z, Jiao Z (2017) Physio-biochemical and molecular mechanism underlying the enhanced heavy metal tolerance in highland barley seedlings pre-treated with low-dose gammairradiation. Sci Rep 7(1):14233

Xu Y, Zhang L, Wang J, Liang D, Xia H, Lv X, Deng Q, Wang X, Luo X, Liao M, Lin L (2022) Gibberellic acid promotes selenium accumulation in Cyphomandra betacea under selenium stress. Front Plant Sci 13:968768

Yang L, Huan Y, Sun J, Lin L, Liao M, Wang Z, Liang D, Xia H, Lv X, Wang J (2021) Effects of exogenous gibberellic acid on growth and cadmium accumulation in Cyphomandra betacea seedlings. Environ Prog Sustain Energy 40:e13655

Yourtchi MS, Bayat HR (2013) Effect of cadmium toxicity on growth, cadmium accumulation, and macronutrient content of durum wheat (Dena CV.). Int J Agric Crop Sci (IJACS) 6(15):1099–1103

Zeid IM (2001) Responses of Phaseolus vulgaris to chromium and cobalt treatments. Biol Plant 44:111–115

Zhang Q, Wang C (2020) Natural and human factors affect the distribution of soil heavy metal pollution: a review. Water Air Soil Pollut 231(7):1–3

Zhang H, Xu Z, Guo K, Huo Y, He G, Sun H, Guan Y, Xu N, Yang W, Sun G (2020) Toxic effects of heavy metal Cd and Zn on chlorophyll, carotenoid metabolism and photosynthetic function in tobacco leaves revealed by physiological and proteomics analysis. Ecotoxicol Environ Saf 202:110856

Chapter 7
Ethylene-Mediated Regulation of Plant Biology Under Heavy Metal Stress: Physiological and Molecular Aspects

Priya Sharma, Lisha Khungar, Simar Malhotra, Sahil Mehta, Bindu Yadav, and Om Prakash Narayan

Abstract Excessive heavy metals (HMs) on agricultural lands pose a significant hazard to plant growth and development, leading to a decline in crop yield and productivity. Plants use diverse strategies to combat the presence of unwanted HMs. Recent research suggests that ethylene has multiple functions in plants, playing a critical role in several physiological processes as well as enhancing HM stress tolerance. The modulation of ethylene levels in plants, achieved through numerous strategies targeting either ethylene production or the ethylene-signaling pathway, has shown encouraging results in bolstering resistance to HM stress. This chapter encompasses multiple facets of ethylene's involvement in plant responses to HM stress. It also explores the intricate interplay between ethylene and various other signaling molecules in the context of adverse HM stress conditions. Additionally, the chapter delves into various approaches for modifying ethylene's actions to enhance HM tolerance. Given our present understanding of ethylene and its regulatory functions, there is a growing belief that optimizing endogenous levels of ethylene under HM stress conditions in plants holds great promise.

Keywords Plants · Ethylene · Heavy metal stress · Ethylene inhibitors and modulators

---

P. Sharma
Department of Botany, University of Delhi, Delhi, India

L. Khungar
Department of Plant Molecular Biology, South Campus, University of Delhi, Delhi, India

S. Malhotra · S. Mehta
Department of Botany, Hansraj College, University of Delhi, Delhi, India

B. Yadav
School of Environmental Sciences, Jawaharlal Nehru University, New Delhi, India

O. P. Narayan (✉)
Division of Chemical Biology and Medicinal Chemistry, University of Texas at Austin, Austin, TX, USA

A. Sharma et al. (eds.), *Hormones and Heavy Metals*, Plant Life and Environment Dynamics, https://doi.org/10.1007/978-981-96-9622-2_7

7.1 Introduction

Plants frequently face various biotic and abiotic stress conditions, including submersion, drought, and extreme temperatures, all of which can significantly impact their growth, development, and yield. In addition to these challenges, the emergence of HM stress has become an increasingly prevalent and urgent issue (Thao et al. 2015). The prevailing issue emanates from a complex interplay of unregulated industrialization, urban expansion, and agricultural practices, including the use of contaminated inputs, viz., metal-contaminated pesticides, fertilizers, and sewage sludge, leading to a pronounced escalation of HM concentrations in the soil (Alloway 2012). Notably, while plants inherently require several different types of HMs for crucial biological functions within their cellular processes (Jogawat et al. 2021), an excessive presence of both non-essential and essential heavy metals necessitates plants to either endure their existence or reluctantly absorb them from the soil (Nies 1999; Clemens et al. 2002; Jogawat et al. 2021). Essential HMs, along with non-essential ones, make their way into root cells, from where they are transported to various tissues and subcellular locations. Subsequently, they accumulate in grains, leaves, and other plant parts, creating health risks when consumed by animals, birds, or humans (Ugulu et al. 2021; Jogawat et al. 2021). Ingesting HMs can lead to severe health issues in humans, including brain damage, liver cirrhosis, dementia, and more (Ugulu et al. 2021). In plants, HMs affect plant growth, physiology, and development, primarily by inducing the production of reactive oxygen species (ROS), which subsequently leads to cell damage and ultimately results in reduced crop growth and yield (Nagajyoti et al. 2010; Gill and Tuteja 2010). Beyond its agricultural repercussions, the gradual accumulation of heavy metal remains in the food chain has been empirically linked to the emergence of significant ecological and health challenges (Malik 2004; Verstraeten et al. 2008). As a result, food crops are being replaced by non-food crops in HM-polluted regions. These non-food crops are specifically chosen for their ability to resist and accumulate metals, ultimately aiming to stabilize and cleanse the soil through a process known as phytoremediation (Vangronsveld et al. 2009; Nguyen et al. 2020; Alengebawy et al. 2021). Therefore, a deep understanding of how plants cope with metal-induced stress is necessary.

HM stress triggers oxidative stress in plants, which is characterized by excessive ROS and methylglyoxal generation. Their elevated levels have detrimental effects on cellular structure maintenance, including lipid peroxidation in the membranes, degradation of biological macromolecules, ion leakage, and DNA strand breakage (Nagajyoti et al. 2010; Gill and Tuteja 2010; Thao et al. 2015). Furthermore, HM stress disrupts numerous biochemical, molecular, and physiological processes. Consequently, plant growth is impeded, leading to a decline in economic yield (Yadav 2010; Hossain et al. 2012; Anjum et al. 2012; Asgher et al. 2015). Plants adopt various approaches to counteract the presence of unwanted HMs. One common response to HM stress is the altered expression of genes involved in ethylene production and ethylene-responsive genes (Maksymiec 2007; Khan and Khan 2014;

Asgher et al. 2015). Studies have shown an increase in ethylene production in plants exposed to toxic levels of HMs, including cadmium (Cd), nickel (Ni), zinc (Zn), copper (Cu), and iron (Fe). Notably, Cd and Cu induce ethylene production by enhancing the activity of 1-aminocyclopropane-1-carboxylic acid (ACC) synthase (ACS), a key enzyme in the ethylene synthesis pathway (Thao et al. 2015; Keunen et al. 2016; Iqbal et al. 2017). While ethylene is recognized for its role in modulating essential physiological processes such as seed germination, nodule and root hair formation, fruit ripening, and plant senescence (Thao et al. 2015; Keunen et al. 2016), its precise function in response to elevated HM stress still needs deep understanding to address a wide range of environmental, agricultural, and health challenges. This chapter explores ethylene's role in plants under HM stress and highlights its integration into the broader signaling cascade activated in metal-exposed plants through experimental evidence.

7.2 Ethylene Biosynthesis and Signaling Under Heavy Metal Stress

Ethylene is a crucial hormone that modulates various stages of plant growth and development, from germination to senescence. It also occupies a pivotal position in how plants react to various stressors. When plants experience stress, ethylene levels increase, acting as a signal for adaptive responses and modulating other signaling pathways (Iqbal et al. 2017). Ethylene synthesis involves a three-step process, commencing with the generation of S-adenosylmethionine (SAM) from the substrate methionine by SAM synthetase. ACC synthase (ACS) then converts SAM into 1-aminocyclopropane-1-carboxylic acid (ACC). (Wang et al. 2002). This is the pathway's rate-limiting step, producing 5′-methylthioadenosine (MTA) and recycling it to methionine via the "Yang cycle." ACC undergoes oxidation via ACC oxidase (ACO) to generate ethylene (Fig. 7.1). Notably, as the primary enzymes in ethylene biosynthesis, both ACS and ACO are governed by multigene families and serve as pivotal regulatory nodes within the ethylene biosynthetic pathway (Wang et al. 2002). Exposure to HM stress triggers an upregulation in the activity of ACO and ACS enzymes, leading to an elevated production of ethylene (Schellingen et al. 2014; Khan et al. 2015).

Ethylene signaling kicks off with a group of receptors embedded in the endoplasmic reticulum (ER). Ethylene, being a volatile molecule, can move freely within the cell, eventually reaching these receptors. In *Arabidopsis thaliana*, there are five genes responsible for high-affinity ethylene receptors, viz., ETHYLENE RESISTANT 1 and 2 (ETR1/2), ETHYLENE RESPONSE SENSOR 1 and 2 (ERS1/2), and ETHYLENE INSENSITIVE 4 (EIN4). In the absence of ethylene, a protein called CONSTITUTIVE TRIPLE RESPONSE1 (CTR1), acting like a Raf-like MAPKKK, interacts with these ethylene receptors. This interaction results in the suppression of a downstream component, EIN2, through direct phosphorylation

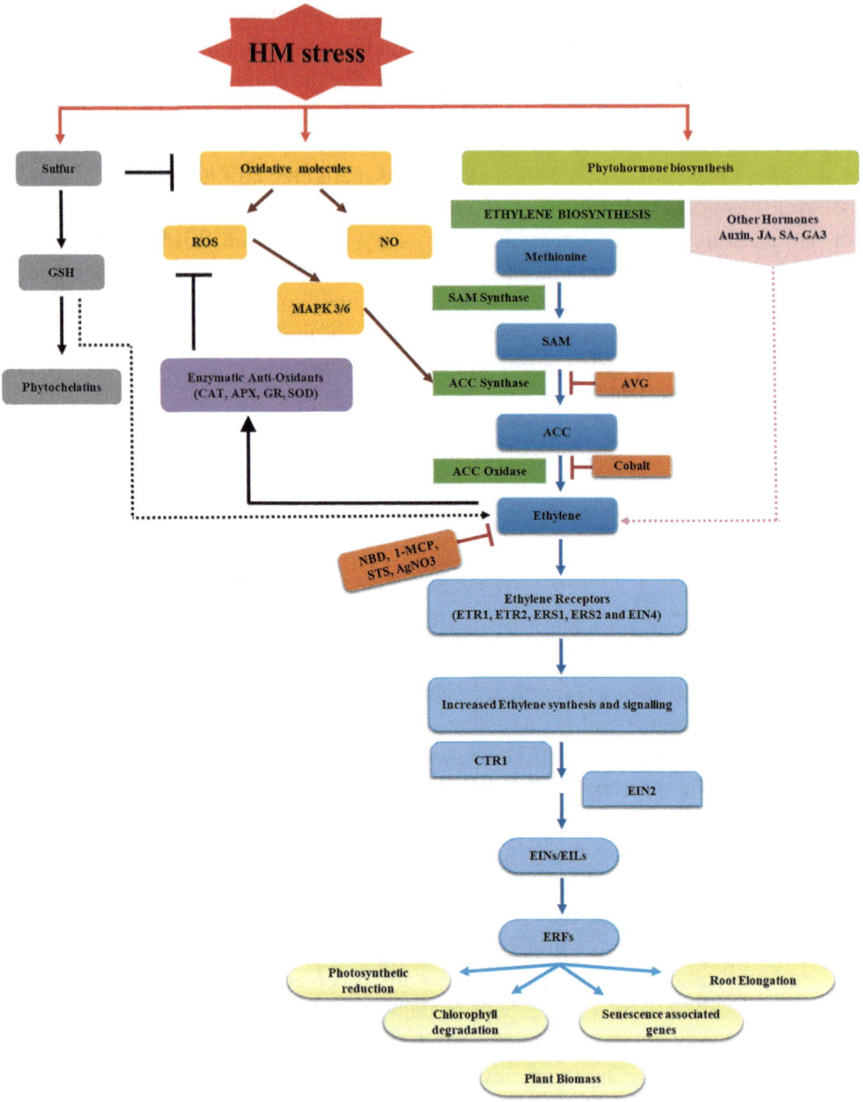

Fig. 7.1 Schematic representation of Ethylene biosynthesis and signaling under HM stress. Exposure of plants to Heavy metals exerts influence on a multitude of signaling pathways. Phytohormone biosynthesis, specifically ethylene biosynthesis, is generally prompted at the enzymatic levels of ACC synthase (ACS) and oxidase (ACO), consequently setting in motion the intricate ethylene-signaling cascade. Further, the enhanced production of reactive oxygen species (ROS) as well as reactive nitrogen species (RNS) serves as a precursor to oxidative signaling pathways. These pathways are notably orchestrated by mitogen-activated protein kinases (MAPK). Lastly, the pivotal role of glutathione (GSH) extends beyond its capacity as an antioxidant; it acts as a foundational element in the biosynthesis of metal-chelating phytochelatins. Metal-stressed plants exhibit intricate interplay among various signaling components. Dashed lines signify potential regulatory effects in response to heavy metal (HM) stress, while the cross symbolizes the release from inhibition

of its cytosolic C-terminal domain, which leads to the deactivation of EIN3 and ETHYLENE-INSENSITIVE3-LIKE1 (EIL1) (Shan et al. 2012). When ethylene binds to these receptors, facilitated by copper ions provided by RESPONSIVE TO ANTAGONIST1 (RAN1), the Cu transporter, CTR1 is inactivated and becomes inactive. This causes cleavage of EIN2's carboxyl terminus; as a result, EIN2 is transported from the endoplasmic reticulum to the nucleus, stabilizing the EIN3 protein and initiating an extensive signaling cascade (Ju et al. 2012). HM stress conditions, such as exposure to cadmium (Cd), predominantly drive ethylene synthesis through an increase in the expression of ACS2 and ACS6 transcripts. (Schellingen et al. 2014) (Table 7.1). An interesting discovery comes from an Arabidopsis mutant that lacks both ACS2 and ACS6 genes. When this mutant was exposed to Cd, it produced less ethylene and exhibited improved leaf biomass. This suggests that ethylene may negatively regulate plant development under HM stress conditions (Schellingen et al. 2014). Further, under HM stress conditions, ACS2 and ACS6, are phosphorylated by MAPKs, increasing their stability and activity. These MAPKs also possess the capability to phosphorylate the WRKY33 TF, which subsequently induces ACS2 and ACS6 gene production by binding to their respective promoters (Li et al. 2012). While most studies have focused on gene expression levels, other regulatory mechanisms like post-transcriptional and post-translational modifications (PTMs) can also impact enzyme activity during HM exposure. Furthermore, the MAPK cascade MKK9-MPK3/6, operating downstream of CTR1, regulates EIN3, a key TF involved in ethylene biosynthesis (Yoo et al. 2008; Xu et al. 2008). Given the current limitations in our understanding of ethylene signaling under HM stress, there is a clear need for further exploration in this pivotal research area.

7.3 Crosstalk of Ethylene with Other Components Under HM Stress Conditions

The molecular mechanisms enabling plants to adapt to various HM stresses are diverse and can vary among different plant species. However, in general, ethylene plays a crucial role in enabling plant adaptation to HM exposure by interacting with other phytohormones and signaling molecules (Thapa et al. 2012; Montero-Palmero et al. 2014a, b). Excessive HM not only affects the production of ethylene but also influences other phytohormones. Apart from ethylene, the levels of several other phytohormones, including salicylic acid (SA), jasmonic acid (JA), and, abscisic acid (ABA) increase, while the levels of gibberellic acid (GA3) and auxin decrease (Metwally et al. 2003; Cánovas et al. 2004; Saini et al. 2021). Brassinosteroids, auxins, SA, GA3, and cytokinin also play roles in antioxidant responses against ROS generated in response to Pb, Cu, and Cd exposure (Saini et al. 2021; Thao et al. 2015).

Table 7.1 Effect of HM stress on ethylene biosynthesis and signaling in plants

Heavy metal	Concentration	Duration	Biological sample	Increased gene expression	Physiological effect	References
Al	10 μM $AlCl_3$	2 and 24 h	*Medicago. truncatula*	ACS and ACO expression	Inhibition of root elongation	Sun et al. (2007)
Al	50 μM AlC_{l3}	0.5, 2, and 12 h	*Arabidopsis thaliana*	ACS and ACO expression	Inhibition of root elongation	Sun et al. (2010)
Al	30 μM $AlCl_3$	24 h	*T. aestivum*	ACS and ACO expression	Inhibition of root elongation	Yu et al. (2016)
As	5 (0.375 mg/L) to 200 μM (15 mg/L) $Na_2HASO_4{\cdot}7H_2O$	3 days	Rice roots	ACS1, ACS2, ACO3, ACO4, ACO5, ACO6, EIN3;3, EIN3;4 and EIN3;5	Root growth inhibition	Huang et al. (2012)
Cu	10 mM CuSO4	48 h	*Nicotiana glutinosa*	*ACO1* and *ACO3* genes	NA	Kim et al. (1998)
Cu	2.5 mM $CuCl_2$	0.5 to 6 h	*B. oleracea*	*ACS and ACO expression*	NA	Jakubowicz et al. (2010)
Cd	5 or 10 μM $CdSO_4$	24 or 72 h	*Arabidopsis thaliana*	*ACS2*, *ACS6*, *ACO2*, *ACO4*, ETR2, ERF1	NA	Schellingen et al. (2014)
Cd	50 μM Cd	2 h	*A.halleri* and *A.thaliana*	ERF1, ERF2 and ERF5	Stunted rosette growth and root growth inhibition in *A.thaliana*	Weber et al. (2006a, b)
Cr	0–400 μM of K_2CrO_4	3 days	Rice	ACS1, ACS2, ACO4, ACO5	Root growth inhibition	Jogawat et al. (2021)
Hg	3 μM $HgCl_2$	3, 6, and 24 h	*Medicago sativa*	ACS, ACO, and ethylene-responsive genes	Root growth inhibition	Montero-Palmero et al. (2014a)

(continued)

Table 7.1 (continued)

Heavy metal	Concentration	Duration	Biological sample	Increased gene expression	Physiological effect	References
Hg	25 μM	1–24 h	*O. sativa*	ACS2, ACO1, ACO2, ACO5, ACO6, 5 MAPKKK, 1 MAPKK and 2 MAPK	Decreased root length	Chen et al. (2014)
Pb	0.5 mM $Pb(NO_3)_2$	2 days	*Arabidopsis thaliana*	*EIN2*	Root growth inhibition	Cao et al. (2014)
Pb	500 mg l^{-1} $Pb(NO_3)_2$	12 days	*Sesbania drummondii*	ACS/ACO gene	Inhibition of root and shoot growth as well as biomass decline	Srivastava et al. (2006)

SA enhances Cd tolerance in wheat and pea and mitigates Cu stress in *Brassica napus* by increasing glutathione (GSH) and boosting metal detoxification (Hayat et al. 2007; Piotrowska-Niczyporuk et al. 2012; Noriega et al. 2012). On the other hand, JA amplifies metal biosorption and ROS in microalgae and Arabidopsis exposed to Cd or Cu, possibly mediated by GSH (Piotrowska-Niczyporuk et al. 2012). Ethylene and JA signaling pathways might intersect through EIN3 and EIL1, the ethylene stabilization factors, to regulate gene expression in Arabidopsis (Song et al. 2014; Zhu et al. 2011). However, other studies have revealed that ethylene and JA regulate ethylene response factors (ERFs) through an EIN3/EIL1-independent post-translational mechanism (Bethke et al. 2009; Van der Does et al. 2013). Interestingly, when Arabidopsis plants were subjected to high levels of cadmium (Cd), the ethylene and JA signaling pathways were activated. This activation led to NITRATE TRANSPORTER 1.8 (NRT1.8) upregulation and NRT1.5 downregulation, which allowed the redistribution of nitrate to the roots in response to stress, thereby improving the plant's ability to respond to Cd stress (Zhang et al. 2014). This suggests that the regulation of gene expression by ethylene and JA in response to HM stress is complex and involves multiple pathways. Further research is needed to gain a deeper insight into the intricate interactions between these signaling molecules as well as their targets and interactors.

The interplay between ethylene and ROS plays an important role in plants' response to HMs. In tomato (*Solanum lycopersicum* L.), ethylene, and hydrogen peroxide (H_2O_2) were found to operate together synergistically, with H_2O_2 playing a significant role in ethylene-induced cell death generated by cadmium (Cd) (Liu et al. 2008). Numerous studies have demonstrated the fact that HMs can induce ROS production and modify antioxidant enzymes' activity such as catalase, peroxidase, superoxide dismutase (SOD), ascorbate peroxidase (APX), and glutathione reductase (GR) in plants (Sun et al. 2010; Montero-Palmero et al. 2014a; Khan et al. 2015). Adjusting the levels of ethylene through ethephon or NBD (see details in next section) can alleviate photosynthetic inhibition and reduce oxidative stress, possibly by enhancing the metabolism of SOD, GR, and APX in mustard plants exposed to nickel (Ni) and zinc (Zn) (Khan and Khan 2014). In some studies, the use of a ROS scavenger (GSH) impeded the activation of specific MAPKs (MAPK3 and MAPK6) triggered by HM-induced ROS accumulation. These MAPKs are known to be involved in ethylene production and signaling, although their precise role remains a matter of debate. This suggests that ROS and ethylene may interact via these MAPKs to regulate plant responses when exposed to HM stressors (Hahn and Harter 2009). The relationship between gibberellins (GA3) and ethylene has been studied using the *gibberellin insensitive ethylene overproducing 2-1* double mutant suggesting that a functional GA3 signaling pathway is necessary for increased ethylene biosynthesis in Arabidopsis (De Grauwe et al. 2008). GA3 and/or sulfur (S) at adequate levels might diminish stress-induced ethylene levels, hence alleviating cadmium (Cd) stress-induced photosynthetic suppression. (Masood and Khan 2013). Sulfur assimilation produces cysteine (Cys), which is essential for both ethylene and glutathione (GSH) production under normal conditions. (Iqbal et al. 2013; De Grauwe et al. 2008). Under HM stress, applying S to Cd-exposed plants was found to regulate stress-induced ethylene, leading to maximal GSH content. This adjustment of ethylene and GSH levels offered efficient protection against oxidative stress and alleviated Cd-induced plant symptoms (Asgher et al. 2015). This suggests that the GSH pathway may act as a checkpoint for crosstalk between ethylene and S in plant responses to HM stress.

A study on aluminum (Al) application in Arabidopsis revealed an increased abundance of ethylene-associated genes, impacting root growth. Al treatment elevated the levels of AUXIN RESISTANT1 (*At*AUX1) and PINFORMED2 (*At*PIN2) transcripts in normal plants, while ethylene inhibitors prevented this effect (Sun et al. 2010). Auxin-insensitive mutants showed reduced root growth inhibition under Al treatment, suggesting ethylene's role in influencing auxin transport (Sun et al. 2010). However, copper (Cu)-induced redistribution of auxin relied on a different gene, PIN1. Surprisingly, the Cu-induced root growth inhibition was similar in ethylene mutants, indicating ethylene was not crucial for this response (Yuan et al. 2013). Thus, plant genes involved in auxin control may act independently or dependently on ethylene signaling, depending on the HM stress type.

Nitric oxide (NO), a key plant signaling molecule, influences various plant responses, including ethylene emission, stress responses, growth, programmed cell death, senescence, and stomatal regulation (Mostofa et al. 2015; Thao et al. 2015). NO's role in heavy metal (HM) tolerance is multifaceted. Exogenous NO enhances plant tolerance to high levels of Ni, Cd, and Al, while endogenous NO may be linked to Cd toxicity (Singh et al. 2008; Kazemi et al. 2010). NO can act as an antioxidant, scavenging ROS and boosting antioxidant enzyme activity in Ni or Cd-treated plants (Kazemi et al. 2010; YUN et al. 2013). Interaction between ethylene, polyamines, NO, MAPKs, and TFs is proposed during stress induced by short-term Cd exposure in *Glycine max* seedlings (Chmielowska-Bąk et al. 2014a). These outcomes suggest NO's involvement in HM-induced ethylene pathways and highlight complex interactions between phytohormones and signaling molecules.

7.4 Approaches to Improve Plant Fitness Under HM Stress

HM stress has emerged as a serious issue due to its detrimental impact on plant productivity and thereby human health (Thapa et al. 2012). A crucial aspect of improving plant HM tolerance involves the manipulation of ethylene biosynthesis, signaling, and response (Thao et al. 2015; Keunen et al. 2016). Numerous investigations have established the effectiveness of adjusting endogenous ethylene levels using ethylene biosynthesis modulators. Ethephon, an external ethylene-releasing compound, is employed to modulate endogenous ethylene production during exposure to cadmium (Cd), nickel (Ni), or zinc (Zn) stress conditions (Masood et al. 2012; Khan and Khan 2014). It reduces HM-induced ethylene, triggering antioxidant system activation and enhanced photosynthesis. Consequently, ethephon-treated plants exhibit greater tolerance to HM stress (Masood et al. 2012; Khan and Khan 2014). Furthermore, sulfur (S) application has shown promise in alleviating Cd stress, primarily by reducing undesirable stress-induced ethylene production in mustard plants (Khan and Khan 2014).

Combining treatments involving GA3 and/or S has been effective in reducing ethylene production induced by Cd exposure while promoting a photosynthetic response to the challenges posed by Cd stress. Moreover, the application of ethylene-action inhibitors has proven effective in alleviating stress symptoms in plants subjected to diverse HM stresses, including aluminum (Al), mercury (Hg), cadmium (Cd), and nickel (Ni) or zinc (Zn) (Sun et al. 2010; Maksymiec 2011; Montero-Palmero et al. 2014b; Khan and Khan 2014). NBD, which inhibits ethylene action by blocking receptors, is commonly employed to reduce the negative effects of ethylene under stress conditions induced by nickel (Ni) and zinc (Zn), particularly by affecting photosystem II activity, enhancing photosynthetic efficiency, nitrogen use efficiency, and bolstering antioxidant capabilities (Sisler and Serek 1997; Khan and Khan 2014). Another ethylene synthesis inhibitor, aminoethoxyvinyl glycine (AVG), reduces ethylene synthesis by inhibiting the ACS enzyme's activity. (Masood et al. 2012). Cobalt (Co), typically beneficial for moderate plant growth, is known to inhibit ethylene production (Chmielowska-Bąk et al. 2014b). Excessive Co-treatment in HM-stressed plants inhibits ACO enzymatic activity and also causes oxidative stress, photosynthesis inhibition, and disruptions in various plant processes (Sun et al. 2010; Chmielowska-Bąk et al. 2014a, 2014b). As a result, using Co as an ethylene inhibitor for research purposes needs to be approached cautiously. These findings highlight the potential of multifaceted treatments to mitigate HM-induced ethylene levels and enhance plant resilience (Thao et al. 2015).

Flowering plant species, targeting the ethylene biosynthesis and signaling pathway components in response to HM stress is the preferred approach. (Tran and Pal 2014). For instance, tomato's ethylene-insensitive never ripe (Nr) mutant has displayed resistance to Cd-induced stress by enhancing antioxidant enzyme activity and influencing intercellular spaces and mesophyll size (Gratão et al. 2009; Monteiro et al. 2011). Alterations in the Nr (LeETR3) sensor domain resulted in its incapacity to respond to either endogenous or exogenous ethylene, highlighting the potential for manipulating ethylene perception in controlling plant responses to HM stress (Table 7.2). Notably, manipulating ethylene signaling-related genes that encode upper ethylene pathway components, situated between the ethylene receptor and EIN2, can often lead to pleiotropic effects (Wuriyanghan et al. 2009; Wang et al. 2013). Furthermore, under Cd stress, Ethylene Response Factor (ERF) TFs are significant regulators of certain stress-related genes (Dal Corso et al. 2010).

Table 7.2 Genetic Approaches to improve HM stress tolerance in plants

HMs	Plant	Mutants	Physiological effect	References
Al	Arabidopsis	*etr1-3* and ein2-1 mutants	Reduced inhibition of root elongation	Sun et al. (2010)
Cd	Arabidopsis	*acs2-1acs6-1* double mutants	Reduced inhibition of leaf biomass	Schellingen et al. (2014)
Cd	Tomato	*Nr* (*LeETR3*) mutant	Reduced root diameter	Gratao et al. (2009)
			Maintenance of pigment content; reduced leaf senescence	Monteiro et al. (2011)
Cd	Arabidopsis	*ein2-1 mutant*	Higher antioxidative activity, less reduction in shoot and root growth	Zhou et al. (2019)
Hg	Arabidopsis	*ein2-5* mutant	Reduced inhibition of root growth	Montero-Palmero et al. (2014b)
Li	Arabidopsis	*etr1-1* and *ein3-3 mutants*	Less sensitive to stress	Bueso et al. (2007)
Pb	Arabidopsis	*ein2-1* mutant	Enhanced antioxidant activity, reduced inhibition of root length; reduced Pb and GSH content	Cao et al. (2014)
Zn	Arabidopsis	*ein2-1 and etr1-3*	Significantly less reduction in rosette biomass, chlorophyll content, and soluble sugars in the mutant and enhanced antioxidant activity compared to the wild type	Khan et al. (2019)

7.5 Conclusion and Perspectives

HM contamination poses a significant hazard to sustainable agriculture on a global scale. Recent research has given insight into the significant role of ethylene in regulating various physiological processes and mediating HM tolerance in plants. Currently, our understanding of how ethylene functions in coping with metal-induced stress and its intricate interactions within the broader signaling network is steadily expanding. Recent findings highlight a close connection between ethylene, cellular redox balance (involving the antioxidant glutathione, GSH), and other plant hormones like gibberellic acid (GA3), salicylic acid (SA), and jasmonic acid (JA). These interactions collectively influence whether a plant becomes more sensitive or tolerant to heavy metals. Understanding the intricate mechanisms of ethylene biology and its interactions with various other signaling molecules will provide valuable insights into developing innovative strategies for enhancing plant stress responses.

Furthermore, the knowledge gained from studying HM stress tolerance can be applied not only to improve crop productivity but also to mitigate the detrimental effects of HM pollution on ecosystems and human health. Thus, this research has far-reaching implications for both agricultural sustainability and environmental conservation efforts.

References

Alengebawy A, Abdelkhalek ST, Qureshi SR, Wang M-Q (2021) Heavy metals and pesticides toxicity in agricultural soil and plants: ecological risks and human health implications. Toxics 9:42. https://doi.org/10.3390/toxics9030042

Alloway BJ (ed) (2012) Heavy metals in soils: trace metals and metalloids in soils and their bioavailability, vol 22. Springer Science & Business Media, Berlin. https://doi.org/10.1007/978-94-007-4470-7

Anjum NA, Ahmad I, Mahmood I, Pacheco M, Duarte AC, Pereira E, Umar S, Ahmad A, Khan NA, Iqbal M, Prasad MNV (2012) Modulation of glutathione and its related enzymes in plants' responses to toxic metals and metalloids—a review. Environ Exp Bot 75:307–324. https://doi.org/10.1016/j.envexpbot.2011.07.002

Asgher M, Khan MIR, Anjum NA, Khan NA (2015) Minimising toxicity of cadmium in plants—role of plant growth regulators. Protoplasma 252:399–413. https://doi.org/10.1007/s00709-014-0710-4

Bethke G, Unthan T, Uhrig JF, Pöschl Y, Gust AA, Scheel D, Lee J (2009) Flg22 regulates the release of an ethylene response factor substrate from MAP kinase 6 in Arabidopsis thaliana via ethylene signaling. Proc Natl Acad Sci 106:8067–8072. https://doi.org/10.1073/pnas.0810206106

Bueso E, Alejandro S, Carbonell P, Perez-Amador MA, Fayos J, Bellés JM, Rodriguez PL, Serrano R (2007) The lithium tolerance of the Arabidopsis cat2 mutant reveals a cross-talk between oxidative stress and ethylene. Plant J 52:1052–1065. https://doi.org/10.1111/j.1365-313x.2007.03305.x

Cánovas D, Vooijs R, Schat H, de Lorenzo V (2004) The role of thiol species in the Hypertolerance of aspergillus sp. P37 to arsenic. J Biol Chem 279:51234–51240. https://doi.org/10.1074/jbc.m408622200

Cao F, Chen F, Sun H, Zhang G, Chen Z-H, Wu F (2014) Genome-wide transcriptome and functional analysis of two contrasting genotypes reveals key genes for cadmium tolerance in barley. BMC Genomics 15:611. https://doi.org/10.1186/1471-2164-15-611

Chen Y-A, Chi W-C, Trinh NN, Huang L-Y, Chen Y-C, Cheng K-T, Huang T-L, Lin C-Y, Huang H-J (2014) Transcriptome profiling and physiological studies reveal a major role for aromatic amino acids in mercury stress tolerance in rice seedlings. PLoS One 9:e95163. https://doi.org/10.1371/journal.pone.0095163

Chmielowska-Bąk J, Gzyl J, Rucińska-Sobkowiak R, Arasimowicz-Jelonek M, Deckert J (2014a) The new insights into cadmium sensing. Front Plant Sci 5:245. https://doi.org/10.3389/fpls.2014.00245

Chmielowska-Bąk J, Lefèvre I, Lutts S, Kulik A, Deckert J (2014b) Effect of cobalt chloride on soybean seedlings subjected to cadmium stress. Acta Soc Bot Pol 83:201–207. https://doi.org/10.5586/asbp.2014.027

Clemens S, Palmgren MG, Krämer U (2002) A long way ahead: understanding and engineering plant metal accumulation. Trends Plant Sci 7(7):309–315. https://doi.org/10.1016/s1360-1385(02)02295-1

Dal Corso G, Farinati S, Furini A (2010) Regulatory networks of cadmium stress in plants. Plant Signal Behav 5(6):663–667. https://doi.org/10.4161/psb.5.6.11425

De Grauwe L, Dugardeyn J, Van Der Straeten D (2008) Novel mechanisms of ethylene-gibberellin crosstalk revealed by their eto2-1double mutant. Plant Signal Behav 3:1113–1115. https://doi.org/10.4161/psb.3.12.7037

Gill SS, Tuteja N (2010) Reactive oxygen species and antioxidant machinery in abiotic stress tolerance in crop plants. Plant Physiol Biochem 48:909–930. https://doi.org/10.1016/j.plaphy.2010.08.016

Gratão PL, Monteiro CC, Rossi ML, Martinelli AP, Peres LEP, Medici LO, Lea PJ, Azevedo RA (2009) Differential ultrastructural changes in tomato hormonal mutants exposed to cadmium. Environ Exp Bot 67:387–394. https://doi.org/10.1016/j.envexpbot.2009.06.017

Hahn A, Harter K (2009) Mitogen-activated protein kinase cascades and ethylene: signaling, biosynthesis, or both?: figure 1. Plant Physiol 149:1207–1210. https://doi.org/10.1104/pp.108.132241

Hayat S, Ali B, Aiman Hasan S, Ahmad A (2007) Brassinosteroid enhanced the level of antioxidants under cadmium stress in Brassica juncea. Environ Exp Bot 60:33–41. https://doi.org/10.1016/j.envexpbot.2006.06.002

Hossain MA, Piyatida P, da Silva JAT, Fujita M (2012) Molecular mechanism of heavy metal toxicity and tolerance in plants: central role of glutathione in detoxification of reactive oxygen species and methylglyoxal and in heavy metal chelation. J Bot 2012:1–37. https://doi.org/10.1155/2012/872875

Huang TL, Nguyen QTT, Fu SF, Lin CY, Chen YC, Huang HJ (2012) Transcriptomic changes and signalling pathways induced by arsenic stress in rice roots. Plant Mol Biol 80:587–608. https://doi.org/10.1007/s11103-012-9969-z

Iqbal N, Masood A, Khan MIR, Asgher M, Fatma M, Khan NA (2013) Cross-talk between sulfur assimilation and ethylene signaling in plants. Plant Signal Behav 8:e22478. https://doi.org/10.4161/psb.22478

Iqbal N, Khan NA, Ferrante A, Trivellini A, Francini A, Khan MIR (2017) Ethylene role in plant growth, development and senescence: interaction with other phytohormones. Front Plant Sci 08:475. https://doi.org/10.3389/fpls.2017.00475

Jakubowicz M, Gałgańska H, Nowak W, Sadowski J (2010) Exogenously induced expression of ethylene biosynthesis, ethylene perception, phospholipase D, and Rboh-oxidase genes in broccoli seedlings. J Exp Bot 61:3475–3491. https://doi.org/10.1093/jxb/erq177

Jogawat A, Yadav B, Chhaya Narayan OP (2021) Metal transporters in organelles and their roles in heavy metal transportation and sequestration mechanisms in plants. Physiol Plant 173(1):259–275

Ju C, Yoon GM, Shemansky JM, Lin DY, Ying ZI, Chang J, Garrett WM, Kessenbrock M, Groth G, Tucker ML, Cooper B, Kieber JJ, Chang C (2012) CTR1 phosphorylates the central regulator EIN2 to control ethylene hormone signaling from the ER membrane to the nucleus in Arabidopsis. Proc Natl Acad Sci 109:19486–19491. https://doi.org/10.1073/pnas.1214848109

Kazemi N, Khavari-Nejad RA, Fahimi H, Saadatmand S, Nejad-Sattari T (2010) Effects of exogenous salicylic acid and nitric oxide on lipid peroxidation and antioxidant enzyme activities in leaves of Brassica napus L. under nickel stress. Sci Hortic 126:402–407. https://doi.org/10.1016/j.scienta.2010.07.037

Keunen E, Schellingen K, Vangronsveld J, Cuypers A (2016) Ethylene and metal stress: small molecule, big impact. Front Plant Sci 7:23. https://doi.org/10.3389/fpls.2016.00023

Khan MIR, Khan NA (2014) Ethylene reverses photosynthetic inhibition by nickel and zinc in mustard through changes in PS II activity, photosynthetic nitrogen use efficiency, and antioxidant metabolism. Protoplasma 251:1007–1019. https://doi.org/10.1007/s00709-014-0610-7

Khan MIR, Nazir F, Asgher M, Per TS, Khan NA (2015) Selenium and sulfur influence ethylene formation and alleviate cadmium-induced oxidative stress by improving proline and glutathione production in wheat. J Plant Physiol 173:9–18. https://doi.org/10.1016/j.jplph.2014.09.011

Khan AR, Wakeel A, Muhammad N, Liu B, Wu M, Liu Y, Ali I, Zaidi SHR, Azhar W, Song G, Wu J, Gan Y (2019) Involvement of ethylene signaling in zinc oxide nanoparticle-mediated biochemical changes in Arabidopsis thaliana leaves. Environ Sci Nano 6:341–355. https://doi.org/10.1039/c8en00971f

Kim YS, Choi D, Lee MM, Lee SH, Kim WT (1998) Biotic and abiotic stress-related expression of 1-aminocyclopropane-l-carboxylate oxidase gene family in Nicotiana glutinosa L. Plant Cell Physiol 39:565–573. https://doi.org/10.1093/oxfordjournals.pcp.a029406

Li G, Meng X, Wang R, Mao G, Han L, Liu Y, Zhang S (2012) Dual-level regulation of ACC synthase activity by MPK3/MPK6 cascade and its downstream WRKY transcription factor during ethylene induction in Arabidopsis. PLoS Genet 8:e1002767. https://doi.org/10.1371/journal.pgen.1002767

Liu K, Shen L, Wang J, Sheng J (2008) Rapid inactivation of chloroplastic ascorbate peroxidase is responsible for oxidative modification to rubisco in tomato (Lycopersicon esculentum) under cadmium stress. J Integr Plant Biol 50:415–426. https://doi.org/10.1111/j.1744-7909.2007.00621.x

Maksymiec W (2007) Signaling responses in plants to heavy metal stress. Acta Physiol Plant 29:177–187. https://doi.org/10.1007/s11738-007-0036-3

Maksymiec W (2011) Effects of jasmonate and some other signaling factors on bean and onion growth during the initial phase of cadmium action. Biol Plant 55:112–118. https://doi.org/10.1007/s10535-011-0015-9

Malik A (2004) Metal bioremediation through growing cells. Environ Int 30(2):261–278. https://doi.org/10.1016/j.envint.2003.08.001

Masood A, Khan N (2013) Ethylene and gibberellic acid interplay in regulation of photosynthetic capacity inhibition by cadmium. J Plant Biochem Physiol 01:3. https://doi.org/10.4172/2329-9029.1000111

Masood A, Iqbal N, Khan NA (2012) Role of ethylene in alleviation of cadmium-induced photosynthetic capacity inhibition by Sulphur in mustard. Plant Cell Environ 35(3):524–533. https://doi.org/10.1111/j.1365-3040.2011.02432.x

Metwally A, Finkemeier I, Georgi M, Dietz K-J (2003) Salicylic acid alleviates the cadmium toxicity in barley seedlings. Plant Physiol 132:272–281. https://doi.org/10.1104/pp.102.018457

Monteiro CC, Carvalho RF, Gratão PL, Carvalho G, Tezotto T, Medici LO, Peres LEP, Azevedo RA (2011) Biochemical responses of the ethylene-insensitive never ripe tomato mutant subjected to cadmium and sodium stresses. Environ Exp Bot 71:306–320. https://doi.org/10.1016/j.envexpbot.2010.12.020

Montero-Palmero MB, Martín-Barranco A, Escobar C, Hernández LE (2014a) Early transcriptional responses to mercury: a role for ethylene in mercury-induced stress. New Phytol 201:116–130. https://doi.org/10.1111/nph.12486

Montero-Palmero MB, Ortega-Villasante C, Escobar C, Hernández LE (2014b) Are plant endogenous factors like ethylene modulators of the early oxidative stress induced by mercury? Front Environ Sci 2:34. https://doi.org/10.3389/fenvs.2014.00034

Mostofa MG, Fujita M, Tran L-SP (2015) Nitric oxide mediates hydrogen peroxide- and salicylic acid-induced salt tolerance in rice (Oryza sativa L.) seedlings. Plant Growth Regul 77:265–277. https://doi.org/10.1007/s10725-015-0061-y

Nagajyoti PC, Lee KD, Sreekanth TVM (2010) Heavy metals, occurrence and toxicity for plants: a review. Environ Chem Lett 8:199–216. https://doi.org/10.1007/s10311-010-0297-8

Nguyen TQ, Sesin V, Kisiala A, Emery RJN (2020) Phytohormonal roles in plant responses to heavy metal stress: implications for using macrophytes in phytoremediation of aquatic ecosystems. Environ Toxicol Chem 40:7–22. https://doi.org/10.1002/etc.4909

Nies DH (1999) Microbial heavy-metal resistance. Appl Microbiol Biotechnol 51:730–750. https://doi.org/10.1007/s002530051457

Noriega G, Caggiano E, Lecube ML, Cruz DS, Batlle A, Tomaro M, Balestrasse KB (2012) The role of salicylic acid in the prevention of oxidative stress elicited by cadmium in soybean plants. Biometals 25:1155–1165. https://doi.org/10.1007/s10534-012-9577-z

Piotrowska-Niczyporuk A, Bajguz A, Zambrzycka E, Godlewska-Żyłkiewicz B (2012) Phytohormones as regulators of heavy metal biosorption and toxicity in the green alga Chlorella vulgaris (Chlorophyceae). Plant Physiol Biochem 52:52–65. https://doi.org/10.1016/j.plaphy.2011.11.009

Saini S, Kaur N, Pati PK (2021) Phytohormones: key players in the modulation of heavy metal stress tolerance in plants. Ecotoxicol Environ Saf 223:112578. https://doi.org/10.1016/j.ecoenv.2021.112578

Schellingen K, Van Der Straeten D, Vandenbussche F, Prinsen E, Remans T, Vangronsveld J, Cuypers A (2014) Cadmium-induced ethylene production and responses in Arabidopsis thaliana rely on ACS2 and ACS6 gene expression. BMC Plant Biol 14:214. https://doi.org/10.1186/s12870-014-0214-6

Shan X, Yan J, Xie D (2012) Comparison of phytohormone signaling mechanisms. Curr Opin Plant Biol 15(1):84–91. https://doi.org/10.1016/j.pbi.2011.09.006

Singh HP, Batish DR, Kaur G, Arora K, Kohli RK (2008) Nitric oxide (as sodium nitroprusside) supplementation ameliorates Cd toxicity in hydroponically grown wheat roots. Environ Exp Bot 63:158–167. https://doi.org/10.1016/j.envexpbot.2007.12.005

Sisler EC, Serek M (1997) Inhibitors of ethylene responses in plants at the receptor level: recent developments. Physiol Plant 100:577–582. https://doi.org/10.1034/j.1399-3054.1997.1000320.x

Song S, Qi T, Wasternack C, Xie D (2014) Jasmonate signaling and crosstalk with gibberellin and ethylene. Curr Opin Plant Biol 21:112–119. https://doi.org/10.1016/j.pbi.2014.07.005

Srivastava AK, Venkatachalam P, Raghothama KG, Sahi SV (2006) Identification of lead-regulated genes by suppression subtractive hybridization in the heavy metal accumulator Sesbania drummondii. Planta 225:1353–1365. https://doi.org/10.1007/s00425-006-0445-3

Sun P, Tian Q-Y, Zhao M-G, Dai X-Y, Huang J-H, Li L-H, Zhang W-H (2007) Aluminum-induced ethylene production is associated with inhibition of root elongation in Lotus japonicus L. Plant Cell Physiol 48:1229–1235. https://doi.org/10.1093/pcp/pcm077

Sun P, Tian Q-Y, Chen J, Zhang W-H (2010) Aluminium-induced inhibition of root elongation in Arabidopsis is mediated by ethylene and auxin. J Exp Bot 61:347–356. https://doi.org/10.1093/jxb/erp306

Thao NP, Khan MIR, Thu NBA, Hoang XLT, Asgher M, Khan NA, Tran L-SP (2015) Role of ethylene and its cross talk with other signaling molecules in plant responses to heavy metal stress. Plant Physiol 169:73–84. https://doi.org/10.1104/pp.15.00663

Thapa G, Sadhukhan A, Panda SK, Sahoo L (2012) Molecular mechanistic model of plant heavy metal tolerance. Biometals 25:489–505. https://doi.org/10.1007/s10534-012-9541-y

Tran LSP, Pal S (2014) Phytohormones: a window to metabolism, signaling and biotechnological applications. Springer, New York. https://doi.org/10.1007/978-1-4939-0491-4

Ugulu I, Akhter P, Iqbal Khan Z, Akhtar M, Ahmad K (2021) Trace metal accumulation in pepper (Capsicum annuum L.) grown using organic fertilizers and health risk assessment from consumption. Food Res Int 140:109992. https://doi.org/10.1016/j.foodres.2020.109992

Van der Does D, Leon-Reyes A, Koornneef A, Van Verk MC, Rodenburg N, Pauwels L, Goossens A, Körbes AP, Memelink J, Ritsema T, Van Wees SCM, Pieterse CMJ (2013) Salicylic acid suppresses jasmonic acid signaling downstream of SCFCOI1-JAZ by targeting GCC promoter motifs via transcription factor ORA59. Plant Cell 25:744–761. https://doi.org/10.1105/tpc.112.108548

Vangronsveld J, Herzig R, Weyens N, Boulet J, Adriaensen K, Ruttens A et al (2009) Phytoremediation of contaminated soils and groundwater: lessons from the field. Environ Sci Pollut Res 16:765–794. https://doi.org/10.1007/s11356-009-0213-6

Verstraeten SV, Aimo L, Oteiza PI (2008) Aluminium and lead: molecular mechanisms of brain toxicity. Arch Toxicol 82:789–802. https://doi.org/10.1007/s00204-008-0345-3

Wang KL-C, Li H, Ecker JR (2002) Ethylene biosynthesis and signaling networks. Plant Cell 14(suppl_1):S131–S151. https://doi.org/10.1105/tpc.001768

Wang Q, Zhang W, Yin Z, Wen C-K (2013) Rice CONSTITUTIVE TRIPLE-RESPONSE2 is involved in the ethylene-receptor signaling and regulation of various aspects of rice growth and development. J Exp Bot 64:4863–4875. https://doi.org/10.1093/jxb/ert272

Weber LM, He J, Bradley B, Haskins K, Anseth KS (2006a) PEG-based hydrogels as an in vitro encapsulation platform for testing controlled β-cell microenvironments. Acta Biomater 2:1–8. https://doi.org/10.1016/j.actbio.2005.10.005

Weber M, Trampczynska A, Clemens S (2006b) Comparative transcriptome analysis of toxic metal responses in Arabidopsis thaliana and the Cd2+-hypertolerant facultative metallophyte Arabidopsis halleri. Plant Cell Environ 29:950–963. https://doi.org/10.1111/j.1365-3040.2005.01479.x

Wuriyanghan H, Zhang B, Cao W-H, Ma B, Lei G, Liu Y-F, Wei W, Wu H-J, Chen L-J, Chen H-W, Cao Y-R, He S-J, Zhang W-K, Wang X-J, Chen S-Y, Zhang J-S (2009) The ethylene receptor ETR2 delays floral transition and affects starch accumulation in rice. Plant Cell 21:1473–1494. https://doi.org/10.1105/tpc.108.065391

Xu J, Li Y, Wang Y, Liu H, Lei L, Yang H, Liu G, Ren D (2008) Activation of MAPK kinase 9 induces ethylene and camalexin biosynthesis and enhances sensitivity to salt stress in Arabidopsis. J Biol Chem 283:26996–27006. https://doi.org/10.1074/jbc.m801392200

Yadav SK (2010) Heavy metals toxicity in plants: an overview on the role of glutathione and phytochelatins in heavy metal stress tolerance of plants. S Afr J Bot 76:167–179. https://doi.org/10.1016/j.sajb.2009.10.007

Yoo S-D, Cho Y-H, Tena G, Xiong Y, Sheen J (2008) Dual control of nuclear EIN3 by bifurcate MAPK cascades in C2H4 signaling. Nature 451:789–795. https://doi.org/10.1038/nature06543

Yu Y, Jin C, Sun C, Wang J, Ye Y, Zhou W, Lu L, Lin X (2016) Inhibition of ethylene production by putrescine alleviates aluminium-induced root inhibition in wheat plants. Sci Rep 6:18888. https://doi.org/10.1038/srep18888

Yuan H-M, Xu H-H, Liu W-C, Lu Y-T (2013) Copper regulates primary root elongation through PIN1-mediated auxin redistribution. Plant Cell Physiol 54:766–778. https://doi.org/10.1093/pcp/pct030

YUN Y, ZHE L, Xing DA (2013) Nitric oxide promotes MPK6-mediated caspase-3-like activation in cadmium-induced Arabidopsis thaliana programmed cell death. Plant Cell Environ 36:1–15. https://doi.org/10.1111/j.1365-3040.2012.02543.x

Zhang G-B, Yi H-Y, Gong J-M (2014) The Arabidopsis ethylene/jasmonic acid-NRT signaling module coordinates nitrate reallocation and the trade-off between growth and environmental adaptation. Plant Cell 26:3984–3998. https://doi.org/10.1105/tpc.114.129296

Zhou Y, Liu C, Li G, Hao L (2019) Ethylene insensitive mutation increases Arabidopsis tolerance to Cd in NPR1-dependent manner. Plant Soil 441:49–69. https://doi.org/10.1007/s11104-019-04080-3

Zhu Z, An F, Feng Y, Li P, Xue L, Jiang Z, Kim JM, To TK, Li W, Zhang X, Yu Q, Dong Z, Chen WQ, Seki M, Zhou JM, Guo H (2011) Derepression of ethylene-stabilized transcription factors (EIN3/EIL1) mediates jasmonate and ethylene signaling synergy in Arabidopsis. Proc Natl Acad Sci 108:12539–12544. https://doi.org/10.1073/pnas.1103959108

Chapter 8
NO-Mediated Regulation of Plant Biology Under Heavy Metal Stress: Physiological and Molecular Aspects

Harleen Kaur, Ravneet Kaur, and Ashish Sharma

Abstract The presence of heavy metals in soil is one of the major catastrophes that limits plant growth and productivity. Heavy metal stress causes excessive and uncontrolled production of Reactive Oxygen Species (ROS). ROS have severe damaging effects in plants, such as lipid peroxidation, damage to proteins, and alteration in various metabolic pathways of plants. Plants exposed to heavy metal stress develop various adaptive strategies. In response to heavy metal (HM) stress, plants release various signaling molecules that initiate a cascade of responses to confer tolerance to plants against heavy metal stress. One such molecule is nitric oxide (NO) that belongs to the class of reactive nitrogen species (RNS). NO is produced within plant cells under normal physiological conditions and is responsible for the regulation of various physiological and biochemical functions of plants. However, under HM stress, NO plays a significant role. NO either directly acts as a signaling molecule, upregulating/downregulating various enzymes, or it acts by activating the antioxidant machinery of plants, or it directly functions as an antioxidant in plants, thereby mitigating the negative effects of HM toxicity. The chapter discloses various pathways of NO synthesis as well as effects of NO (exogenous and endogenous) on plant machinery under HM stress. The chapter further discusses how NO helps in alleviating the negative effects of HM toxicity and counteracting the detrimental effects of HMs on plants.

Keywords Abiotic stress · Hormones · Nitric acid · Plant stress · Stress biology · Toxicity

---

H. Kaur · R. Kaur · A. Sharma (✉)
Department of Botany and Environment Studies, DAV University, Jalandhar, Punjab, India
e-mail: ashish10210@davuniversity.org

A. Sharma et al. (eds.), *Hormones and Heavy Metals*, Plant Life and Environment Dynamics, https://doi.org/10.1007/978-981-96-9622-2_8

8.1 Nitric Oxide

Nitric oxide is a nitrogen-based chief reactive nitrogen species (RNS), whose reactivity is due to the existence of the unpaired electron, due to which it occurs in three interconvertible forms within the plant cell, viz. NO^- (nitroxyl anion), $NO^•$ (free radical) and NO^+ (nitrosonium cation) (Khan et al. 2014). Being produced within plants subjected to environmental stress, RNS are known to play a crucial role in redox signaling within plant cells as well as to activate the plant antioxidant defense system.

NO is an endogenous, ubiquitous, multifunctional, bioactive signaling molecule involved in regulating various plant processes such as growth and development, aging, flowering, resistance to diseases, stress tolerance, etc., at developmental, physiological, and biochemical levels (Yu et al. 2014; Serrano et al. 2015; Sanz et al. 2015; Astier et al. 2018; Costa-Broseta et al. 2018; Umbreen et al. 2018). The first report of the generation of NO by plants was given by Klepper in 1975. It is generated within plants via inducible enzymatic or non-enzymatic channels. NO can have both beneficial (cytoprotective) and harmful (cytotoxic) effects on plants depending upon various factors such as intracellular concentration, site of production, interaction with other cellular products, and NO and Reactive Oxygen Species (ROS) balance (Correa-Aragunde et al. 2015).

According to Fancy et al. (2017), nitric oxide (NO) is involved in combating abiotic stress factors, such as extreme temperatures, ozone exposure, water deficiency, high salinity, and heavy metal (HM) stress. Two mechanisms have been proposed by Radi et al. (1991), explaining how NO can reduce stress in plants. First, NO acts as a signaling molecule, triggering signal transduction pathway and resulting in the expression of stress-related genes. Secondly, NO can directly function as an antioxidant, thereby acting as ROS scavengers in plant cells by stimulating the activity of many enzymes (Groß et al. 2013; Lindermayr and Durner 2015), thereby indirectly conferring tolerance to plants under stress (Mengel et al. 2013; Duan et al. 2015). For example, supplying Sodium Nitroprusside (SNP), an NO supplement to plants under metal or salt stress, resulted in enhanced plant growth and productivity and increased photosynthesis, thereby providing metal or salt tolerance to plants (Manai et al. 2014; Bai et al. 2015; Fatma et al. 2016).

Crosstalk between ROS and the NO signal pathways helps in controlling the over-responsive reaction, the automatic death of cells, and the aging of leaves, thus behaving as plant protectors and enhancing stress tolerance in plants (Wang et al. 2013; Serrano et al. 2015). When the concentration of ROS becomes toxic to a plant, NO may act as a detoxifier and minimize any detrimental effects by activating the antioxidant machinery of the plant (Lipton et al. 1993).

8.2 Synthesis of NO

During metal stress conditions, endogenous levels of NO increase within plants. NO being a gaseous agent can also be perceived from plant's surrounding environment along with its endogenous origin. Within a plant cell, biosynthesis of NO occurs in parallel to ROS levels in cellular organelles, such as chloroplast, mitochondria, peroxisomes, as well as cytoplasm. NO synthesis in plants occurs via various pathways (Fig. 8.1). The well-documented reductive pathway for NO synthesis is the nitrate reductase (NR) pathways (Gupta et al. 2011). The key enzymes involved in the biosynthesis of NO are NO synthase (NOS). Three different isoforms of NOS, viz., endothelial (eNOS), neuronal (nNOS), and inducible (iNOS), have been identified in animals (Alderton et al. 2001).

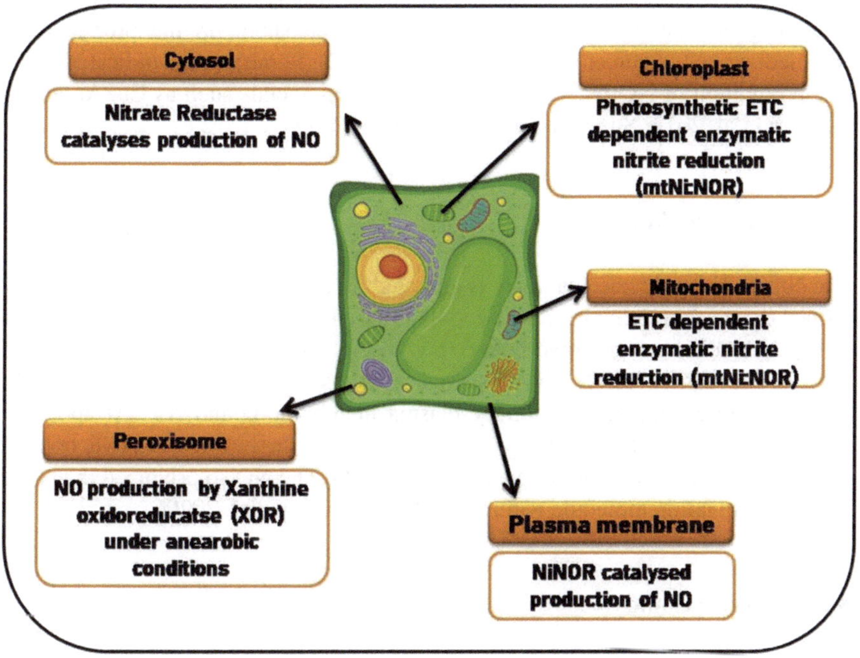

Fig. 8.1 Various pathways for the synthesis of nitric oxide in plant cell

8.2.1 *Pathways for NO Biosynthesis*

8.2.1.1 Cytoplasm

Nitrate Reductase (NR) pathway operates in the cytosol of plant cells. NR is a cytosolic enzyme encoded by NIA1 and NIA2 genes (Gupta et al. 2011) that catalyzes the reduction of nitrate to nitrite at the expenditure of NAD(P)H. During this, one electron transfers from NAD(P)H to nitrite, resulting in NO formation (Planchet and Kaiser 2006) via the reaction:

$$\mathrm{NAD(P)H + 3H_2O^+ + 2NO^{2-} \rightarrow NAD(P)^+ + 2NO + 5H_2O.}$$

Under unstressed conditions, NR activity accounts for up to only 1% of the nitrate reduction capacity which is sufficient to carry out NO-based activities within the cell. However, exposure to abiotic stress such as salinity, drought, cold, and heavy metal triggers the plant defense machinery which shoot up the production of NR-mediated NO (Sun et al. 2012; Tan et al. 2012). Some reports indicated excessive production of NO by NR in response to heavy metals such as Cu, Al, and As (Kolbert et al. 2012; Sun et al. 2014; Xue and Yi 2017).

8.2.1.2 Plasma Membrane

The Nitrite:NO reductase pathway is operated by plasma membrane bound protein termed as the nitrate:NO reductase (NiNOR) enzyme (Chamizo-Ampudia et al. 2017). NiNOR is a membrane-bound root-specific enzyme that occurs in fractions (StoÈhr and Ullrich 2002). Nitrite serves as a substrate for NiNOR enzyme and is derived from a coupled reaction by plasma membrane-bound NR. At apoplastic acidic pH (6.1), while working with full capability, the enzyme NiNOR converts nitrite produced by membrane-bound NR into NO. The NO so produced then functions as a signaling molecule within the plant (Shivaraj et al. 2020).

8.2.1.3 Peroxisome

The xanthine oxidoreductase (XOR) pathway is responsible for ample production of NO in plants. Xanthine oxidoreductase/xanthine oxidase (XOD)/xanthine dehydrogenase (XDH) is an enzyme containing molybdenum, which is present in peroxisomes and is widely distributed (Corpas et al. 2008). It exists in two interchangeable forms: the superoxide-producing xanthine oxidase (form O, EC 1.1.3.22) and xanthine dehydrogenase (form D, EC 1.1.1.204) (Palma et al. 2002). In peroxisomes of pea plants, Xanthine oxidoreductase is predominantly in the xanthine oxidase form, with only 30% existing as xanthine dehydrogenase (Sandalio et al. 1988; Corpas et al. 1997; del Rıo et al. 2004).

XOR regulated the conversion of NO^{2-} to NO. During this process, NADH/xanthine acts as an e^- donor (Corpas et al. 2008). XOR works differently under both aerobic and anaerobic conditions. Depending upon the oxygen availability, XOR can initiate ROS signaling or NO signaling. Under aerobic conditions, superoxides are produced by the activity of XOR, which further dismutes to H_2O_2. In anaerobic conditions, XOD activity invariably leads to the production of NO (Shivaraj et al. 2020).

8.2.1.4 Mitochondrial Pathway

When oxygen level declines below 20 μM, NiR-mediated NO production occurs in mitochondria of both plants and animals (Shiva 2010). In mitochondria, electron transport complex (ETC) dependent enzyme mediated nitrite reduction (mtNi:NOR) occurs at complex I, II and III, present on the mitochondrial membrane. In this reaction, NR uses nitrite and NADPH as substrate and e^- donor respectively, for the production of NO (Sahay and Gupta 2017).

8.2.1.5 Chloroplast Pathway

As reported by Sahay and Gupta in 2017, in chloroplast, NO synthesis occurs via two pathways:

(i) NR-based nitrite reduction (NiR): In this, nitrite (NO^{2-}) gets converted into NO.
(ii) NOSLE-based pathways: The NOSLE activity utilizes L-Arginine as a substrate and NADPH as an electron donor, operating independently of the presence of Ca^{2+}/Calmodulin (CaM). The occurrence of this reaction was identified through the transformation of L-Arginine into Citrulline and the generation of nitric oxide (NO).

8.3 NO Signal Transduction Under Heavy Metal Stress

In a plant cell, NO displays dual functions: During unstressed conditions, the endogenous level of NO is kept extremely low through the elimination of excessive NO from the plant cell. Apart from interrupting the ROS-initiated chain in plant cell, NO also causes upregulation of gene expression involved in providing tolerance to plants (Beligni and Lamattina 1999). Under stressed conditions, NO production increases abruptly resulting in nitrosative stress causing cell death (Fig. 8.2).

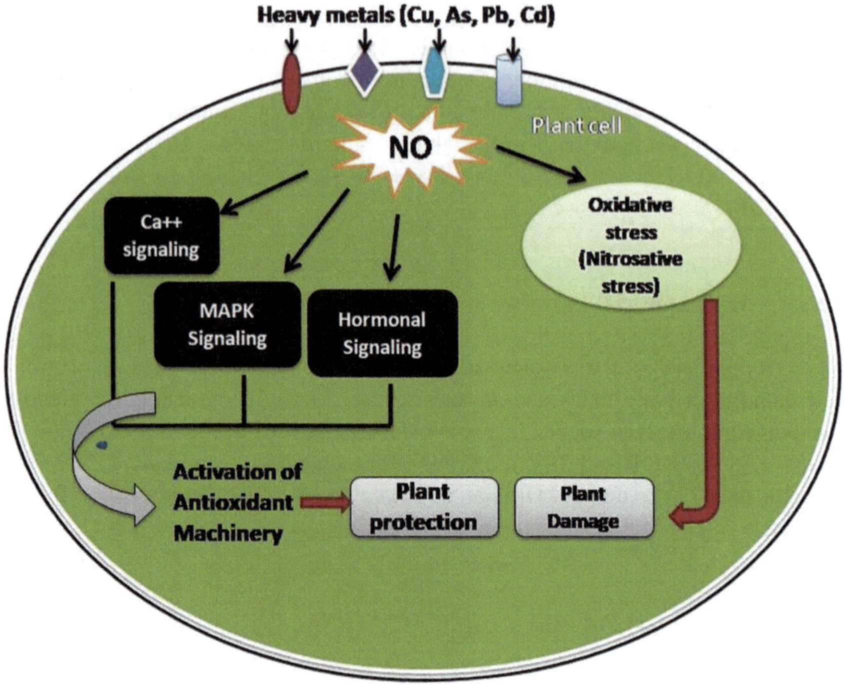

Fig. 8.2 Signaling mechanism of nitric oxide in alleviating heavy metal stress in plants

8.3.1 The Signal Transduction of NO Exists in Multiple Pathways

Endogenous NO metabolism and its mechanism of action occurs through NO-mediated post-transitional modifications (NO-PTMs). Although various PTMs have been identified so far, two well-characterized PTMs governed by NO under normal unstressed environment as well as under stressed (abiotic and biotic) environments are S-Nitrosylation and tyrosine nitration (Greenacre and Ischiropoulos 2001; Corpas et al. 2009).

S-Nitrosylation is almost ubiquitous among living organisms and numerous proteins are known to undergo the process of S-Nitrosylation, meant for regulating their activity in plants exposed to abiotic stresses such as chilling stress, salinity stress, high light stress, and HM stress (Anand and Stamler 2012; Zhang et al. 2018). S-Nitrosylation, a rapid, reversible, redox-based PTM, being observed as a NO-mediated regulatory process in which NO gets attached to specific cysteine thiols by a covalent bond resulting in formation of S-Nitrosothiols (SNOs), a key regulators in the plant response to abiotic stress (Fancy et al. 2017) and, interestingly, the above said process can occur independent of the production level of NO (Feechan et al. 2005; Chaki et al. 2011; Yun et al. 2016).

Amongst various SNOs, one such SNO that act as a signaling molecule is GSNO, a low molecular weight (LMW) GSNO, being considered as a storehouse of NO with a lifetime of hours in aqueous solution (Floryszak-Wieczorek et al. 2006) as well as a vital NO-mediated signaling molecule (Broniowska and Hogg 2012; Mur et al. 2013). Antioxidant capability of the cell is mainly dependent on the presence of low molecular-weight antioxidants. Out of range of small, low molecular-weight thiols present in cells (Hancock and Whiteman 2018) but one of the most imperative is the highly abundant redox-sensitive tripeptide molecule glutathione (Schafer and Buettner 2001). GSNO is formed by the covalent attachment of NO to the glutathione (GSH: Glutamic acid, Cysteine, and Glycine) leading to the formation of S-nitrosoglutathione (GSNO). In a similar way, tyrosine nitration (Corpas et al. 2009) occurs in plants in which NO gets covalently attached to a tyrosine residue (Greenacre and Ischiropoulos 2001).

One of the targets of elevated NO levels in plant cells are calcium ions (Ca^{2+}) (Fig. 8.2). Increase in the levels of endogenous NO causes activation of Ca^{2+} channels resulting in elevated cytosolic Ca^{2+} levels by activating Ca^{2+} channels. NO modulates that NO may exert its effects by acting through cyclic guanosine monophosphate (cGMP) and cyclic adenosine diphosphate ribose (cADPR) to modulate intracellular calcium (Ca^{2+})-permeable channels. This modulation leads to an increase in free cytosolic calcium levels within cells. Pharmacological studies investigating variations in Ca^{2+} levels have suggested that NO has the potential to activate both plasma membrane and intracellular Ca^{2+}-permeable channels (Garcia-Mata et al. 2003; Lecourieux et al. 2005; Vandelle et al. 2006). Changes in cytosolic-free Ca^{2+} concentrations occur in response to various stress stimuli, initiating complex interactions and signal transduction pathways (Rudd and Franklin-Tong 2001).

The chemical signal initiated by a transient increase in calcium concentration is converted into a biological response through various highly sensitive calcium-sensing proteins. Plants contain several calcium-sensing proteins, such as Calmodulins (CaMs), CaM like proteins (CMLs), Calcineurin B-like proteins (CBLs), and Ca^{2+}-dependent protein kinases (CDPKs). Such proteins initiate a myriad of downstream signaling pathways in response to binding to Ca^{2+} molecule (Luan et al. 2002; Sanders et al. 2002; Dodd et al. 2010). Exogenous application of Ca^{2+} also results in enhancement in the activity of several antioxidant enzymes, viz. superoxide dismutase (SOD), ascorbate peroxidase (APX), and glutathione reductase (GR) thereby imparting tolerance to the plants under abiotic stresses (Ahmad et al. 2015).

Endogenous NO initiates signal transduction via interactions with signal transducing proteins such as Mitogen Activated protein kinases (MAPK) (Kumar and Klessig 2000; Capone et al. 2004). NO mediates the activation of a 46-kDa MAPK via CDPK and CaM to transfer signals (Lamotte et al. 2005). MAPKs are enzymes classified as serine/threonine kinases. They constitute a three-tier phospho-relay signaling module that is evolutionarily conserved across the plant kingdom (Jonak et al. 2002). The interconnected MAPK module transmits various environmental signals, and the activated MAPKs induce a multitude of cellular and physiological responses by triggering the activation of transcription factors, along with cytosolic proteins (Šamajová et al. 2013). Consequently, this signaling pathway influences crucial cellular functions like differentiation, growth, and various developmental processes, as well as the response to environmental stress tolerance (Lin and Aarts 2012). The transcriptional activation of these protein kinases in rice underscores the regulatory role of MAPKs in the development, defense mechanisms, and signaling responses to stress in plants (Kim et al. 2003).

Amongst different abiotic stresses studied so far, it has been observed that HM stress displays a profound effect on MAPK signaling pathways. Entry of HM into the plant cell caused excessive production of ROS, thereby altering NO-induced cell signaling, causing activation of MAPKs (Jonak et al. 2004; Smeets et al. 2013; Jalmi and Sinha 2015). Another study revealed that MAPK signaling pathways are activated in plants subjected to HM stress, such as Cu stress and Cd stress (Jalmi et al. 2018).

In plants, NO is known to play an imperative role in auxin signaling. The signaling pathway of auxin (AUX) can be activated through NO-mediated S-nitrosylation of transport inhibitor response 1 (TIR1), which is an AUX receptor, at the cysteine (Cys) residue. S-nitrosylation of TIR1 enhances the interaction between TIR1 and Auxin/Indole-3-acetic acid (IAA), thereby promoting the degradation of Auxin/IAA complexes. This mechanism was elucidated by Terrile et al. (2012).

Studies revealed that in *Cucumis sativus,* AUX prompts the production of endogenous NO, and in turn, endogenous NO enhanced the AUX-mediated adventitious rooting (Pagnussat et al. 2002). In *Arabidopsis*, Cadmium (Cd) exposure leads to decline in AUX by reducing polar auxin transporters, hence resulting in inhibition in root meristem growth (Yuan and Huang 2016).

On the other hand, ethylene (ETH) and NO have antagonistic effect on each other. An inverse relationship was observed between ETH and NO in *T. repens*, in which Cd stress caused enhanced ETH level and the level of ETH was decreased by exogenous supply of SNP. The possible reason behind this was inhibition of NO-mediated inhibition of methionine adenosyltransferase (MAT) through S-nitrosylation, thereby arresting ETH biosynthesis (Lindermayr et al. 2006).

In *Arabidopsis*, exogenous application of NO reduces the endogenous levels of gibberellins as NO downregulates the expression of GA3 oxidase, thereby affecting the gibberellin biosynthetic pathways (Lozano-Juste and León 2011).

8.4 Cross Talk Between NO and ROS

In response to HM stress, ROS levels in plants increase. These ROS such as OH or O^{2-} are formed as a result of the reduction of molecular oxygen. The ROS produced and endogenous NO undergo cross talk (Fig. 8.3) and this cross talk takes place during two phases: (i) biosynthesis phase and (ii) regulation of expression of genes or enzymatic activity (Qiao et al. 2014). The balance between levels of ROS and NO and their interaction plays an imperative role in providing tolerance to plants against

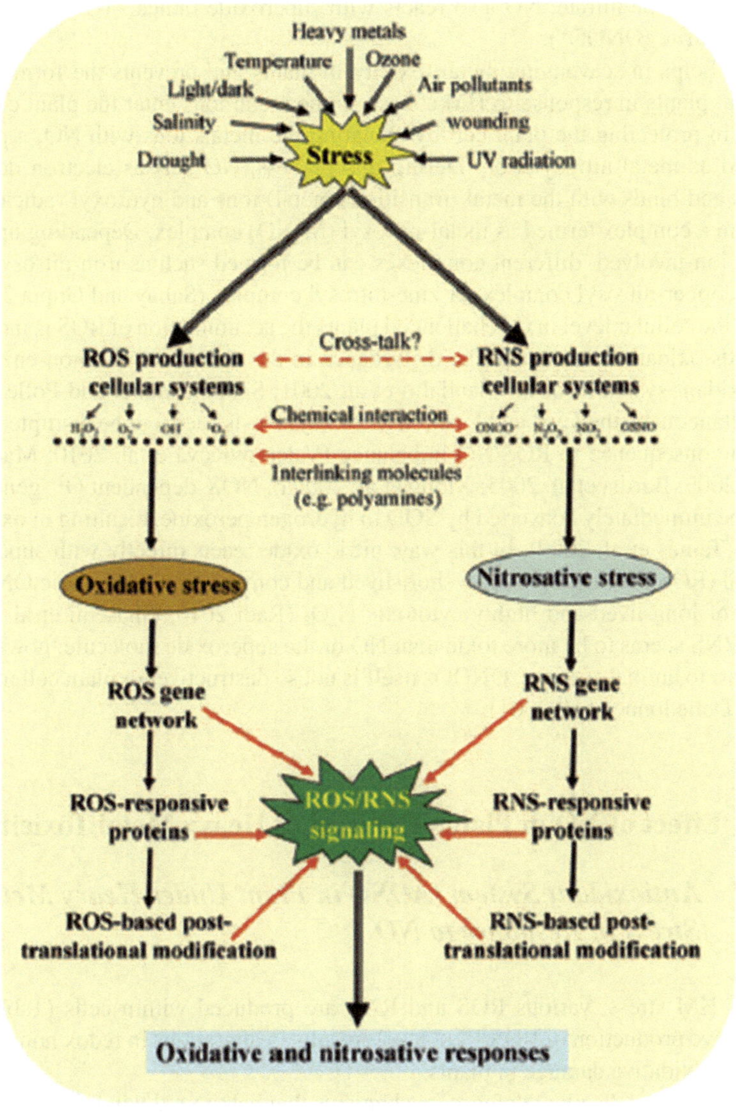

Fig. 8.3 A framework model for the oxidative and nitrosative signaling cross-talk in response to abiotic stress in plants. (Source: Molassiotis and Fotopoulos 2011)

HM stress or any abiotic stress (Pető et al. 2013; Wang et al. 2014; Feigl et al. 2015; Silveira et al. 2015; Thao et al. 2015; Farnese et al. 2016). NO reacts with proteins in different ways: (1) through metals present in the protein giving metal nitrosyl; (2) through sulfhydryl groups to render a process of S-Nitrosylation, and (3) by adding a nitro ($-NO_2$) group in the nitration process. Cysteine, methionine, tryptophan, and tyrosine can be nitrate. NO also reacts with superoxide radicals (O^{2-}) to generate peroxynitrite ($ONOO^-$).

NO helps in scavenging metal toxicity in plants and prevents the formation of ROS in plants in response to HM stress. As the metal ions enter the plant cell, NO helps in protecting the plant cell by chelating the metals ions with NO, a process termed as metal-nitrosylation. During this process, NO acts as electron donor, it reacts and binds with the metal (transitions metal) ions and hydroxyl radical (OH) to form a complex termed as metal-nitrosyl (M-NO) complex. Depending upon the metal ion involved; different complexes can be formed such as iron-nitrosyl complex, copper-nitrosyl complex, or zinc-nitrosyl complex (Sahay and Gupta 2017).

At the cellular level in Cd challenged plants the accumulation of ROS is increased towards oxidative stress, and the disturbance in the enzymatic and non-enzymatic antioxidant system follows (Sandalio et al. 2001; Schutzendubel and Polle 2002). Simultaneously, the level and location of NO synthesis seems to be disrupted, leading in consequence to ROS/NO imbalance (Valentovičová et al. 2010; Mahmood et al. 2009; Bartha et al. 2005; Xiong et al. 2009a). NOX-dependent O^{2-} generation may be immediately converted by SOD to hydrogen peroxide, resulting in oxidative burst (Tamás et al. 2009). In this way, nitric oxide reacts directly with superoxide radical (ROS precursor) to form short-lived and comparatively less toxic ONOO in place of long-lived and highly cytotoxic H_2O_2 (Radi 2013; Ghasemi et al. 2014). This RNS seems to be more toxic than NO or the superoxide molecule; however, in contrast to animal systems, $ONOO^-$ itself is not so destructive for plant cell metabolism (Delledonne et al. 2001).

8.5 Effect of NO in Plants Exposed to Heavy Metal Toxicity

8.5.1 Antioxidant System (AOSs) in Plant Under Heavy Metal Stress in Response to NO

Under HM stress, various ROS and RNS are produced within cells (Table 8.1). Excessive production of ROS/RNS levels results in alterations in redox homeostasis causing oxidative damage in plants.

This triggers the plant defense mechanisms that help in reducing the rate of ROS generation as well as helps in protecting the cell components from ROS injury hence, developing tolerance in plants against HM stress. The antioxidant system (AOSs) helps in maintaining the balance between ROS generation and its scavenging with the help of antioxidant enzymes (Gill et al. 2012; Sandalio et al. 2012). Regulation of antioxidant enzyme activity can be done via binding or modifying the metal ions. In heme-containing enzymes, NO functions as an iron ligand, hence causing either activation or deactivation of enzyme. It has been observed in *Antiaria toxicaris*, NO upregulated the activity of ROS scavanging enzymes viz. CAT, APX due to enhanced S-nitrosylation in *Antiaria toxicaris* (Bai et al. 2011).

Table 8.1 Main Reactive Oxygen Species (ROS) and Reactive Nitrogen Species (RNS) found in plant cells

Free radicals	Non-radicals
Reactive oxygen species	*Reactive oxygen species*
Superoxide, $O^{\bullet -}{}_2$	Hydrogen peroxide, H_2O_2
Alkoxyl, $RO^\bullet$	Hypochlorous acid, HOCl
Hydroxyl, OH^-	Ozone, O_3
Peroxyl, $ROO^\bullet$	Peroxynitrite, $ONOO^-$
Hydroperoxyl, $HO_2^\bullet$	Singlet oxygen, $^1O_2$
Reactive nitrogen species	*Reactive nitrogen species*
Nitric oxide, $NO^\bullet$	Nitrous acid, HNO_2
Nitric dioxide, $NO_2^\bullet$	Nitrosonium cation, NO^+
Nitrate radical, $NO_3^\bullet$	Nitrosyl anion, NO^-
	Peroxynitrite, $ONOO^-$
	Alkylperoxynitrites: ROONO
	Dinitrogen trioxide, N_2O_3

Source: Farnese et al. (2016) and Rahman et al. (2012)

The mechanism entails an elevation in root exudation, limitation, and accumulation of HMs within susceptible plant tissues, as well as processes such as chelation, metal binding, and sequestration. All these mechanisms are believed to be directed with the help of the plant sensory system, consisting of enzymatic and non-enzymatic antioxidants compounds. Enzymatic plant defense compounds that are mainly produced in cell organelles such as chloroplast, mitochondria and peroxisome include SOD (EC 1.15.1.1), guaiacol peroxidase (POX- EC1.11.1.7), CAT (EC 1.11.1.6), GR (EC 1.6.4.2), APX (EC 1.11.1.11), mono- and dehydro ascorbate reductase (MDHAR- EC 1.6.5.4, DHAR- EC 1.8.5.1), etc. Non-enzymatic compounds involve glutathione (GSH), phytochelatins (PCs), α-tocopherol (vitamin E), carotenoids, ascorbate (AsA), proline (Pro), etc. (Sahay and Gupta 2017).

In wheat plants subjected to copper stress, pretreatment of seeds with NO helps in improving seed germination as well as upregulates CAT activity, thereby protecting plants from oxidative stress (Hu et al. 2007). In HM/salt stressed lupin (*Lupinus* sp.) plants, NO treatment caused increased SOD activity, thereby providing protection against stress (Kopyra and Gwóźdź 2003). In *Arabidopsis* plants subjected to copper (Cu) stress, NO helps in providing tolerance to plants by undergoing ROS metabolism (Pető et al. 2013). NO upregulated the level and activity of POX, hence providing protection to plants against oxidative stress in wheat roots (Singh et al. 2008) and chickpea (Kumari et al. 2010) under Cd stress, and in chickpea (Sheokand et al. 2010). It was reported by Ahmad et al. (2018) that exogenous supplementation of SNP causes enhanced activity of antioxidant enzymes, hence reducing the levels of ROS in tomato seedlings exposed to Cd toxicity.

NO is believed to have antioxidant capacity by virtue of which it causes detoxification of Cd within plant cells. During Cd stress, CAT and GOX could be S-nitrosylated, and the S-nitrosylation levels could undergo alteration in response to Cd stress (Ortega-Galisteo et al. 2012). Interestingly, an elevated activity of antioxidant enzymes (CAT, SOD, APX, and GR) was observed in tomato plants under Cd stress, activity of which was further enhanced by addition of NO (Ahmad et al. 2018).

Under Arsenic stress (As), exogenous supply of SNP (a donor of NO) causes activation of defense machinery in plants that helps in ROS scavenging, thereby reducing the damage caused by As. A study revealed elevated expression of the antioxidant enzymes SOD, CAT, APX, and GR in rice under As stress, while NO could confront the stimulation to ameliorate the toxic effect of As (Praveen and Gupta 2018). On the other hand, instead of activating the plant AOSs, NO could directly be involved in amelioration of toxicity in plants caused by HM stress, probably due to the antioxidant role of NO to alleviate ROS from plants exposed to heavy metal stress. Such an observation was made after a study revealed that in water hyacinth grown in As excess environment, the level of antioxidants declined after supplementation with SNP, indicating direct involvement of NO in alleviating As induced toxicity (Wei et al. 2020).

In lead (Pb)-stressed wheat plants, oxidative stress caused due to excess of ROS can be counteracted by increased activities of antioxidant enzymes (Kaur et al. 2015). It was observed that plants exhibit an innate ability to combat oxidative stress due to enhanced levels of ROS by activation of various antioxidant enzymes (Sytar et al. 2013). Similarly, in cowpea, Pb toxicity resulted in an enhancement in the level of antioxidant enzymes the levels of which were reduced by exogenous application of NO, thereby reducing Pb stress in plants (Sadeghipour 2016).

8.5.2 Effect of Nitric Oxide on Plant Physiology Under Heavy Metal Stress

HM toxicity within plants causes disruption in normal metabolic pathways of plants negatively affecting the growth, photosynthesis, and reproduction of plants, thereby the plants show notable changes in morphology, physiology, cellular and biochemical functions of plants (Sharma and Dietz 2009). Being present in soil, HM first interacts with the root system of plants, thus roots are the foremost organ that get affected by HM toxicity. For instance, oxidative and glutathione depletion was observed in Alfalfa roots exposed to Cd and Hg (Mercury) stress (Ortega-Villasante et al. 2005).

8.5.3 Effect of Nitric Oxide on Lipid Peroxidation Under Heavy Metal Stress

The generation of ROS in plants leads to the oxidation of lipids present in cell membranes, as well as proteins and enzymes, consequently affecting the proper functioning of plant cells (Mittler 2002). Malondialdehyde (MDA), a byproduct of lipid peroxidation, serves as an indicator for assessing the extent of lipid peroxidation within plant tissues. Numerous studies have demonstrated an elevation in MDA content in plants subjected to heavy metal stress. Interestingly, the application of SNP has been shown to result in a reduction in the levels of MDA. This suggests that SNP application may have a protective effect on plant cells by mitigating oxidative damage, as evidenced by the decreased levels of MDA. Research showed that arsenic stress caused a significant increase in MDA contents. However, exogenous supply of SNP (a NO donor) before exposure to As stress remarkably reduced MDA contents in tall fescue leaves (Jin et al. 2010).

8.5.4 Effect of Nitric Oxide on Seed Germination Under Heavy Metal Stress

Seed germination is considered as the most significant and important phase during plant development. Presence/supplementation of NO can mitigate the negative effects of HM toxicity during seed germination. In *Catharanthus roseus* seeds exposed to Cd stress, treatment with SNP resulted in elevated seed germination as well as seedling growth through interaction with melatonin (Nabaei and Amooaghaie 2019). The prompted seed germination can be attributed to the stimulation of activity of α-amylase and protease enzyme, and upregulation of the H^+ -ATPase activity, hence increasing the starch mobilizing rate (Maurice et al. 2016).

8.5.5 Effect of Nitric Oxide on Photosynthesis Under Heavy Metal Stress

It has been observed that HM stress significantly affects photosynthesis in the plant by affecting photosynthetic pigments accumulation, water-oxidizing complex of PhotoSystem II (PSII) due to replacement of Ca^{2+} with Mn^{2+} ions in the reaction center, thereby affecting the electron transport chain (Parmar et al. 2013).

In plants under Aluminium (Al) stress HM (Wang and Yang 2005; Yang et al. 2012; Tiwari et al. 2019), As (Singh et al. 2009), Cd (Laspina et al. 2005; Groppa et al. 2008; Gong et al. 2017; Per et al. 2017; Ahmad et al. 2018; Chen et al. 2018), Cu (Dong et al. 2014; Khairy et al. 2016), nickel (Ni) (Rizwan et al. 2017) and Pb

(Bai et al. 2015), exogenous supply of NO improves photosynthesis rate as well as plant growth hence, causing amelioration of HM.

Research revealed that the photosynthesis rate was enhanced in Cd-exposed *Cucumis sativus* plants after treatment with NO. In addition to this, NO reduced the damage caused by HM toxicity on light harvesting complex (LHC)I and (LHC)II (Gong et al. 2017). In Cd stressed *Brassica juncea* plants, NO supplementation helped improve the chlorophyll content, Rubisco activity chlorophyll fluorescence and other photosynthetic parameters (Per et al. 2017).

In *Nicotiana tabacum* exposed to Cu stress, supplementation of NO enhanced Rubisco activase activity (Khairy et al. 2016). Similarly, in *Lolium perenne,* photosynthetic pigments, photosynthetic rate, and ultimately plant growth were improved after NO supplementation (Dong et al. 2014).

8.5.6 Effect of Nitric Oxide on Plant Productivity Under Heavy Metal Stress

Supplementation of metal stressed plants with NO exhibited reduced levels of metals in above ground biomass (shoot) (Yang et al. 2012; Per et al. 2017). Exogenous application of NO in the form of SNP functions to influence the cell wall components, causing cell wall expansion, thereby upregulating the plant growth in heavy metal-stressed plants (Wang et al. 2016; Akram et al. 2018).

Exogenous supply of NO mitigates HM (Cd) stress in rice plants by increasing accumulation of Cd in the cell wall due to enhanced levels of root cell wall components, viz. pectin and hemicellulose (Xiong et al. 2009a, b). It was observed that tall fescue plants grown under arsenic had reduced dry weight, while the negative impact of HM was reversed by exogenous application of NO (Jin et al. 2010).

An experiment showed that plant biomass was significantly reduced on exposure to As stress. However, the application of SNP mitigates the detrimental effects of As toxicity and counteracts As-induced stress (Singh et al. 2009). Exogenous supplementation of SNP helps in the protection of Ni stressed *B. napus* plants by restricting the translocation of Ni to aboveground tissues (Kazemi et al. 2010).

8.5.7 Effect of Nitric Oxide on Non-antioxidant System Under Heavy Metal Stress

It was observed that osmolyte level, such as Pro content, increases in plants subjected to Pd stress, and exogenous application of SNP has an additive effect on Pro. Similar results were reported by Zhang et al. (2008) that NO application causes enhancement in Pro levels in algae exposed to Cu stress. The primary mechanism underlying the elevation of Pro levels in response to HM stress involves the activation of D-pyroline-5-carboxylate synthetase (P5CS), the enzyme responsible for Pro biosynthesis. This process also entails the upregulation of P5CS expression in copper-stressed algae.

8.5.8 *Effect of Nitric Oxide on Phytohormones Under Heavy Metal Stress*

In studies on *Medicago truncatula* exposed to Cd stress, the introduction of NO resulted in the inhibition of indole-3-acetic acid (IAA) oxidase activity. This inhibition led to a decrease in the degradation of AUXs, revealing an intriguing interplay between NO and AUXs in enhancing Cd tolerance (Xu et al. 2010). Similarly, the application of SNP led to an increase in the IAA content in wheat (*Triticum aestivum*) and rye (*Secale cereale*) plants exposed to Al toxicity, indicating a beneficial interaction between AUXs and NO during HM stress (He et al. 2012).

Likewise, there exists a beneficial interaction between Gibberellins (GA) and NO, where the provision of SNP encouraged the accumulation of GA. This, in turn, facilitated the growth of apical roots, enhancing tolerance to Al toxicity in wheat roots (He et al. 2012).

Recently, there has been a suggestion of a potential connection between NO and ETH mediated by MAPKs during plant reactions to HM stress (Thao et al. 2015). Furthermore, it is extensively recorded that the external application of NO mitigates the harmful effects of excessive ETH production under abiotic stresses. This is primarily attributed to NO's ability to inhibit the activities of 1-aminocyclopropane-1-carboxylic acid synthase (ACS) or 1-aminocyclopropane-1-carboxylic acid oxidase (ACO), thereby preventing the excessive formation of ETH (Manjunatha et al. 2010).

The external application of NO also confers resistance to rice against As-induced toxicity and plays a protective role against As-induced oxidative stress. NO operates downstream of abscisic acid (ABA) in the physiological processes that regulate

Table 8.2 Effect of NO on plants subjected to heavy metal stress

Heavy metal stress	Plant species	Responses of NO in plants	References
Chromium (Cr)	Tomato	Improved activity of α-Amylase and Proline	Khan et al. (2021)
Cu	Rice seedlings	Counteraction of Cu induced increase in GSH and proline content	Mostofa et al. (2014)
Cd	*T. angustifolia*	Enhancement in the innate chelation ability of cell wall	Zhao et al. (2016)
Ni	*Brassica napus*	Reduction in translocation of Ni from root to shoot and increase in antioxidative enzymatic activities	Kazemi et al. (2010)
Cu	Wheat	Improved seed germination	Hu et al. (2007)

(continued)

Table 8.2 (continued)

Heavy metal stress	Plant species	Responses of NO in plants	References
As	*Oryza sativa* roots	Enhancement in chlorophyll content and plant growth	Singh et al. (2016)
Cu	*Panax ginseng*	Upregulation of antioxidative enzyme activities, thereby reducing the Cu toxicity	–
Cd	*Triticum aestivum*	Reduced electrolyte leakage, decline in MDA and H_2O_2 content, as well as reduction in antioxidant enzyme activity	Singh et al. (2008)
Cd	*Medicago truncatula*	Improved proline and Glutathione content as well as enhancement in plant growth rate	Xu et al. (2010)
Cd	*Oryza sativa*	Reduction in MDA and H_2O_2 content as well as increase in chlorophyll and protein content	Hsu and Kao (2004)
Pb	Pepper (*Capsicum annuum* L.)	Alleviation of heavy metal-induced oxidative stress	Kaya et al. (2019)
Cd	*Glycine max* L. Cv. Navico	Enhancement of plant growth, increased antioxidant enzyme activity and reduction in the level of H_2O_2 and O_2	Kopyra et al. (2006)
Cd	*Arabidopsis*	Repression in auxin accumulation	Yuan and Huang (2016)
As	*Pistia stratiotes*	Improved antioxidative enzymatic activity and improvement in photosynthesis	Farnese et al. (2013)
Pb	*Vigna unguiculata* L.	Increased photosynthetic and antioxidant activity	Sadeghipour and Aghaei (2013)
Ni	*Eleusine coracana*	Mitigation of oxidative stress via improvement in antioxidant enzyme activity	Viswanath et al. (2016)
Cd	*Solanum lycopersicum*	Mitigation of oxidative stress via improvement in antioxidant enzyme activity	Ahmad et al. (2018)
Cd	*Helianthus annuus* (L.)	Increased SOD and catalase activity	Laspina et al. (2005)
Cu	*Vicia faba* L.	The Cu-induced NO in vascular bundles was associated primarily with the presence of the induced NADPH-d activity	Zou et al. (2012)
Ni	*Oryza sativa* L., cv. yangliangyou 6	Increased level of Peroxidase and Catalase in both root and shoot	Rizwan et al. (2017)
Ni	*Phaseolus vulgaris* L.	Mitigation of oxidative stress in plants	Mihailovic and Drazic (2011)

stomatal transpiration (Neill et al. 2003; Desikan et al. 2002; Neill et al. 2008). The main driving force responsible for the translocation of HMs from the root shoot is transpiration stream (Salt et al. 1995). Reports suggest that exogenous supplementation of NO inhibits the transpiration rate of plants, hence decreases HM translocation from root to shoot (Xiong et al. 2009a, b). Table 8.2 provides an overview of the NO roles in plants under heavy metal stress.

8.6 Conclusion

Heavy metal toxicity in plants induces oxidative stress in plants by the production of ROS in excess amount. Elevated levels of ROS adversely affect various plant components and disrupt plant metabolism. NO has the ability to mitigate the negative effects of heavy metal stress in plants. Under HM toxicity, high levels of NO (exogenous/endogenous) provide protection to plants by activating antioxidant and non-antioxidant enzymes in plants, thereby relieving the plants from phytotoxicity. NO is a signaling molecule that functions to activate various defense-related genes and various plant pathways that are meant to provide tolerance to plants under stress. NO also interacts with phytohormones affecting and regulating various physiological responses of plants. In general, NO reduces the toxicity and accumulation of heavy metals in plants.

References

Ahmad A, Hadi F, Ali N (2015) Effective phytoextraction of cadmium (Cd) with increasing concentration of total phenolics and free proline in Cannabis sativa (L) plant under various treatments of fertilizers, plant growth regulators and sodium salt. Int J Phytoremediation 17(1):56–65

Ahmad P, Ahanger MA, Alyemeni MN, Wijaya L, Alam P (2018) Exogenous application of nitric oxide modulates osmolyte metabolism, antioxidants, enzymes of ascorbate-glutathione cycle and promotes growth under cadmium stress in tomato. Protoplasma 255(1):79–93

Akram NA, Iqbal M, Muhammad A, Ashraf M, Al-Qurainy F, Shafiq S (2018) Aminolevulinic acid and nitric oxide regulate oxidative defense and secondary metabolisms in canola (Brassica napus L.) under drought stress. Protoplasma 255(1):163–174

Alderton WK, Cooper CE, Knowles RG (2001) Nitric oxide synthases: structure, function and inhibition. Biochem J 357(3):593–615

Anand P, Stamler JS (2012) Enzymatic mechanisms regulating protein S-nitrosylation: implications in health and disease. J Mol Med 90(3):233–244

Astier J, Gross I, Durner J (2018) Nitric oxide production in plants: an update. J Exp Bot 69(14):3401–3411

Bai X, Yang L, Tian M, Chen J, Shi J, Yang Y, Hu X (2011) Nitric oxide enhances desiccation tolerance of recalcitrant Antiaris toxicaria seeds via protein S-nitrosylation and carbonylation. PLoS One 6(6):e20714

Bai XY, Dong YJ, Wang QH, Xu LL, Kong J, Liu S (2015) Effects of lead and nitric oxide on photosynthesis, antioxidative ability, and mineral element content of perennial ryegrass. Biol Plant 59(1):163–170

Bartha B, Kolbert Z, Erdei L (2005) Nitric oxide production induced by heavy metals in Brassica juncea L. Czern. and Pisum sativum L. Acta Biol Szeged 49(1–2):9–12

Beligni M, Lamattina L (1999) Is nitric oxide toxic or protective? Trends Plant Sci 4(8):299–300

Broniowska KA, Hogg N (2012) The chemical biology of S-nitrosothiols. Antioxid Redox Signal 17(7):969–980

Capone R, Tiwari BS, Levine A (2004) Rapid transmission of oxidative and nitrosative stress signals from roots to shoots in Arabidopsis. Plant Physiol Biochem 42(5):425–428

Chaki M, Valderrama R, Fernández-Ocaña AM, Carreras A, Gómez-Rodríguez MV, López-Jaramillo JAVIER et al (2011) High temperature triggers the metabolism of S-nitrosothiols in sunflower mediating a process of nitrosative stress which provokes the inhibition of ferredoxin–NADP reductase by tyrosine nitration. Plant Cell Environ 34(11):1803–1818

Chamizo-Ampudia A, Sanz-Luque E, Llamas A, Galvan A, Fernandez E (2017) Nitrate reductase regulates plant nitric oxide homeostasis. Trends Plant Sci 22(2):163–174

Chen W, Dong Y, Hu G, Bai X (2018) Effects of exogenous nitric oxide on cadmium toxicity and antioxidative system in perennial ryegrass. J Soil Sci Plant Nutr 18(1):29–143

Pagnussat GC, Simontacchi M, Puntarulo S, Lamattina L (2002) Nitric oxide is required for root organogenesis. Plant Physiol 129(3):954–956

Corpas FJ, de la Colina C, Sánchez-Rasero F, Luis A (1997) A role for leaf peroxisomes in the catabolism of purines. J Plant Physiol 151(2):246–250

Corpas FJ, Palma JM, Sandalio LM, Valderrama R, Barroso JB, Luis A (2008) Peroxisomal xanthine oxidoreductase: characterization of the enzyme from pea (Pisum sativum L.) leaves. J Plant Physiol 165(13):1319–1330

Corpas FJ, Chaki M, Leterrier M, Barroso JB (2009) Protein tyrosine nitration: a new challenge in plants. Plant Signal Behav 4(10):920–923

Correa-Aragunde N, Foresi N, Lamattina L (2015) Nitric oxide is a ubiquitous signal for maintaining redox balance in plant cells: regulation of ascorbate peroxidase as a case study. J Exp Bot 66(10):2913–2921

Costa-Broseta Á, Perea-Resa C, Castillo MC, Ruíz MF, Salinas J, León J (2018) Nitric oxide controls constitutive freezing tolerance in Arabidopsis by attenuating the levels of osmoprotectants, stress-related hormones and anthocyanins. Sci Rep 8(1):1–10

del Rıo LA, Corpas FJ, Barroso JB (2004) Nitric oxide and nitric oxide synthase activity in plants. Phytochemistry 65(7):783–792

Delledonne M, Zeier J, Marocco A, Lamb C (2001) Signal interactions between nitric oxide and reactive oxygen intermediates in the plant hypersensitive disease resistance response. Proc Natl Acad Sci 98(23):13454–13459

Desikan R, Griffiths R, Hancock J, Neill S (2002) A new role for an old enzyme: nitrate reductase-mediated nitric oxide generation is required for abscisic acid-induced stomatal closure in Arabidopsis thaliana. Proc Natl Acad Sci 99(25):16314–16318

Dodd AN, Kudla J, Sanders D (2010) The language of calcium signaling. Annu Rev Plant Biol 61:593–620

Dong Y, Xu L, Wang Q, Fan Z, Kong J, Bai X (2014) Effects of exogenous nitric oxide on photosynthesis, antioxidative ability, and mineral element contents of perennial ryegrass under copper stress. J Plant Interact 9(1):402–411

Duan X, Li X, Ding F, Zhao J, Guo A, Zhang L et al (2015) Interaction of nitric oxide and reactive oxygen species and associated regulation of root growth in wheat seedlings under zinc stress. Ecotoxicol Environ Saf 113:95–102

Fancy NN, Bahlmann AK, Loake GJ (2017) Nitric oxide function in plant abiotic stress. Plant Cell Environ 40(4):462–472

Farnese FS, de Oliveira JA, Gusman GS, Leão GA, Ribeiro C, Siman LI, Cambraia J (2013) Plant responses to arsenic: the role of nitric oxide. Water Air Soil Pollut 224:1–11

Farnese FS, Menezes-Silva PE, Gusman GS, Oliveira JA (2016) When bad guys become good ones: the key role of reactive oxygen species and nitric oxide in the plant responses to abiotic stress. Front Plant Sci 7:471

Fatma M, Masood A, Per TS, Rasheed F, Khan NA (2016) Interplay between nitric oxide and sulfur assimilation in salt tolerance in plants. Crop J 4(3):153–161

Feechan A, Kwon E, Yun BW, Wang Y, Pallas JA, Loake GJ (2005) A central role for S-nitrosothiols in plant disease resistance. Proc Natl Acad Sci 102(22):8054–8059

Feigl G, Lehotai N, Molnár A, Ördög A, Rodríguez-Ruiz M, Palma JM, Corpas FJ, Erdei L, Kolbert Z (2015) Zinc induces distinct changes in the metabolism of reactive oxygen and nitrogen species (ROS and RNS) in the roots of two Brassica species with different sensitivity to zinc stress. Ann Bot 116(4):613–625

Floryszak-Wieczorek J, Milczarek G, Arasimowicz M, Ciszewski A (2006) Do nitric oxide donors mimic endogenous NO-related response in plants? Planta 224(6):1363–1372

Garcia-Mata C, Gay R, Sokolovski S, Hills A, Lamattina L, Blatt MR (2003) Nitric oxide regulates K+ and Cl-channels in guard cells through a subset of abscisic acid-evoked signaling pathways. Proc Natl Acad Sci 100(19):11116–11121

Ghasemi H, Rostampour F, Ranjbar A (2014) The role of oxidative stress in metals toxicity/mitochondrial dysfunction as a key player. Galen Med J 3(1):2–13

Gill SS, Peter Singh L, Gill R, Tuteja N (2012) Generation and scavenging of reactive oxygen species in plants under stress. In: Improving crop resistance to abiotic stress, pp 49–70. Wiley-Blackwell, Germany

Gong B, Nie W, Yan Y, Gao Z, Shi Q (2017) Unravelling cadmium toxicity and nitric oxide induced tolerance in Cucumis sativus: insight into regulatory mechanisms using proteomics. J Hazard Mater 336:202–213

Greenacre SA, Ischiropoulos H (2001) Tyrosine nitration: localisation, quantification, consequences for protein function and signal transduction. Free Radic Res 34(6):541–581

Groppa MD, Rosales EP, Iannone MF, Benavides MP (2008) Nitric oxide, polyamines and Cd-induced phytotoxicity in wheat roots. Phytochemistry 69(14):2609–2615

Groß F, Durner J, Gaupels F (2013) Nitric oxide, antioxidants and prooxidants in plant defense responses. Front Plant Sci 4:419

Gupta KJ, Fernie AR, Kaiser WM, van Dongen JT (2011) On the origins of nitric oxide. Trends Plant Sci 16(3):160–168

Hancock JT, Whiteman M (2018) Cellular redox environment and its influence on redox signalling molecules. Reactive Oxygen Species 5(14)

He HY, He LF, Gu MH, Li XF (2012) Nitric oxide improves aluminum tolerance by regulating hormonal equilibrium in the root apices of rye and wheat. Plant Sci 183:123–130

Hsu YT, Kao CH (2004) Cadmium toxicity is reduced by nitric oxide in rice leaves. Plant Growth Regul 42:227–238

Hu KD, Hu LY, Li YH, Zhang FQ, Zhang H (2007) Protective roles of nitric oxide on germination and antioxidant metabolism in wheat seeds under copper stress. Plant Growth Regul 53(3):173–183

Jalmi SK, Sinha AK (2015) ROS mediated MAPK signaling in abiotic and biotic stress-striking similarities and differences. Front Plant Sci 6:769

Jalmi SK, Bhagat PK, Verma D, Noryang S, Tayyeba S, Singh K et al (2018) Traversing the links between heavy metal stress and plant signaling. Front Plant Sci 9:12

Jin JW, Xu YF, Huang YF (2010) Protective effect of nitric oxide against arsenic-induced oxidative damage in tall fescue leaves. Afr J Biotechnol 9(11):1619–1627

Jonak C, Ökrész L, Bögre L, Hirt H (2002) Complexity, cross talk and integration of plant MAP kinase signalling. Curr Opin Plant Biol 5(5):415–424

Jonak C, Nakagami H, Hirt H (2004) Heavy metal stress. Activation of distinct mitogen-activated protein kinase pathways by copper and cadmium. Plant Physiol 136(2):3276–3283

Kaur G, Singh HP, Batish DR, Mahajan P, Kohli RK, Rishi V (2015) Exogenous nitric oxide (NO) interferes with lead (Pb)-induced toxicity by detoxifying reactive oxygen species in hydroponically grown wheat (Triticum aestivum) roots. PLoS One 10(9):e0138713

Kaya C, Akram NA, Sürücü A, Ashraf M (2019) Alleviating effect of nitric oxide on oxidative stress and antioxidant defence system in pepper (Capsicum annuum L.) plants exposed to cadmium and lead toxicity applied separately or in combination. Sci Hortic 255:52–60

Kazemi N, Khavari-Nejad RA, Fahimi H, Saadatmand S, Nejad-Sattari T (2010) Effects of exogenous salicylic acid and nitric oxide on lipid peroxidation and antioxidant enzyme activities in leaves of Brassica napus L. under nickel stress. Sci Hortic 126(3):402–407

Khairy AIH, Oh MJ, Lee SM, Roh KS (2016) Nitric oxide overcomes Cd and Cu toxicity in in vitro-grown tobacco plants through increasing contents and activities of rubisco and rubisco activase. Biochim Open 2:41–51

Khan MN, Mobin M, Mohammad F, Corpas FJ (eds) (2014) Nitric oxide in plants: metabolism and role in stress physiology. Springer International Publishing, Cham

Khan MN, Alamri S, Al-Amri AA, Alsubaie QD, Al-Munqedi B, Ali HM et al (2021) Effect of nitric oxide on seed germination and seedling development of tomato under chromium toxicity. J Plant Growth Regul 40(6):2358–2370

Kim JA, Agrawal GK, Rakwal R, Han KS, Kim KN, Yun CH, Heu S, Park SY, Lee YH, Jwa NS (2003) Molecular cloning and mRNA expression analysis of a novel rice (Oryzasativa L.) MAPK kinase kinase, OsEDR1, an ortholog of ArabidopsisAtEDR1, reveal its role in defense/stress signalling pathways and development. Biochem Biophys Res Commun 300(4):868–876

Kolbert Z, Pető A, Lehotai N, Feigl G, Erdei L (2012) Long-term copper (Cu^{2+}) exposure impacts on auxin, nitric oxide (NO) metabolism and morphology of Arabidopsis thaliana L. Plant Growth Regul 68(2):151–159

Kopyra M, Gwóźdź EA (2003) Nitric oxide stimulates seed germination and counteracts the inhibitory effect of heavy metals and salinity on root growth of Lupinus luteus. Plant Physiol Biochem 41(11–12):1011–1017

Kopyra M, Stachoń-Wilk M, Gwóźdź EA (2006) Effects of exogenous nitric oxide on the anti-oxidant capacity of cadmium-treated soybean cell suspension. Acta Physiol Plant 28:525–536

Kumar D, Klessig DF (2000) Differential induction of tobacco MAP kinases by the defense signals nitric oxide, salicylic acid, ethylene, and jasmonic acid. Mol Plant-Microbe Interact 13(3):347–351

Kumari A, Sheokand S, Swaraj K (2010) Nitric oxide induced alleviation of toxic effects of short term and long term Cd stress on growth, oxidative metabolism and Cd accumulation in Chickpea. Braz J Plant Physiol 22(4):271–284

Lamotte O, Courtois C, Barnavon L, Pugin A, Wendehenne D (2005) Nitric oxide in plants: the biosynthesis and cell signalling properties of a fascinating molecule. Planta 221(1):1–4

Laspina NV, Groppa MD, Tomaro ML, Benavides MP (2005) Nitric oxide protects sunflower leaves against Cd-induced oxidative stress. Plant Sci 169(2):323–330

Lecourieux D, Lamotte O, Bourque S, Wendehenne D, Mazars C, Ranjeva R, Pugin A (2005) Proteinaceous and oligosaccharidic elicitors induce different calcium signatures in the nucleus of tobacco cells. Cell Calcium 38(6):527–538

Lin YF, Aarts MG (2012) The molecular mechanism of zinc and cadmium stress response in plants. Cell Mol Life Sci 69(19):3187–3206

Lindermayr C, Durner J (2015) Interplay of reactive oxygen species and nitric oxide: nitric oxide coordinates reactive oxygen species homeostasis. Plant Physiol 167(4):1209–1210

Lindermayr C, Saalbach G, Bahnweg G, Durner J (2006) Differential inhibition of Arabidopsis methionine adenosyltransferases by protein S-nitrosylation. J Biol Chem 281(7):4285–4291

Lipton SA, Choi YB, Pan ZH, Lei SZ, Chen HSV, Sucher NJ et al (1993) A redox-based mechanism for the neuroprotective and neurodestructive effects of nitric oxide and related nitroso-compounds. Nature 364(6438):626–632

Lozano-Juste J, León J (2011) Nitric oxide regulates DELLA content and PIF expression to promote photomorphogenesis in Arabidopsis. Plant Physiol 156(3):1410–1423

Luan SKJ, Rodriguez-Concepcion M, Yalovsky S, Gruissem S (2002) CaMs and CBLs: calcium sensors for specific signal-response coupling in plants. Plant Cell 14:S389–S400

Mahmood T, Gupta KJ, Kaiser WM (2009) Cadmium stress stimulates nitric oxide production by wheat roots. Pak J Bot 41(3):6

Manai J, Kalai T, Gouia H, Corpas FJ (2014) Exogenous nitric oxide (NO) ameliorates salinity-induced oxidative stress in tomato (Solanum lycopersicum) plants. J Soil Sci Plant Nutr 14(2):433–446

Manjunatha G, Lokesh V, Neelwarne B (2010) Nitric oxide in fruit ripening: trends and opportunities. Biotechnol Adv 28(4):489–499

Maurice N, Ping CY, Miaomiao Q, Constantine U, Bo Y, Qi KY (2016) Effects of exogenous nitric oxide on germination and carbohydrates mobilization in alfalfa seedlings under cadmium stress. Int J Environ Sci Technol 4:2337–2350

Mengel A, Chaki M, Shekariesfahlan A, Lindermayr C (2013) Effect of nitric oxide on gene transcription–S-nitrosylation of nuclear proteins. Front Plant Sci 4:293

Mihailovic N, Drazic G (2011) Incomplete alleviation of nickel toxicity in bean by nitric oxide supplementation. Plant Soil Environ 57(8):396–401

Mittler R (2002) Oxidative stress, antioxidants and stress tolerance. Trends Plant Sci 7(9):405–410

Molassiotis A, Fotopoulos V (2011) Oxidative and nitrosative signaling in plants: two branches in the same tree? Plant Signal Behav 6(2):210–214

Mostofa MG, Seraj ZI, Fujita M (2014) Exogenous sodium nitroprusside and glutathione alleviate copper toxicity by reducing copper uptake and oxidative damage in rice (Oryza sativa L.) seedlings. Protoplasma 251(6):1373–1386

Mur LA, Mandon J, Persijn S, Cristescu SM, Moshkov IE, Novikova GV et al (2013) Nitric oxide in plants: an assessment of the current state of knowledge. AoB Plants 5:pls052

Nabaei M, Amooaghaie R (2019) Interactive effect of melatonin and sodium nitroprusside on seed germination and seedling growth of Catharanthus roseus under cadmium stress. Russ J Plant Physiol 66(1):128–139

Neill SJ, Desikan R, Hancock JT (2003) Nitric oxide signalling in plants. New Phytol 159(1):11–35

Neill S, Barros R, Bright J, Desikan R, Hancock J, Harrison J, Morris P, Ribeiro D, Wilson I (2008) Nitric oxide, stomatal closure, and abiotic stress. J Exp Bot 59(2):165–176

Ortega-Galisteo AP, Rodríguez-Serrano M, Pazmiño DM, Gupta DK, Sandalio LM, Romero-Puertas MC (2012) S-Nitrosylated proteins in pea (Pisum sativum L.) leaf peroxisomes: changes under abiotic stress. J Exp Bot 63(5):2089–2103

Ortega-Villasante C, Rellán-Alvarez R, Del Campo FF, Carpena-Ruiz RO, Hernández LE (2005) Cellular damage induced by cadmium and mercury in Medicago sativa. J Exp Bot 56(418):2239–2251

Palma JM, Sandalio LM, Corpas FJ, Romero-Puertas MC, McCarthy I, del Río LA (2002) Plant proteases, protein degradation, and oxidative stress: role of peroxisomes. Plant Physiol Biochem 40(6–8):521–530

Parmar P, Kumari N, Sharma V (2013) Structural and functional alterations in photosynthetic apparatus of plants under cadmium stress. Bot Stud 54(1):1–6

Per TS, Masood A, Khan NA (2017) Nitric oxide improves S-assimilation and GSH production to prevent inhibitory effects of cadmium stress on photosynthesis in mustard (Brassica juncea L.). Nitric Oxide 68:111–124

Pető A, Lehotai N, Feigl G, Tugyi N, Ördög A, Gémes K et al (2013) Nitric oxide contributes to copper tolerance by influencing ROS metabolism in Arabidopsis. Plant Cell Rep 32(12):1913–1923

Planchet E, Kaiser WM (2006) Nitric oxide (NO) detection by DAF fluorescence and chemiluminescence: a comparison using abiotic and biotic NO sources. J Exp Bot 57(12):3043–3055

Praveen A, Gupta M (2018) Nitric oxide confronts arsenic stimulated oxidative stress and root architecture through distinct gene expression of auxin transporters, nutrient related genes and modulates biochemical responses in Oryza sativa L. Environ Pollut 240:950–962

Qiao W, Li C, Fan LM (2014) Cross-talk between nitric oxide and hydrogen peroxide in plant responses to abiotic stresses. Environ Exp Bot 100:84–93

Radi R (2013) Protein tyrosine nitration: biochemical mechanisms and structural basis of functional effects. Acc Chem Res 46(2):550–559

Radi R, Beckman JS, Bush KM, Freeman BA (1991) Peroxynitrite-induced membrane lipid peroxidation: the cytotoxic potential of superoxide and nitric oxide. Arch Biochem Biophys 288(2):481–487

Rahman T, Hosen I, Islam MM, Shekhar HU (2012) Oxidative stress and human health. Adv Biosci Biotechnol 3(07):997–1019

Rizwan M, Imtiaz M, Dai Z, Mehmood S, Adeel M, Liu J, Tu S (2017) Nickel stressed responses of rice in Ni subcellular distribution, antioxidant production, and osmolyte accumulation. Environ Sci Pollut Res 24(25):20587–20598

Rudd JJ, Franklin-Tong VE (2001) Unravelling response-specificity in Ca^{2+} signalling pathways in plant cells. New Phytol 151(1):7–33

Sadeghipour O (2016) Pretreatment with nitric oxide reduces lead toxicity in cowpea (Vigna unguiculata [L.] Walp.). Arch Biol Sci 68(1):165–175

Sadeghipour O, Aghaei P (2013) Improving the growth of cowpea (Vigna unguiculata L. Walp.) by magnetized water. J Biodivers Environ Sci 3(1):37–43

Sahay S, Gupta M (2017) An update on nitric oxide and its benign role in plant responses under metal stress. Nitric Oxide 67:39–52

Salt DE, Prince RC, Pickering IJ, Raskin I (1995) Mechanisms of cadmium mobility and accumulation in Indian mustard. Plant Physiol 109(4):1427–1433

Šamajová O, Plíhal O, Al-Yousif M, Hirt H, Šamaj J (2013) Improvement of stress tolerance in plants by genetic manipulation of mitogen-activated protein kinases. Biotechnol Adv 31(1):118–128

Sandalio LM, Fernández VM, Rupérez FL, Del Río LA (1988) Superoxide free radicals are produced in glyoxysomes. Plant Physiol 87(1):1–4

Sandalio LM, Dalurzo HC, Gomez M, Romero-Puertas MC, Del Rio LA (2001) Cadmium-induced changes in the growth and oxidative metabolism of pea plants. J Exp Bot 52(364):2115–2126

Sandalio LM, Rodríguez-Serrano M, Gupta DK, Archilla A, Romero-Puertas MC, Río LAD (2012) Reactive oxygen species and nitric oxide in plants under cadmium stress: from toxicity to signaling. In: Environmental adaptations and stress tolerance of plants in the era of climate change. Springer, New York, pp 199–215

Sanders D, Pelloux J, Brownlee C, Harper JF (2002) Calcium at the crossroads of signaling. Plant Cell 14(Suppl 1):S401–S417

Sanz L, Albertos P, Mateos I, Sánchez-Vicente I, Lechón T, Fernández-Marcos M, Lorenzo O (2015) Nitric oxide (NO) and phytohormones crosstalk during early plant development. J Exp Bot 66(10):2857–2868

Schafer FQ, Buettner GR (2001) Redox environment of the cell as viewed through the redox state of the glutathione disulfide/glutathione couple. Free Radic Biol Med 30(11):1191–1212

Schutzendubel A, Polle A (2002) Plant responses to abiotic stresses: heavy metal-induced oxidative stress and protection by mycorrhization. J Exp Bot 53(372):1351–1365

Serrano I, Romero-Puertas MC, Sandalio LM, Olmedilla A (2015) The role of reactive oxygen species and nitric oxide in programmed cell death associated with self-incompatibility. J Exp Bot 66(10):2869–2876

Sharma SS, Dietz KJ (2009) The relationship between metal toxicity and cellular redox imbalance. Trends Plant Sci 14(1):43–50

Sheokand S, Bhankar V, Sawhney V (2010) Ameliorative effect of exogenous nitric oxide on oxidative metabolism in NaCl treated chickpea plants. Braz J Plant Physiol 22:81–90

Shiva S (2010) Mitochondria as metabolizers and targets of nitrite. Nitric Oxide 22(2):64–74

Shivaraj SM, Vats S, Bhat JA, Dhakte P, Goyal V, Khatri P et al (2020) Nitric oxide and hydrogen sulfide crosstalk during heavy metal stress in plants. Physiol Plant 168(2):437–455

Silveira NM, de Oliveira JA, Ribeiro C, Canatto RA, Siman L, Cambraia J, Farnese F (2015) Nitric oxide attenuates oxidative stress induced by arsenic in lettuce (Lactuca sativa) leaves. Water Air Soil Pollut 226(11):1–9

Singh HP, Batish DR, Kaur G, Arora K, Kohli RK (2008) Nitric oxide (as sodium nitroprusside) supplementation ameliorates Cd toxicity in hydroponically grown wheat roots. Environ Exp Bot 63(1–3):158–167

Singh HP, Kaur S, Batish DR, Sharma VP, Sharma N, Kohli RK (2009) Nitric oxide alleviates arsenic toxicity by reducing oxidative damage in the roots of Oryza sativa (rice). Nitric Oxide 20(4):289–297

Singh AP, Dixit G, Kumar A, Mishra S, Singh PK, Dwivedi S et al (2016) Nitric oxide alleviated arsenic toxicity by modulation of antioxidants and thiol metabolism in rice (Oryza sativa L.). Front Plant Sci 6:1272

Smeets K, Opdenakker K, Remans T, Forzani C, Hirt H, Vangronsveld J, Cuypers A (2013) The role of the kinase OXI1 in cadmium-and copper-induced molecular responses in A rabidopsis thaliana. Plant Cell Environ 36(6):1228–1238

StoÈhr C, Ullrich WR (2002) Generation and possible roles of NO in plant roots and their apoplastic space. J Exp Bot 53(379):2293–2303

Sun YF, Liu SB, Meng FL, Liu JY, Jin Z, Kong LT, Liu JH (2012) Metal oxide nanostructures and their gas sensing properties: a review. Sensors 12(3):2610–2631

Sun C, Lu L, Liu L, Liu W, Yu Y, Liu X et al (2014) Nitrate reductase-mediated early nitric oxide burst alleviates oxidative damage induced by aluminum through enhancement of antioxidant defenses in roots of wheat (Triticum aestivum). New Phytol 201(4):1240–1250

Sytar O, Kumar A, Latowski D, Kuczynska P, Strzałka K, Prasad MNV (2013) Heavy metal-induced oxidative damage, defense reactions, and detoxification mechanisms in plants. Acta Physiol Plant 35(4):985–999

Tamás L, Valentovičová K, Halušková Ľ, Huttová J, Mistrík I (2009) Effect of cadmium on the distribution of hydroxyl radical, superoxide and hydrogen peroxide in barley root tip. Protoplasma 236(1):67–72

Tan DX, Hardeland R, Manchester LC, Korkmaz A, Ma S, Rosales-Corral S, Reiter RJ (2012) Functional roles of melatonin in plants, and perspectives in nutritional and agricultural science. J Exp Bot 63(2):577–597

Terrile MC, París R, Calderón-Villalobos LI, Iglesias MJ, Lamattina L, Estelle M, Casalongué CA (2012) Nitric oxide influences auxin signaling through S-nitrosylation of the Arabidopsis TRANSPORT INHIBITOR RESPONSE 1 auxin receptor. Plant J 70(3):492–500

Thao NP, Khan MIR, Thu NBA, Hoang XLT, Asgher M, Khan NA, Tran LSP (2015) Role of ethylene and its cross talk with other signaling molecules in plant responses to heavy metal stress. Plant Physiol 169(1):73–84

Tiwari S, Verma N, Singh VP, Prasad SM (2019) Nitric oxide ameliorates aluminium toxicity in Anabaena PCC 7120: regulation of aluminium accumulation, exopolysaccharides secretion, photosynthesis and oxidative stress markers. Environ Exp Bot 161:218–227

Umbreen S, Lubega J, Cui B, Pan Q, Jiang J, Loake GJ (2018) Specificity in nitric oxide signalling. J Exp Bot 69(14):3439–3448

Valentovičová K, Halušková LU, Huttová J, Mistrík I, Tamás L (2010) Effect of cadmium on diaphorase activity and nitric oxide production in barley root tips. J Plant Physiol 167(1):10–14

Vandelle E, Poinssot B, Wendehenne D, Bentéjac M, Pugin A (2006) Integrated signaling network involving calcium, nitric oxide, and active oxygen species but not mitogen-activated protein kinases in BcPG1-elicited grapevine defenses. Mol Plant-Microbe Interact 19(4):429–440

Wang YS, Yang ZM (2005) Nitric oxide reduces aluminum toxicity by preventing oxidative stress in the roots of Cassia tora L. Plant Cell Physiol 46(12):1915–1923

Wang Y, Loake GJ, Chu C (2013) Cross-talk of nitric oxide and reactive oxygen species in plant programed cell death. Front Plant Sci 4:314

Wang L, Su H, Han L, Wang C, Sun Y, Liu F (2014) Differential expression profiles of poplar MAP kinase kinases in response to abiotic stresses and plant hormones, and overexpression of PtMKK4 improves the drought tolerance of poplar. Gene 545(1):141–148

Wang WW, Bai XY, Dong YJ, Chen WF, Song YL, Tian XY (2016) Effects of application of exogenous NO on the physiological characteristics of perennial ryegrass grown in Cd-contaminated soil. J Soil Sci Plant Nutr 16(3):731–744

Wei L, Zhang J, Wang C, Liao W (2020) Recent progress in the knowledge on the alleviating effect of nitric oxide on heavy metal stress in plants. Plant Physiol Biochem 147:161–171

Xiong J, An L, Lu H, Zhu C (2009a) Exogenous nitric oxide enhances cadmium tolerance of rice by increasing pectin and hemicellulose contents in root cell wall. Planta 230:755–765

Xiong J, Lu H, Lu K, Duan Y, An L, Zhu C (2009b) Cadmium decreases crown root number by decreasing endogenous nitric oxide, which is indispensable for crown root primordia initiation in rice seedlings. Planta 230(4):599–610

Xu J, Wang W, Yin H, Liu X, Sun H, Mi Q (2010) Exogenous nitric oxide improves antioxidative capacity and reduces auxin degradation in roots of Medicago truncatula seedlings under cadmium stress. Plant Soil 326(1):321–330

Xue M, Yi H (2017) Nitric oxide signaling is involved in arsenic-induced guard cell death in Vicia faba L.(Fabaceae). Rev Bras Bot 40(3):635–642

Yang LT, Qi YP, Chen LS, Sang W, Lin XJ, Wu YL, Yang CJ (2012) Nitric oxide protects sour pummelo (Citrus grandis) seedlings against aluminum-induced inhibition of growth and photosynthesis. Environ Exp Bot 82:1–13

Yu M, Lamattina L, Spoel SH, Loake GJ (2014) Nitric oxide function in plant biology: a redox cue in deconvolution. New Phytol 202(4):1142–1156

Yuan HM, Huang X (2016) Inhibition of root meristem growth by cadmium involves nitric oxide-mediated repression of auxin accumulation and signalling in Arabidopsis. Plant Cell Environ 39(1):120–135

Yun BW, Skelly MJ, Yin M, Yu M, Mun BG, Lee SU et al (2016) Nitric oxide and S-nitrosoglutathione function additively during plant immunity. New Phytol 211(2):516–526

Zhang LP, Mehta SK, Liu ZP, Yang ZM (2008) Copper-induced proline synthesis is associated with nitric oxide generation in Chlamydomonas reinhardtii. Plant Cell Physiol 49(3):411–419

Zhang X, Rui H, Zhang F, Hu Z, Xia Y, Shen Z (2018) Overexpression of a functional Vicia sativa PCS1 homolog increases cadmium tolerance and phytochelatins synthesis in Arabidopsis. Front Plant Sci 9:107

Zhao H, Jin Q, Wang Y, Chu L, Li X, Xu Y (2016) Effects of nitric oxide on alleviating cadmium stress in Typha angustifolia. Plant Growth Regul 78(2):243–251

Zou T, Zheng LP, Yuan HY, Yuan YF, Wang JW (2012) The nitric oxide production and NADPH-diaphorase activity in root tips of *Vicia faba* L. under copper toxicity. Plant Omics 5(2):115–121

Chapter 9
Jasmonic Acid-Mediated Regulation of Plant Biology Under Heavy Metal Stress: Physiological and Molecular Aspects

Ekhlaque A. Khan, Sudha Yadav, and Amarendra N. Misra

Abstract Plants, like sessile organisms, must withstand a variety of environmental stressors. Because of increased industrialization and other anthropogenic activities, heavy metal pollution from heavy metals like cadmium, arsenic, and lead is the greatest abiotic stress for plants in the contemporary scenario, posing a substantial risk to animal and human health. There is a high demand for less harmful metal deposition in plant-based foods. Plant hormones are essential to cope with biotic and abiotic stress in plants. Among all the hormones, jasmonic acid (JA) and jasmonates (JAs), which act as its precursors and derivatives, play an essential role in mediating plant responses and defense against biotic and abiotic stress and have attracted much attention in research. Although there are several reviews on JA, this review is based on the regulation of JA in plants against stress and synthesis of jasmonic acid, metabolic regulation, and signaling. We review current developments in understanding the roles and processes of JAs in plant responses to abiotic stimuli like heavy metals, salt, drought, cold, and heat, as well as biotic stresses. This chapter will cover the physiological and molecular approaches of jasmonic acid and jasmonate derivatives (JAs) under heavy metal stress. Finally, we highlight present difficulties and potential in JA research in plant stress responses.

Keywords Hormones · Jasmonic acid · Plant physiology · Stress biology · Toxicity

E. A. Khan (✉)
Department of Biotechnology, Chaudhary Bansi Lal University, Bhiwani, Haryana, India

S. Yadav
Department of Botany, Chaudhary Bansi Lal University, Bhiwani, Haryana, India

A. N. Misra
Office of the Vice Chancellor, Jagdguru Kripalu University, Banara, Cuttack, Odisha, India

A. Sharma et al. (eds.), *Hormones and Heavy Metals*, Plant Life and Environment Dynamics, https://doi.org/10.1007/978-981-96-9622-2_9

9.1 Introduction

Both biotic and abiotic stresses affect the plants. The primary major abiotic factors that affect plant growth and development are temperature, soil nutrient level, salt, water, light, and ozone (Isah 2019). Fluctuations in the level of any of these factors affect the normal function and plant physiology. This abiotic stress disturbs the various biological processes like homeostasis of ROS (reactive oxygen species), composition of membrane lipid, electrolyte maintenance, electron transport system, and integrity of organelles (You and Chan 2015).

Unlike animals, plants cannot move to a suitable environment to avoid stress conditions. Therefore, plants evolve mechanisms to withstand unfavorable conditions by modifying physiological and developmental processes. In recent times, changes in the world's climate may increase the frequency of abiotic stress, posing a threat to food production all over the world. It is predicted that more than 50% of the world's farming land will be under drought stress by 2050 (Marthandan et al. 2020), indicating the need for abiotic stress-tolerant crop production.

In plants, several plant hormones are linked with metabolic and various physiological functions (Sirhindi et al. 2020). Plant hormones perform various functions in plants like development, growth, and response to various types of stresses by inducing a signaling cascade. Auxin, cytokinin, abscisic acid, gibberellin, ethylene, brassinosteroid, jasmonic acid, and salicylic acid are phytohormones that have the potential to protect plants from various abiotic stresses (Zaid and Mohammad 2018).

Among all the phytohormones, jasmonic acid (JA) has been widely studied in recent years. Jasmonic acid and its derivatives like methyl jasmonate (MeJA), JA-glucosyl ester, jasmonyl isoleucine (JA-Ile), jasmonoyl-amino acid, cis-jasmone, lactones of 12-hydroxy-JA-IIe, JA-Ile methyl ester, 12-carboxy-JA-IIe, JA-Ile glucosyl ester, 12-hydroxyjasmonic acid sulfate (12-HSO4-JA), 12-O-glucosyl-JA-IIe, and 12-O-glucosyl-JA are collectively named as jasmonates (JAs). These are all formed by cyclopentanones and are collectively called oxylipins, which are oxidized lipid family members (Wasternack and Feussner 2018). By enzymatic process by alpha-dioxygenases or lipoxygenases and via non-enzymatic process by polyunsaturated fatty acids autoxidation, biologically active signaling molecule oxylipins are synthesized (Göbel and Feussner 2009).

Jasmonic acid and its derivatives are major growth and plant stress regulators, also have a crucial role in seed germination, callus growth, senescence, and primary root growth (Fahad et al. 2015). It is an essential signal molecule in both abiotic stress and biotic stress signaling in plants (Per et al. 2018; Ali and Baek 2020). For some jasmonates, fungus is the source, and petals of jasmine (*Jasminum grandiflorum*) are the source for methyl jasmonate (MeJA), a methyl ester form (Avanci et al. 2010). JA mainly resides in reproductive tissues and flowers of plants and in low amounts in mature leaves and roots. JA also regulates vegetative growth, anthocyanin biosynthesis, stamen, cell cycle regulation, and trichome growth, as well as Rubisco biosynthesis inhibition, glucose transport, fruit ripening, and uptake of nitrogen and phosphorus (Wasternack 2007; Browse 2005).

JA, as a signaling molecule, regulates the various gene expressions that are involved in specific defense mechanisms and abiotic stresses like heavy metal, salt, drought, low temperature, and micronutrient toxicity. Due to various human-made activities like smelting, mining, and agricultural and industrial activities, various toxic heavy metals are disposed of in large amounts into the ecosystem (Deng et al. 2021). Metalloids and heavy metals contain elements with a density higher than 5 g cm^{-3} (Tchounwou et al. 2012). Heavy metals and metalloids like mercury (Hg), lead (Pb), arsenic (As), and cadmium (Cd) have severe effects on the environment and human health (Järup 2003). There is an increase in diseases and mortality due to slow poisoning of As and Cd (Clemens and Ma 2016). The main route in humans is through food ingestion, and the primary source of this can be understood by the food chain (Rojas-Rueda et al. 2021). So there is a need to generate such food crops that have less toxic elements in the edible part of the plant.

We have to select new crop varieties for generating new cultivars with lower accumulation potential (Zhao and Wang 2020). Using genetically engineered or naturally present hyperaccumulating species for phytoremediation can be a source for food safety in the future (Deng et al. 2021). Jasmonates act as primary signaling molecules for defense, which play an important role in growth and also respond to several developmental and environmental factors (Yu et al. 2019b). For effective plant growth, spatial and temporal regulation of jasmonate signaling is very important, and jasmonic acid (JA) synthesis is induced as a defense response in plants against insect attacks (Yan et al. 2018).

For long-distance and cellular translocation of minerals, membrane-located transporters play a crucial role in plants for deposition and detoxification of harmful heavy metalloids and metals (Sharma et al. 2021; Tang and Zhao 2020). A member of the ZIP (zinc-regulated transporter) family, known as AtIRT1 (Arabidopsis thaliana iron-regulatedtransporter 1), is the primary transporter responsible for cadmium (Cd) uptake in Arabidopsis (Lux et al. 2011). The transport of Cd from roots to shoots is facilitated by heavy metal ATPases (HMAs) located in the root pericycle, specifically AtHMA4 and AtHMA2 (Mills et al. 2005; Hussain et al. 2004). For Cd sequestration to the vacuole from cytosol, its homolog AtHMA3 is present in the vacuole membrane (Chao et al. 2012).

9.2 Biosynthesis of Jasmonic Acid

Jasmonates are synthesized from the octadecanoid pathway, which is a defense signaling pathway in plants (Wasternack and Feussner 2018). The biosynthesis pathway initiates from α-linolenic acid (α-LeA), secreted from the chloroplast membrane component, i.e., galactolipid, by the action of galactolipases named phospholipase A1 (PLA1). Defective in anther dehiscence 1 (DAD1) and its homologs DAD1-like lipases (DALL) and Dongle (DGL) encode the chloroplast-targeted galactolipases (Rudus et al. 2014). In the next step, in (plastid) chloroplast enzyme, 13-lipoxygenases (13-LOXs) convert α-LeA into 13-HPOT (13S-hydroperoxyoctadecatrienoic acid).

After that, 13-HPOT is converted into cis-(+)-12-oxophytodienoic acid (OPDA) in a 2-step process catalyzed by AOSs (allene oxide synthases) (Sivasankar et al. 2000) and AOCs (allene oxide cyclases) (Ziegler et al. 2000). Until this, all steps occur in the plastid (chloroplast). For the next step, OPDA must move to the peroxisome for ß-oxidation and reduction in the side chain of the carboxylic acid in higher plants (Howe and Yoshida 2019). The movement of OPDA from the chloroplast membrane localized transporter OPDA (12-oxophytodienoic acid), known as JASSY (Guan et al. 2019), and influx into peroxisome is mediated by ABC transporter 1 (PXA1), which is a member of the ATP-binding cassette (ABC) transporter family in Arabidopsis (AtABCD1; Theodoulou et al. 2005). OPDA is converted into OPC-8 [3-oxo-2-(2-pentenyl)-cyclopentane-1-octanoic acid] by OPRs (OPDA reductases) in peroxisomes (Schaller et al. 2000). In the next step, OPC-8 is converted into (+)-7-iso-JA via three successive reactions of β-oxidation catalyzed by three enzymes. In the first step of β-oxidation, the enzyme is ACX (acyl-CoA oxidase); in the second step, MFP (multifunctional protein); and in the third step, KAT (L-3-ketoacyl CoA thiolase) (Yang et al. 2020; Huang et al. 2017). In the next step, peroxisomal (+)-7-iso-JA is translocated to the cytoplasm for further modifications by forming conjugate with Ile (isoleucine) amino acid to form (+)-7-iso-JA-Ile, the most bioactive JA compound, by the action of enzyme JA-Ile synthesizing enzyme (JAR1), GH3 family member (Staswick and Tiryaki 2004). Finally, JA-Ile from the cytoplasm translocates to the nucleus through ABC transporter JAT1 and gets involved in the JA signaling pathway (Li et al. 2017a). Cytochrome P450 subfamily of CYP94 proteins, including CYP94B3, CYP94B2, CYP94B1, and CYP94C1, mediate oxidative inactivation from JA-Ile to 12-hydroxyJA-Ile (12OH-JA-Ile) (Bruckhoff et al. 2016). For the inactivation and JA hydroxylation, jasmonate-induced oxygenase (JOX) enzymes have been discovered (Caarls et al. 2017). In most of the land plants, methyl jasmonate (MeJA) is also identified during development and stress responses. This methyl jasmonate (MeJA) is synthesized by JMT (jasmonic acid carboxyl methyltransferase) via a methyl esterification reaction (Seo et al. 2001). For storage and transport of JA, modifications through O-glycosylation and sulfation are required (Wasternack and Feussner 2018).

9.3 JA Signaling Pathway

JA after epimerization is converted into JA-Ile, which gets deposited in stressed leaves' cytoplasm under stress conditions. From the cytoplasm, JA-Ile translocated to adjacent leaves and to the nucleus for responses to defense (Li et al. 2017b). This subcellular localization of JAs in *Arabidopsis thaliana* (At) is controlled by jasmonic acid transporter protein 1 (AtJAT1, also named AtABCG16) (Wang et al. 2019). So, in plant cells under abiotic stress conditions, JA-Ile or JA transport to the apoplast and nucleus from the cytoplasm is controlled by JAT1.

JA application results in a rapid increase in gene expression which are involved in the defense system of plants, showing the involvement of the JA-specific pathway

(Hashimoto et al. 2004). Repressor protein JAZ (JASMONATE ZIM-DOMAIN PROTEIN), receptor COI1 (CORONATINE INSENSITIVE 1), and MYC2 transcription factor or JA are mainly involved in the JA-specific signaling pathway, which regulates plant response and tolerance for stress due to environmental conditions (Wasternack and Strnad 2018; Ali and Baek 2020). JAZs undergo ligase-mediated proteolytic ubiquitinization by the action of JA receptor COI1 (F-box protein) after interaction with JA-Ile. After ubiquitination of JAZs, MYC2 transcription factor becomes free and binds with G-box (50-CACGTG-30) in the promoter region, and further JA-responsive gene expression occurs. JA response is suppressed in jasmonate insensitive 1 (JIN1) mutants in *Arabidopsis* (Berger et al. 1996). Further studies showed that transcription factor MYC2 is encoded by the JIN1 gene (Lorenzo et al. 2004).

SCFCOI1-JAZ protein complex encoding genes are responsible for JA signaling, discovered in rice through published transcriptomic datasets. In the rice genome, a total of 15 JAZs have been identified (Ye et al. 2009). Six gene transcripts of OsJAZs (OsJAZ5/6/9/10/11/12) were upregulated in rice roots induced by arsenic stress (Huang et al. 2012).

JA results in a reduction in heavy metals and metalloids accumulation and their detoxification through antioxidant function, transport system, sequestration, and chelation process. In the JA regulatory mechanism, calcium signaling is involved, which plays a crucial role in signal transmission generated by stress caused by metalloids and heavy metals to plant physiological processes (Zhang et al. 2020b). Furthermore, exogenous application of auxin can delay senescence of leaves, which is balanced out by the application of JA exogenously. Here, in the JA–auxin signaling pathway, IAA29, JAZ8, and JAZ4 genes are negative regulators (Jiang et al. 2014). JAZs suppress the JA signaling pathway, and *Arabidopsis* WRKY57 acts as a node of convergence for auxin-mediated and jasmonic acid signaling in leaf senescence induced by jasmonic acid (Jiang et al. 2014). In Fig. 9.1, we have discussed the JA signaling pathway.

9.4 Role of Jasmonic Acid in Plants

In plants, jasmonic acid plays several important roles in regulating both biotic stress and abiotic stress, such as salt, drought, heavy metal stress, and temperature fluctuations, at both the molecular and physiological levels. For example, tomato (*Solanum lycopersicum*) gets tolerance for TYLCV (yellow leaf curl virus) by increasing the expression of SIMAPK3 gene, which controls jasmonic acid and salicylic acid signaling (Li et al. 2017b). Similarly, plants like Yam plants (*Dioscorea batatas Decne*) and Jerusalem artichoke plants (*Helianthus tuberosus* L.) get JA-induced tolerance for abiotic stresses. JA also plays an important role in inducing in vitro tuberization (Koda and Kikuta 1991).

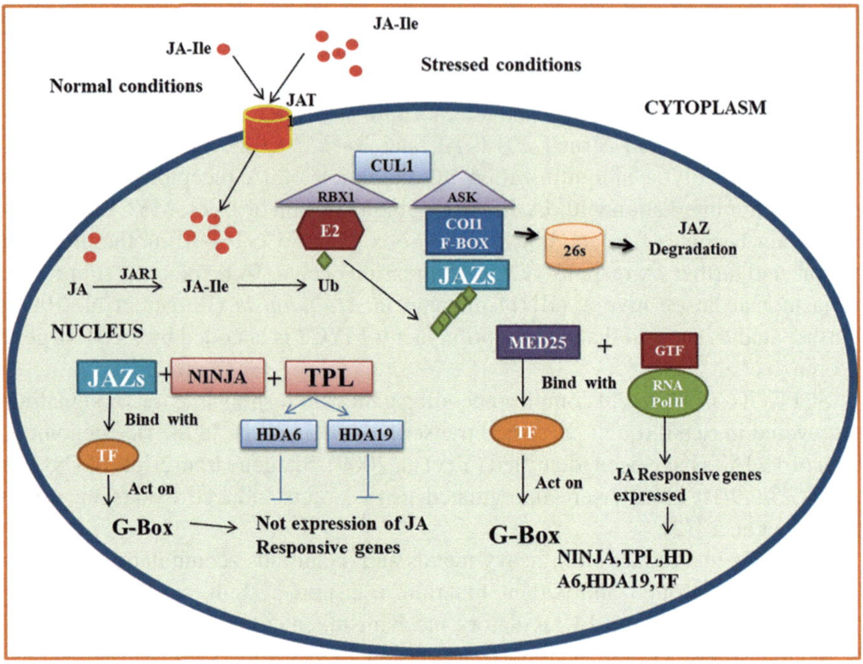

Fig. 9.1 Signaling pathway of jasmonic acid

9.4.1 Drought Stress

Water deficiency is one of the most serious stresses in plants, affecting the overall growth and development and leading to a decline in both crop quality and quantity. One of the causes of water stress to plants is stomatal opening, which regulates CO_2 and water exchange between the environment and the plant (Schroeder et al. 2001). So, plants can improve their drought resistance power by controlling water loss by stomata via closing of stomata (Yadav et al. 2019). Studies show that drought stress prevents the conversion of OPDA into JA; it results in accumulation of OPDA, which causes a decrease in stomatal aperture, thus providing drought resistance (Daszkowska-Golec and Szarejko 2013; Savchenko et al. 2014). It is observed that LOX genes play a crucial role in drought resistance, as these genes are essential for 12-OPDA production. Under water scarcity, LOX6 increases the synthesis of 12-OPDA, which causes stomatal closure in the presence of ABA hormone (Sarwat and Tuteja 2017). In rice, a basic HLH (helix-loop-helix) protein (OsbHLH148) plays a role in drought tolerance by interacting with OsJAZ1, which results in activation of an essential drought resistance providing TF named OsDREB1 (Seo et al. 2011).

9.4.2 Salt Stress

Salt stress is a major problem mainly in semi-arid and arid regions (Fahad et al. 2015). Plants evolved complex mechanisms to regulate physiological, biochemical, and molecular processes to withstand salt stress. In plants, high salinity causes activation of JA signaling and production of JA. Salt stress activates the JA signaling pathway in *Arabidopsis*, which results in inhibition of cell elongation in the zone of elongation (Valenzuela et al. 2016). JA may also work as a signal molecule that induces JA synthesis to provide salt stress and promote the synthesis of flavonoid in leaves of *Dendrobium officinale* (Zhang et al. 2021). In rice, salt stress results in oxidative damage by reactive oxygen species (ROS), but in a mutant with OPDA deficiency, which is linked to the synthesis of JA, there is an increase in the scavenging activity for ROS (Hazman et al. 2015). These studies showed that JA plays a crucial role in salt tolerance in plants. JA-related genes might also play a crucial role in tolerance to salt stress. Like in *Arabidopsis*, the JA synthesis enzyme, LOX3, is excessively induced under salt stress and increases tolerance to salt (Ding et al. 2016).

9.4.3 Temperature Stress

Extreme temperature (very low/very high) exposure results in stress to plants and affects the normal functions in plants (Hatfield and Prueger 2015), and JA plays an important function in the recovery of symptoms caused due to extreme temperature (Zhao et al. 2013). Cold stress in plants is a main threat that inhibits the plant to fully grow and develop, which results in reduction in crop yield all over the world (Yadav 2009). Low temperature conditions result in oxidative stress in plants. To overcome this stress, plants activate antioxidant activity to normalize plant functions. Antioxidants provide tolerance to low temperature stress, aid in cold acclimatization, and regulate redox homeostasis of cells (Khan et al. 2015). Application of MeJA exogenously results in cold tolerance in loquat fruit (Cai et al. 2011) and Arabidopsis (Hu et al. 2013). Upregulation of the JA signaling key regulator gene MYC and a rise in JA synthesis precursor molecule, i.e., α-linolenic acid, occur in cold-resistant *Camellia japonica* (Li et al. 2016). The ICE-CBF pathway, which induces CBF gene expression, plays important functions in cold stress responses in plants. Normally, ICE1 and ICE2 factors bind with the CBF gene promoter region at the CANNTG sequence. These transcription factors (TFs), with JAZ1 and JAZ4, also result in ICE-CBF pathway inhibition. During cold stress conditions, JA-Ile formation occurs, which causes degradation of JAZ factors through the 26S proteasome, which causes ICE-CBF pathway activation because JAZ degradation activates the ICE1 and ICE2 (Chen et al. 2019). In rubber tree *(Hevea brasiliensis)*, exogenous application of MeJA removes the JAZ protein suppression on ICE2 transcription factor, which activates the CBF (Crepeat binding factor) pathway involving genes COR47, CBF1, and CBF2. An increase in CBF1, CBF2, and COR47 gene expression results in acclimatization to cold stress in rubber trees (Chen et al. 2019). Cold-resistant varieties of rice store more GSH than non-resistant varieties (Yu et al. 2020). GSH, along with JA, regulates plant development processes and various stress signaling pathways (Per et al. 2018).

9.4.4 Heat Stress

There is a continuous increase in the temperature of the environment due to the emission of greenhouse gases. Heat stress severely affects the growth, development, crop yield, and physiological aspects, which generate ROS, which cause oxidative stress in plants (Hasanuzzaman et al. 2013; Sarwar et al. 2018). Several Studies show that the application of JA helps to reduce the effects of heat stress through activation of detoxification mechanisms and oxidative defense in plants (Sharma and Laxmi 2016).

9.4.5 Ozone Stress

Ozone stress in plants caused lesion formation and programmed cell death. In an ozone-treated wild-type Arabidopsis plant, JA level is significantly increased (Rao et al. 2000). Ozone-induced increase in programmed cell death could be prevented by exogenous application of MeJA (Kanna et al. 2003). Exogenous application of MeJA results in a decrease in ozone sensitivity in tomato (*L. esculentum*) and hybrid poplar (*Populus maximowiczii* × *P. trichocarpa*) (Rao et al. 2000). Further studies showed that ozone stress in tomato plants activates the JA pathway, which upregulates the emission of several volatile compounds that provide protection to the plant from natural enemies (Cui et al. 2016).

9.4.6 Light Stress

There are very less reports available for the effect of light and JA synthesis on plant development and growth. UVB light treatment in *Brassica* and *Nicotiana* species induced the formation of JA, which enhanced the defense system of plants (Mewis et al. 2012; Svyatyna and Riemann 2012). In tomato *(L. esculentum)*, blue light-directed light morphogenesis (Gupta et al. 2014) and *A. thaliana* and red light/far-red light-directed photo morphogenesis in rice *(O. sativa)* and *A. thaliana* are linked with JA synthesis (Mewis et al. 2012).

9.4.7 *Carbon Dioxide Stress*

In lima bean (*Phaseolus lunatus*), at high concentrations of CO_2 (500, 700, and 1000 ppm), a rise in cis-JA and MeJA levels is observed. A rise in CO_2 concentration to 750 ppm increases the defense process in tomato by inducing SA and JA signaling pathways against nematodes (Sun et al. 2011). An increase in CO_2 level works differently in various plant species, like in tobacco, increased CO_2 level activates JA synthesis and defense mechanisms, but in rice, it decreases the JA synthesis (Lu et al. 2018).

9.5 Heavy Metal Stress in Plants

These days, heavy metal contamination in agriculture has become a problem worldwide due to an increase in industrialization and human-generated activities like mining (Noriega et al. 2012).

Cu and Zn are essential micronutrients used by plants to perform metal-dependent enzymes and proteins activity. However, if the metal accumulation reaches above the threshold value, it causes metal and oxidative stress. These stresses result in a change in the metabolism of the plant and a decrease in transpiration and photosynthesis. Studies showed that heavy metal stress results in plant physiology and morphology (Qureshi et al. 2016). High levels of lead (Pb) cause the synthesis of reactive oxygen species (ROS), which induce oxidative stress both directly and indirectly in plants (Reddy et al. 2005).

9.6 Interaction of Jasmonic Acid with Heavy Metal Stress

Nowadays, heavy metal waste such as copper (Cu), lead (Pb), cadmium (Cd), and mercury (Hg) from industrial sources or the environment is creating harmful conditions for plant development and growth when their levels exceed threshold concentrations (Qin et al. 2021). JA plays an important role in regulating the metal stress in plants. Recent studies show that Cd treatment causes a Cd stress condition in the roots of Arabidopsis, resulting in rapid regulation of JA synthesis gene expression endogenously. Application of MeJA exogenously decreases the Cd concentration in tissues of shoot and root, and it also stops the expression of heavy metal ATPase (AtHMA4), iron-regulated transporter (AtIRT1), and heavy metal ATPase (AtHMA2) genes (Lei et al. 2020). We can say that the expression of genes responsible for Cd uptake and translocation of Cd to long distances is reduced by JAs, resulting in a decrease in root cell and relaxing the plant from Cd stress. Cd stress, which affects oxidative metabolism in roots of *Pisum sativum*, results in a rise in JAs level (Rodríguez-Serrano et al. 2006). There are various studies that have shown that JA application exogenously can reduce the harmful effects on plants due to heavy metals by controlling antioxidant defense system functions, inducing secondary metabolites, and increasing chlorophyll content. Application 25 μM JA exogenously improve the Cd toxicity tolerance in rapeseed (*Brassica napus*) (Ali et al. 2018). A twofold rise in JA level was observed at the cellular level when exposed to Cd for 48 h at 50 mg L^{-1} (Yan et al. 2013). In *Glycine max* (L.), to overcome cadmium stress, 0.01 mM exogenous application of JA is required, which increases the antioxidant response (Noriega et al. 2012). The negative results of decreased root growth and the level of chlorophyll b due to Cd stress in pepper can be eliminated by JA application exogenously (Yan et al. 2013).

In Cd-hyperaccumulator *Noccaea praecox*, exposure to mechanical stress or fungal infection causes a rise in the endogenous level of JAs; it shows that Cd induced a rise in JA level in metal hyperaccumulators in biotic and abiotic stress conditions (Llugany et al. 2013). Harmful effects of both Cu and Cd in *Avicennia marina* can be decreased by the application of MeJA or JA (Yan et al. 2015a). JA application decreases the negative effect of Ni stress through increasing the activity of osmolytes, which act as antioxidant enzymes in soybean seedlings (Sirhindi et al. 2016). A rise in JA levels is observed in Ni-treated *Daphne jasmine*, a woody shrub (Wiszniewska et al. 2018).

Cu stress in wheat is decreased by the application of JA exogenously, which increased the ROS scavenging related gene expression of glutathione S-transferase (GST) (Li et al. 2013). JAs concentration rises in Cu treated *Phaseolus coccineus*. In rice leaves, Cu treatment results in an increase in JAs level (Rakwal et al. 1996). JA application effectively relieves the plant from Cu stress-induced growth blockage of alfalfa by a rise in activity of antioxidant enzyme, MDA, leaf chlorophyll content, and hydrogen peroxide (Dai et al. 2020).

Similarly, in tomato seedlings, lead (Pb) toxicity stress can be reduced by exogenous application of JA, which results in a rise in concentration of secondary metabolites like flavonoids, anthocyanin, organic acids, metal-chelating compounds, total phenols, and polyphenols (Bali et al. 2019b). Pb toxicity can also be reduced by suppressing the expression of P-type ATPase transporter and RBO gene (Bali et al. 2019a). In rice plants, Pb toxicity causes disturbance in antioxidant enzyme activity and induces peroxidation of lipid (Verma and Dubey 2003). Studies on plants (*Lolium perenne* and *Trifolium repens*) show that Pb stress causes changes in the ratio of unsaturated to saturated fatty acids and composition of fatty acids (Bidar et al. 2008).

Pb stress in maize, which results in a decrease in growth, can be eliminated by JA treatment along with SA and proline (Sofy et al. 2020). Lead toxicity in *Wolffia arrhizal*, which is increased with the application of 100 μM of JA, causes a decrease in the concentration of monosaccharides, carotenoids, chlorophyll, and fresh weight. These symptoms can be recovered by exogenous application of 0.1 μM JA, which decreases the accumulation and absorption of lead and increases the activity of antioxidant enzymes and non-enzymatic antioxidants such as GSH and AsA (Piotrowska et al. 2009).

Treatment of metalloid arsenite [As(III)] for 8 hours in rice results in a rapid rise in the level of JA-Ile in roots (Ronzan et al. 2019). In rice, 0.25, 0.5, and 1 μM MeJA concentration elevates As(III) toxic effects (Mousavi et al. 2020). The pretreated roots of rice plants with 0.5–5 μM concentration of JA show elongation and significantly lower root inhibition by As(V) in comparison to non-treated rice plants (Wang et al. 2018).

High boron levels are harmful for plant development and growth in wheat (Triticum aestivum) (Gunes et al. 2007), apple root stock (Molassiotis et al. 2006), barley (Inal et al. 2009), and tomato plants.

Boron toxicity in plants can be eliminated by the exogenous application of MeJA. In sweet wormwood (*Artemisia annua*), boron toxicity can be recovered by MeJA by decreasing lipid peroxidation level and inducing the antioxidant enzymes (CAT, SOD, and POD) synthesis (Aftab et al. 2011).

9.7 Molecular Mechanism of Jasmonic Acid During Heavy Metal Stress

Here, we will discuss the toxic metal-induced JA biosynthesis upregulated gene at the molecular level. The expression of genes AtAOS, AtLOX3, and AtLOX4 rises rapidly in roots treated with Cd for 1 hour (Lei et al. 2020) in *Arabidopsis*. Gene knockout of AtAOS in *Arabidopsis* results in shorter length of root and more severe chlorosis symptoms with Cd stress (Lei et al. 2020). JA has a role in metal tolerance in tomato crops. In tomato, the spr2 mutant is a JA-deficient mutant (Sánchez-Hernández et al. 2006). Under Cd stress, spr2 mutant plants have a more severe effect on water loss and inhibition of growth than wild-type plants (Zhao et al. 2016). Mutant suppressor of gene spr2 (prosystemin-mediated responses 2), with the absence of activity of chloroplast FAD (fatty acid desaturase), results in a decrease in biomass and an increase in Cd deposition due to severe decrease in JA in tomato (Zhao et al. 2016).

Studies showed that ZIP homolog AtIRT1 is essential for Cd uptake in most of the studied green plants (Hu et al. 2020). In Arabidopsis, mutants that lack the activity of MYC2 (atmyc2) reduce the ALCOHOL DEHYDROGENASE 1 (ADH1) and DEHYDRATION 22 (RD22) gene expression. Here, the JA+-responsive gene named MYC2 is controlled by ABA (Abe et al. 2003). In rice, OsbHLH148 gene expression is upregulated by ABA and JA, which is a key regulating transcription factor for drought response and JA-responsive genes like OsJAZs (Seo et al. 2011).

In rice, Cu stress increases the JA level by increasing gene expression linked to JA synthesis enzymes like LOXs, phospholipase, AOS,12-Oxo-PDA reductase (OPR) (Lin et al. 2013). Cu and Cd treatment in rice results in a rise in expression of OsAOCs, OsDADs, OsAOSs, and OsLOXs in the overall plant (Tan et al. 2017; Lin et al. 2013). Similarly, transcriptomic analysis shows the activation of JA production and signaling pathways in rice under stress of As(V) metal (Huang et al.

2012). Arsenic exposure in rice results in an increase in expression levels of seven genes, OsAIM1, OsDAD1;3, OsLOX2;1, OsAOS1, OsLOX2;3, OsAOS2, and OsDAD1;2, which have a role in JA biosynthesis (Huang et al. 2012). In rice, coleoptile photomorphogenesis 2 (cpm2), which is a jasmonate synthesis mutant, inhibited the translocation of As(III) from root to shoot and a decrease in adventitious roots count (Ronzan et al. 2019).

Overexpression of GhAOS-activated JA production in cotton (*Gossypium hirsutum*) provides tolerance for heavy metal stress (Wang et al. 2015). In transgenic plants with overexpressing GhAOS, the survival rate is almost two-fold higher than wild plants under Cu-stress conditions. In wheat, overexpression of the TaAOS gene results in increased tolerance to Zn metal stress (Liu et al. 2014). In tobacco (*Nicotiana tabacum*) under zinc (Zn) stress conditions, the chlorophyll level in plants overexpressing TaAOS was three times higher than in wild-type plants. Jasmonic acid (JA) mitigates detoxification of lead stress in tomato by decreasing the expression of P-type ATPase metal transporter genes and RBO genes, as well as by promoting the formation of secondary metabolites (Bali et al. 2019b).

Studies showed that the JA biosynthesis mutant, named aos, was hypersensitive to cadmium (Cd) in Arabidopsis. In aos mutant plants, Cd induces an albino phenotype, and Cd accumulation in cells is more severe in comparison to wild plants. In lox1-1 mutant Arabidopsis is less tolerant to Cd stress than wild plants (Lei et al. 2020), and in lox1-1 mutated plants, Cd also induces an increase in H_2O_2 accumulation (Keunen et al. 2013). In plants, a mutant for JA signaling, jar1-1, is hypersensitive to heavy metal stress and severely affected by JA response. Selenium (Sel) stress causes root growth inhibition in Arabidopsis, and this inhibitory effect is more severe in jar1-1 mutant than wild plant (Van Hoewyk et al. 2008).

In cotton (Gossypium hirsutum), overexpression of gene GhAOS induced JA biosynthesis and increased resistance to heavy metal stress (Wang et al. 2015). Overexpression of the GhAOS gene in Cu treated plants increases the survival rate twofold more than wild plants. Similarly, in wheat, the TaAOS gene increases plant resistance to Zn heavy metal stress (Liu et al. 2014).

9.8 Physiological Aspects of Jasmonic Acid During Heavy Metal Stress

JA, along with auxin, influences the various crucial processes in plants like root growth, development of flowers, seed development and germination, senescence, formation of tuber, and seedling development (Xu et al. 2020). Physiological processes like tendril coiling, secondary metabolites synthesis, and stem cell elongation are controlled by both JA and auxin during biotic and abiotic stress in Arabidopsis (Saniewski et al. 2002).

Expression of jasmonate signaling repressor gene JAZ1/TIFY10A is stimulated by auxin in *Arabidopsis* (Grunewald et al. 2009). JA works as an inhibitor for root growth and also decreases the activity of the meristem (Jang et al. 2017). MYC2, which is a basic HLH transcription factor, regulates the interaction between phytohormone auxin and JA by binding with the promoter of the PLT gene, which is an auxin-responsive gene, causing regulation in cell division and niche of stem cells. MYC2 binding with PLT gene results in suppression of its expression and thus inhibits meristematic activity of root in *Arabidopsis* (Chen et al. 2011).

In *Arabidopsis*, transcription factor MYC2 directly suppresses the expression of plethora during jasmonate-mediated modulation in the niche of root stem cells. In rice, this interaction influences the development of the root system induced due to soil pollutant exposure. In *Arabidopsis*, JA-responsive ethylene response factor 109 (ERF109) induces JA–auxin interaction, which controls lateral root growth (Cai et al. 2014).

JA, along with auxin, also controls floral organ development like cellular elongation of stamen filament, anther dehiscence, maturation of nectar, and petal expansion (Reeves et al. 2012). These developments are regulated by ARF6 and ARF8 (Auxin response factors 6 and 8), which induce the JA production by regulating MYB21 and MYB24 TFs for JA-responsive genes to induce floral organ development in *Aquilegia* (Zhang et al. 2020a). JA regulates senescence of leaves, which induces severe leaf yellowing, cell death, and low chlorophyll content.

JA also plays a role in cell division, seed germination, fruit set, and development of root and flower, along with gibberelins (GAs). Under cold stress conditions, GA signaling is inhibited, which causes inhibition of plant growth and accumulation of DELLA proteins (Achard et al. 2008). DELLA proteins bind with JAZ proteins, which cause activation of JA-responsive genes (Wingler et al. 2020). Extreme uptake and deposition of cadmium (Cd) heavy metal inhibit the growth and development of plants (Haider et al. 2022). The Cd stress disturbs functions in plants, like the inhibition of mineral nutrients in the photosynthesis system and water uptake (Baryla et al. 2001).

Methyl jasmonate (MeJA) and JA both control the various physiological and biochemical processes (Yu et al. 2019a; Bukhat et al. 2021) in plants, like photosynthesis machinery and stomatal movement (Yan et al. 2015b). MeJA relieves Cd stress in *Solanum nigrum* by controlling Cd uptake and by using antioxidative activity (Yan et al. 2015b). MeJA works as a stimulator for photosynthesis pigments and photosystem II (PSII) photochemistry (Qiu et al. 2020; Attaran et al. 2014). A study on *Brassica juncea* by using transmission electron microscopy showed that MeJA plays an important role in reducing Cd toxicity by protecting the chloroplast structure (Per et al. 2016).

In *Kandelia obovata,* exogenous application of MJ (Methyl Jasmonate) regulates endogenous JA level, decreases transpiration rate, controls closure of stomata, inhibits Cd uptake, and decreases photosynthetic damage (Chen et al. 2014). Under drought stress conditions, *Allium sativum* L., after application of MJ, results in root, bulb, and shoot growth (Bideshki and Arvin 2013).

JA application to Pigeon pea results in protein accumulation (carotenoid and total chlorophyll) and a decrease in copper stress on seedling growth (Gallego et al. 2012). In Cd-treated *Capsicum frutescens,* application of MJ at a concentration of 0.1 μM results in increased chlorophyll content of leaves, dry weight, and APX and

CAT activity (Yan et al. 2013). In *Brassica napus,* exogenous application of MJ reduces the arsenic-induced oxidative stress and increases photosynthesis and overall growth of the plant (Farooqi et al. 2012). In tomato, application of JA and NO exogenously reduced NaCl toxicity by regulating osmolyte synthesis, accumulation of metabolites, and antioxidant metabolism (Ahmad et al. 2018).

In soybean, JA regulates the root and leaf morphogenesis and regulates plant fertility (Xue and Zhang 2007). In *Arabidopsis,* the allene oxide synthase (AOS)-knockout mutation causes male sterility due to decrease in JA production. Expression of several genes related to stamen tissue development is also controlled by JA in *Arabidopsis* (Song et al. 2011).

Lead (Pb) is a heavy, non-essential, and toxic element for plants, which gets absorbed through shoots, roots, or leaves. In plants, it causes alteration in membrane structure, mineral nutrition, hormonal status, and also inhibition of enzymatic activities, resulting in chlorosis and decreased growth (Küpper 2017). Alteration in enzyme activity due to Pb stress results in seed germination inhibition in *Eichhornia crassipes* and *Sporobolus alterniflorus* (Malar et al. 2016).

In privet seedlings, lead stress affects cell structure, physiology, and growth (Zhou et al. 2018). ROS induced in plants due to high Pb concentration causes the oxidative stress directly or indirectly in rice (Verma and Dubey 2003). Lead also induced alteration in antioxidant metabolism in Bengal gram and horse gram (Reddy et al. 2005).

9.9 Conclusions and Future Directions

JAs have been the focus of significant and rigorous research related to stress-induced regulation of plant development and defense because they are the primary hormones controlling biotic and abiotic stress responses. JA synthesis, signaling, and metabolic pathways have been steadily defined using model plants (e.g., Arabidopsis and rice) over the years, offering essential references for other plants. Despite the fact that biotic and abiotic stressors have had a significant negative economic influence on the production of certain essential agricultural crops and economic trees, little research has been done. To present, the use of JA in agricultural output has been restricted. Previous research on JAs has generally focused on JA functional mechanisms within plant cells, while the tissue, organ, individual, and inter-individual levels remain unknown. Because JAs, particularly OPDA and its derivatives, are mobile and can be transported to different parts of plants to function in plant defense systems, future research on JA transmission and its applications is required. JAs are important plant natural compounds for the plant defense system. The environmental factors in which the plant grows influence JA biosynthesis. To generate stress-resistant crop cultivars, molecular biology techniques such as gene editing can be utilized to change the genetic regulation of JA synthesis. Exogenous JAs are also commonly used as anti-stress protective agents by increasing plant tolerance. The evolution of nano-biotechnology will allow for the creation of nanocarriers for the quick and efficient delivery of exogenous JAs to specified plant cells.

References

Abe H et al (2003) Arabidopsis AtMYC2 (bHLH) and AtMYB2 (MYB) function as transcriptional activators in abscisic acid signaling. Plant Cell 15(1):63–78. https://doi.org/10.1105/tpc.006130

Achard P et al (2008) The cold-inducible CBF1 factor-dependent signaling pathway modulates the accumulation of the growth-repressing DELLA proteins via its effect on gibberellin metabolism. Plant Cell 20(8):2117–2129. https://doi.org/10.1105/tpc.108.058941

Aftab T et al (2011) Methyl jasmonate counteracts boron toxicity by preventing oxidative stress and regulating antioxidant enzyme activities and artemisinin biosynthesis in *Artemisia annua* L. Protoplasma 248(3):601–612. https://doi.org/10.1007/s00709-010-0218-5

Ahmad P et al (2018) Mitigation of sodium chloride toxicity in *Solanum lycopersicum* L. by supplementation of jasmonic acid and nitric oxide. J Plant Interact 13(1):64–72. https://doi.org/10.1080/17429145.2017.1420830

Ali MS, Baek KH (2020) Jasmonic acid signaling pathway in response to abiotic stresses in plants. Int J Mol Sci. https://doi.org/10.3390/ijms21020621

Ali E et al (2018) Role of jasmonic acid in improving tolerance of rapeseed (*Brassica napus* L.) to Cd toxicity. J Zhejiang Univ Sci B:130–146. https://doi.org/10.1631/jzus.B1700191

Attaran E et al (2014) Temporal dynamics of growth and photosynthesis suppression in response to jasmonate signaling. Plant Physiol 165(3):1302–1314. https://doi.org/10.1104/pp.114.239004

Avanci NC et al (2010) Jasmonates are phytohormones with multiple functions, including plant defense and reproduction. Genet Mol Res 9(1):484–505. https://doi.org/10.4238/vol9-1gmr754

Bali S, Jamwal VL, Kaur P et al (2019a) Role of P-type ATPase metal transporters and plant immunity induced by jasmonic acid against Lead (Pb) toxicity in tomato. Ecotoxicol Environ Saf 174:283–294. https://doi.org/10.1016/j.ecoenv.2019.02.084

Bali S, Jamwal VL, Kohli SK et al (2019b) Jasmonic acid application triggers detoxification of lead (Pb) toxicity in tomato through the modifications of secondary metabolites and gene expression. Chemosphere 235:734–748. https://doi.org/10.1016/j.chemosphere.2019.06.188

Baryla A et al (2001) Leaf chlorosis in oilseed rape plants (*Brassica napus*) grown on cadmium-polluted soil: causes and consequences for photosynthesis and growth. Planta 212(5–6):696–709. https://doi.org/10.1007/s004250000439

Berger S, Bell E, Mullet JE (1996) Two methyl jasmonate-insensitive mutants show altered expression of AtVsp in response to methyl jasmonate and wounding. Plant Physiol 111(2):525–531. https://doi.org/10.1104/pp.111.2.525

Bidar G et al (2008) Changes in fatty acid composition and content of two plants (*Lolium perenne* and *Trifolium repens*) grown during 6 and 18 months in a metal (Pb, Cd, Zn) contaminated field. Water Air Soil Pollut 192(1–4):281–291. https://doi.org/10.1007/s11270-008-9655-6

Bideshki A, Arvin MJ, Darini M (2013) Interactive effects of Indole-3-butyric acid (IBA) and salicylic acid (SA) on growth parameters, bulb yield and allicin contents of garlic (Allium sativum) under drought stress in field. Int J Agron Plant Prod 4(2):271–279. http://www.ijappjournal.com

Browse J (2005) Jasmonate: an oxylipin signal with many roles in plants. Vitam Horm 72:431–456. https://doi.org/10.1016/S0083-6729(05)72012-4

Bruckhoff V et al (2016) Functional characterization of CYP94-genes and identification of a novel jasmonate catabolite in flowers. PLoS One 11(7). https://doi.org/10.1371/journal.pone.0159875

Bukhat S et al (2021) Jasmonates: debatable role in temperature stress tolerance. In: Plant growth regulators for climate-smart agriculture, pp 45–62. https://doi.org/10.1201/9781003109013-4

Caarls L et al (2017) Arabidopsis JASMONATE-INDUCED OXYGENASES down-regulate plant immunity by hydroxylation and inactivation of the hormone jasmonic acid. Proc Natl Acad Sci USA 114(24):6388–6393. https://doi.org/10.1073/pnas.1701101114

Cai Y et al (2011) MeJA regulates enzymes involved in ascorbic acid and glutathione metabolism and improves chilling tolerance in loquat fruit. Postharvest Biol Technol 59(3):324–326. https://doi.org/10.1016/j.postharvbio.2010.08.020

Cai XT et al (2014) Arabidopsis ERF109 mediates cross-talk between jasmonic acid and auxin biosynthesis during lateral root formation. Nat Commun 5. https://doi.org/10.1038/ncomms6833

Chao DY et al (2012) Genome-wide association studies identify heavy metal ATPase3 as the primary determinant of natural variation in leaf cadmium in *Arabidopsis thaliana*. PLoS Genet 8(9). https://doi.org/10.1371/journal.pgen.1002923

Chen Q et al (2011) The basic helix-loop-helix transcription factor myc2 directly represses plethora expression during jasmonate-mediated modulation of the root stem cell niche in Arabidopsis. Plant Cell 23(9):3335–3352. https://doi.org/10.1105/tpc.111.089870

Chen J, Yan Z, Li X (2014) Effect of methyl jasmonate on cadmium uptake and antioxidative capacity in *Kandelia obovata* seedlings under cadmium stress. Ecotoxicol Environ Saf 104(1):349–356. https://doi.org/10.1016/j.ecoenv.2014.01.022

Chen WJ et al (2019) The ICE-like transcription factor HbICE2 is involved in jasmonate-regulated cold tolerance in the rubber tree (*Hevea brasiliensis*). Plant Cell Rep 38(6):699–714. https://doi.org/10.1007/s00299-019-02398-x

Clemens S, Ma JF (2016) Toxic heavy metal and metalloid accumulation in crop plants and foods. Annu Rev Plant Biol:489–512. https://doi.org/10.1146/annurev-arplant-043015-112301

Cui H et al (2016) Elevated O3 increases volatile organic compounds via jasmonic acid pathway that promote the preference of parasitoid *Encarsia formosa* for tomato plants. Plant Sci 253:243–250. https://doi.org/10.1016/j.plantsci.2016.09.019

Dai H et al (2020) Exogenous jasmonic acid decreased Cu accumulation by alfalfa and improved its photosynthetic pigments and antioxidant system. Ecotoxicol Environ Saf 190. https://doi.org/10.1016/j.ecoenv.2020.110176

Daszkowska-Golec A, Szarejko I (2013) Open or close the gate - stomata action under the control of phytohormones in drought stress conditions. Front Plant Sci. https://doi.org/10.3389/fpls.2013.00138

Deng F et al (2021) Metalloid hazards: from plant molecular evolution to mitigation strategies. J Hazard Mater 409. https://doi.org/10.1016/j.jhazmat.2020.124495

Ding H et al (2016) Jasmonate complements the function of Arabidopsis lipoxygenase3 in salinity stress response. Plant Sci 244:1–7. https://doi.org/10.1016/j.plantsci.2015.11.009

Fahad S, Hussain S, Matloob A et al (2015) Phytohormones and plant responses to salinity stress: a review. In: Plant growth regulation, pp 391–404. https://doi.org/10.1007/s10725-014-0013-y

Farooqi AA, Butt G, Razzaq Z (2012) Algae extracts and methyl jasmonate anti-cancer activities in prostate cancer: choreographers of "the dance macabre". Cancer Cell Int. https://doi.org/10.1186/1475-2867-12-50

Gallego SM et al (2012) Unravelling cadmium toxicity and tolerance in plants: insight into regulatory mechanisms. In: Environmental and experimental botany, pp 33–46. https://doi.org/10.1016/j.envexpbot.2012.04.006

Göbel C, Feussner I (2009) Methods for the analysis of oxylipins in plants. Phytochemistry 70:1485–1503. https://doi.org/10.1016/j.phytochem.2009.07.040

Grunewald W et al (2009) Expression of the Arabidopsis jasmonate signalling repressor JAZ1/TIFY10A is stimulated by auxin. EMBO Rep 10(8):923–928. https://doi.org/10.1038/embor.2009.103

Guan L et al (2019) JASSY, a chloroplast outer membrane protein required for jasmonate biosynthesis. Proc Natl Acad Sci USA 116(21):10568–10575. https://doi.org/10.1073/pnas.1900482116

Gunes A et al (2007) Silicon increases boron tolerance and reduces oxidative damage of wheat grown in soil with excess boron. Biol Plant 51(3):571–574. https://doi.org/10.1007/s10535-007-0125-6

Gupta N, Prasad VBR, Chattopadhyay S (2014) LeMYC2 acts as a negative regulator of blue light mediated photomorphogenic growth, and promotes the growth of adult tomato plants. BMC Plant Biol 14(1). https://doi.org/10.1186/1471-2229-14-38

Haider FU et al (2022) Integrated application of thiourea and biochar improves maize growth, antioxidant activity and reduces cadmium bioavailability in cadmium-contaminated soil. Front Plant Sci 12. https://doi.org/10.3389/fpls.2021.809322

Hasanuzzaman M et al (2013) Physiological, biochemical, and molecular mechanisms of heat stress tolerance in plants. Int J Mol Sci:9643–9684. https://doi.org/10.3390/ijms14059643

Hashimoto M et al (2004) A novel rice PR10 protein, RSOsPR10, specifically induced in roots by biotic and abiotic stresses, possibly via the jasmonic acid signaling pathway. Plant Cell Physiol 45(5):550–559. https://doi.org/10.1093/pcp/pch063

Hatfield JL, Prueger JH (2015) Temperature extremes: effect on plant growth and development. Weather Clim Extrem 10:4–10. https://doi.org/10.1016/j.wace.2015.08.001

Hazman M et al (2015) Increased tolerance to salt stress in OPDA-deficient rice ALLENE OXIDE CYCLASE mutants is linked to an increased ROS-scavenging activity. J Exp Bot 66(11):3339–3352. https://doi.org/10.1093/jxb/erv142

Howe GA, Yoshida Y (2019) Evolutionary origin of JAZ proteins and jasmonate signaling. In: Molecular plant, pp 153–155. https://doi.org/10.1016/j.molp.2019.01.015

Hu Y et al (2013) Jasmonate regulates the INDUCER OF CBF expression-C-repeat binding factor/dre binding factor1 Cascade and freezing tolerance in Arabidopsis. Plant Cell 25(8):2907–2924. https://doi.org/10.1105/tpc.113.112631

Hu B et al (2020) Evolution of abscisic acid signaling for stress responses to toxic metals and metalloids. Front Plant Sci. https://doi.org/10.3389/fpls.2020.00909

Huang TL et al (2012) Transcriptomic changes and signalling pathways induced by arsenic stress in rice roots. Plant Mol Biol 80(6):587–608. https://doi.org/10.1007/s11103-012-9969-z

Huang H et al (2017) Jasmonate action in plant growth and development. J Exp Bot:1349–1359. https://doi.org/10.1093/jxb/erw495

Hussain D et al (2004) P-type ATPase heavy metal transporters with roles in essential zinc homeostasis in Arabidopsis. Plant Cell 16(5):1327–1339. https://doi.org/10.1105/tpc.020487

Inal A, Pilbeam DJ, Gunes A (2009) Silicon increases tolerance to boron toxicity and reduces oxidative damage in barley. J Plant Nutr 32(1):112–128. https://doi.org/10.1080/01904160802533767

Isah T (2019) Stress and defense responses in plant secondary metabolites production. Biol Res:39. https://doi.org/10.1186/s40659-019-0246-3

Jang G et al (2017) Antagonistic interaction between jasmonic acid and cytokinin in xylem development. Sci Rep 7(1). https://doi.org/10.1038/s41598-017-10634-1

Järup L (2003) Hazards of heavy metal contamination. Br Med Bull:167–182. https://doi.org/10.1093/bmb/ldg032

Jiang Y et al (2014) Arabidopsis WRKY57 functions as a node of convergence for jasmonic acid- and auxin-mediated signaling in jasmonic acid-induced leaf senescence. Plant Cell 26(1):230–245. https://doi.org/10.1105/tpc.113.117838

Kanna M et al (2003) Isolation of an ozone-sensitive and jasmonate-semi-insensitive Arabidopsis mutant (oji1). Plant Cell Physiol 44(12):1301–1310. https://doi.org/10.1093/pcp/pcg157

Keunen E et al (2013) A mutant of the Arabidopsis thaliana LIPOXYGENASE1 gene shows altered signalling and oxidative stress related responses after cadmium exposure. Plant Physiol Biochem 63:272–280. https://doi.org/10.1016/j.plaphy.2012.12.005

Khan MIR et al (2015) Salicylic acid-induced abiotic stress tolerance and underlying mechanisms in plants. Front Plant Sci 6(June). https://doi.org/10.3389/fpls.2015.00462

Koda Y, Kikuta Y (1991) Possible involvement of jasmonic acid in tuberization of yam plants. Plant Cell Physiol 32(5):629–633. https://doi.org/10.1093/oxfordjournals.pcp.a078125

Küpper H (2017) Lead toxicity in plants. In: Lead: its effects on environment and health, pp 491–500. https://doi.org/10.1515/9783110434330-015

Lei GJ et al (2020) Jasmonic acid alleviates cadmium toxicity in Arabidopsis via suppression of cadmium uptake and translocation. J Integr Plant Biol 62(2):218–227. https://doi.org/10.1111/jipb.12801

Li G et al (2013) Proteomic analysis of leaves and roots of common wheat (*Triticum aestivum* L.) under copper-stress conditions. J Proteome Res 12(11):4846–4861. https://doi.org/10.1021/pr4008283

Li Q et al (2016) RNA-seq based transcriptomic analysis uncovers α-linolenic acid and jasmonic acid biosynthesis pathways respond to cold acclimation in *Camellia japonica*. Sci Rep 6. https://doi.org/10.1038/srep36463

Li Q et al (2017a) Transporter-mediated nuclear entry of jasmonoyl-isoleucine is essential for jasmonate signaling. Mol Plant 10(5):695–708. https://doi.org/10.1016/j.molp.2017.01.010

Li Y et al (2017b) SlMAPK3 enhances tolerance to tomato yellow leaf curl virus (TYLCV) by regulating salicylic acid and jasmonic acid signaling in tomato (*Solanum lycopersicum*). PLoS One 12(2). https://doi.org/10.1371/journal.pone.0172466

Lin CY et al (2013) Comparison of early transcriptome responses to copper and cadmium in rice roots. Plant Mol Biol 81(4–5):507–522. https://doi.org/10.1007/s11103-013-0020-9

Liu HH et al (2014) Improved zinc tolerance of tobacco by transgenic expression of an allene oxide synthase gene from hexaploid wheat. Acta Physiol Plant 36(9):2433–2440. https://doi.org/10.1007/s11738-014-1616-7

Llugany M et al (2013) Endogenous jasmonic and salicylic acids levels in the Cd-hyperaccumulator *Noccaea* (Thlaspi) *praecox* exposed to fungal infection and/or mechanical stress. Plant Cell Rep 32(8):1243–1249. https://doi.org/10.1007/s00299-013-1427-0

Lorenzo O et al (2004) JASMONATE-INSENSITIVE1 encodes a MYC transcription factor essential to discriminate between different jasmonate-regulated defense responses in Arabidopsis. Plant Cell 16(7):1938–1950. https://doi.org/10.1105/tpc.022319

Lu C et al (2018) Elevated CO_2 differentially affects tobacco and rice defense against lepidopteran larvae via the jasmonic acid signaling pathway. J Integr Plant Biol 60(5):412–431. https://doi.org/10.1111/jipb.12633

Lux A et al (2011) Root responses to cadmium in the rhizosphere: a review. J Exp Bot:21–37. https://doi.org/10.1093/jxb/erq281

Malar S et al (2016) Lead heavy metal toxicity induced changes on growth and antioxidative enzymes level in water hyacinths [*Eichhornia crassipes* (Mart.)]. Bot Stud 55(1). https://doi.org/10.1186/S40529-014-0054-6

Marthandan V et al (2020) Seed priming: a feasible strategy to enhance drought tolerance in crop plants. Int J Mol Sci:1–23. https://doi.org/10.3390/ijms21218258

Mewis I et al (2012) UV-B irradiation changes specifically the secondary metabolite profile in broccoli sprouts: induced signaling overlaps with defense response to biotic stressors. Plant Cell Physiol 53(9):1546–1560. https://doi.org/10.1093/pcp/pcs096

Mills RF et al (2005) The plant P1B-type ATPase AtHMA4 transports Zn and Cd and plays a role in detoxification of transition metals supplied at elevated levels. FEBS Lett 579(3):783–791. https://doi.org/10.1016/j.febslet.2004.12.040

Molassiotis A et al (2006) Boron-induced oxidative damage and antioxidant and nucleolytic responses in shoot tips culture of the apple rootstock EM 9 (*Malus domestica* Borkh). Environ Exp Bot 56(1):54–62. https://doi.org/10.1016/j.envexpbot.2005.01.002

Mousavi SR et al (2020) Methyl jasmonate alleviates arsenic toxicity in rice. Plant Cell Rep 39(8):1041–1060. https://doi.org/10.1007/s00299-020-02547-7

Noriega G et al (2012) Heme oxygenase is involved in the protection exerted by jasmonic acid against cadmium stress in soybean roots. J Plant Growth Regul 31(1):79–89. https://doi.org/10.1007/s00344-011-9221-0

Per TS et al (2016) Methyl jasmonate alleviates cadmium-induced photosynthetic damages through increased S-assimilation and glutathione production in mustard. Front Plant Sci 7(December 2016). https://doi.org/10.3389/fpls.2016.01933

Per TS et al (2018) Jasmonates in plants under abiotic stresses: crosstalk with other phytohormones matters. Environ Exp Bot:104–120. https://doi.org/10.1016/j.envexpbot.2017.11.004

Piotrowska A et al (2009) Jasmonic acid as modulator of lead toxicity in aquatic plant *Wolffia arrhiza* (Lemnaceae). Environ Exp Bot 66(3):507–513. https://doi.org/10.1016/j.envexpbot.2009.03.019

Qin G et al (2021) Soil heavy metal pollution and food safety in China: effects, sources and removing technology. Chemosphere. https://doi.org/10.1016/j.chemosphere.2020.129205

Qiu X et al (2020) Effects of exogenous methyl jasmonate on the synthesis of endogenous jasmonates and the regulation of photosynthesis in citrus. Physiol Plant 170(3):398–414. https://doi.org/10.1111/ppl.13170

Qureshi AS et al (2016) Evaluating heavy metal accumulation and potential health risks in vegetables irrigated with treated wastewater. Chemosphere 163:54–61. https://doi.org/10.1016/j.chemosphere.2016.07.073

Rakwal R, Tamogami S, Kodama O (1996) Role of jasmonic acid as a signaling molecule in copper chloride-elicited rice phytoalexin production. Biosci Biotechnol Biochem 60(6):1046–1048. https://doi.org/10.1271/bbb.60.1046

Rao MV et al (2000) Jasmonic acid signaling modulates ozone-induced hypersensitive cell death. Plant Cell 12(9):1633–1646. https://doi.org/10.1105/tpc.12.9.1633

Reddy AM et al (2005) Lead induced changes in antioxidant metabolism of horsegram (*Macrotyloma uniflorum* (Lam.) Verdc.) and bengal gram (*Cicer arietinum* L.). Chemosphere 60(1):97–104. https://doi.org/10.1016/j.chemosphere.2004.11.092

Reeves PH et al (2012) A regulatory network for coordinated flower maturation. PLoS Genet 8(2). https://doi.org/10.1371/journal.pgen.1002506

Rodríguez-Serrano M et al (2006) Cadmium effect on oxidative metabolism of pea (*Pisum sativum* L.) roots. Imaging of reactive oxygen species and nitric oxide accumulation in vivo. Plant Cell Environ 29(8):1532–1544. https://doi.org/10.1111/j.1365-3040.2006.01531.x

Rojas-Rueda D et al (2021) Environmental risk factors and health: an umbrella review of meta-analyses. Int J Environ Res Public Health:1–38. https://doi.org/10.3390/ijerph18020704

Ronzan M et al (2019) Auxin-jasmonate crosstalk in *Oryza sativa* L. root system formation after cadmium and/or arsenic exposure. Environ Exp Bot 165:59–69. https://doi.org/10.1016/j.envexpbot.2019.05.013

Ruduś I et al (2014) Wound-induced expression of DEFECTIVE IN ANTHER DEHISCENCE1 and DAD1-like lipase genes is mediated by both CORONATINE INSENSITIVE1-dependent and independent pathways in *Arabidopsis thaliana*. Plant Cell Rep 33(6):849–860. https://doi.org/10.1007/s00299-013-1561-8

Sánchez-Hernández C, López MG, Délano-Frier JP (2006) Reduced levels of volatile emissions in jasmonate-deficient spr2 tomato mutants favour oviposition by insect herbivores. Plant Cell Environ 29(4):546–557. https://doi.org/10.1111/j.1365-3040.2005.01431.x

Saniewski M, Ueda J, Miyamoto K (2002) Relationships between jasmonates and auxin in regulation of some physiological processes in higher plants. Acta Physiol Plant:211–220. https://doi.org/10.1007/s11738-002-0013-9

Sarwar M et al (2018) Exogenously applied growth regulators protect the cotton crop from heat-induced injury by modulating plant defense mechanism. Sci Rep 8(1). https://doi.org/10.1038/s41598-018-35420-5

Sarwat M, Tuteja N (2017) Hormonal signaling to control stomatal movement during drought stress. Plant Gene 11:143–153. https://doi.org/10.1016/j.plgene.2017.07.007

Savchenko T et al (2014) Functional convergence of oxylipin and abscisic acid pathways controls stomatal closure in response to drought. Plant Physiol 164(3):1151–1160. https://doi.org/10.1104/pp.113.234310

Schaller F et al (2000) 12-oxophytodienoate reductase 3 (OPR3) is the isoenzyme involved in jasmonate biosynthesis. Planta 210(6):979–984. https://doi.org/10.1007/s004250050706

Schroeder JI, Kwak JM, Allen GJ (2001) Engineering drought hardiness in plants. Nature 410(6826):327–330. Available at: http://www.nature.com/nature/journal/v410/n6826/abs/410327a0.html

Seo HS et al (2001) Jasmonic acid carboxyl methyltransferase: a key enzyme for jasmonate-regulated plant responses. Proc Natl Acad Sci USA 98(8):4788–4793. https://doi.org/10.1073/pnas.081557298
Seo JS et al (2011) OsbHLH148, a basic helix-loop-helix protein, interacts with OsJAZ proteins in a jasmonate signaling pathway leading to drought tolerance in rice. Plant J 65(6):907–921. https://doi.org/10.1111/j.1365-313X.2010.04477.x
Sharma M, Laxmi A (2016) Jasmonates: emerging players in controlling temperature stress tolerance. Front Plant Sci. https://doi.org/10.3389/fpls.2015.01129
Sharma SS, Kumar V, Dietz KJ (2021) Emerging trends in metalloid-dependent signaling in plants. Trends Plant Sci:452–471. https://doi.org/10.1016/j.tplants.2020.11.003
Sirhindi G et al (2016) Jasmonic acid modulates the physio-biochemical attributes, antioxidant enzyme activity, and gene expression in *Glycine max* under nickel toxicity. Front Plant Sci 7(May 2016). https://doi.org/10.3389/fpls.2016.00591
Sirhindi G et al (2020) Jasmonic acid and methyl jasmonate modulate growth, photosynthetic activity and expression of photosystem II subunit genes in *Brassica oleracea* L. Sci Rep 10(1). https://doi.org/10.1038/s41598-020-65309-1
Sivasankar S, Sheldrick B, Rothstein SJ (2000) Expression of allene oxide synthase determines defense gene activation in tomato. Plant Physiol 122(4):1335–1342. https://doi.org/10.1104/pp.122.4.1335
Sofy MR et al (2020) Minimizing adverse effects of Pb on maize plants by combined treatment with jasmonic, salicylic acids and proline. Agronomy 10(5). https://doi.org/10.3390/agronomy10050699
Song S et al (2011) The jasmonate-ZIM domain proteins interact with the R2R3-MYB transcription factors MYB21 and MYB24 to affect jasmonate-regulated stamen development in Arabidopsis. Plant Cell 23(3):1000–1013. https://doi.org/10.1105/tpc.111.083089
Staswick PE, Tiryaki I (2004) The oxylipin signal jasmonic acid is activated by an enzyme that conjugate it to isoleucine in Arabidopsis W inside box sign. Plant Cell 16(8):2117–2127. https://doi.org/10.1105/tpc.104.023549
Sun Y et al (2011) Elevated CO_2 influences nematode-induced defense responses of tomato genotypes differing in the JA pathway. PLoS One 6(5). https://doi.org/10.1371/journal.pone.0019751
Svyatyna K, Riemann M (2012) Light-dependent regulation of the jasmonate pathway. Protoplasma:137–145. https://doi.org/10.1007/s00709-012-0409-3
Tan M et al (2017) Co-expression network analysis of the transcriptomes of rice roots exposed to various cadmium stresses reveals universal cadmium-responsive genes. BMC Plant Biol 17(1). https://doi.org/10.1186/s12870-017-1143-y
Tang Z, Zhao FJ (2020) The roles of membrane transporters in arsenic uptake, translocation and detoxification in plants. Crit Rev Environ Sci Technol:1–36. https://doi.org/10.1080/10643389.2020.1795053
Tchounwou PB et al (2012) Heavy metal toxicity and the environment. Exp Suppl 101:133–164. https://doi.org/10.1007/978-3-7643-8340-4_6
Theodoulou FL et al (2005) Jasmonic acid levels are reduced in COMATOSE ATP-binding cassette transporter mutants. Implications for transport of jasmonate precursors into peroxisomes. Plant Physiol 137(3):835–840. https://doi.org/10.1104/pp.105.059352
Valenzuela CE et al (2016) Salt stress response triggers activation of the jasmonate signaling pathway leading to inhibition of cell elongation in Arabidopsis primary root. J Exp Bot 67(14):4209–4220. https://doi.org/10.1093/jxb/erw202
Van Hoewyk D et al (2008) Transcriptome analyses give insights into selenium-stress responses and selenium tolerance mechanisms in Arabidopsis. Physiol Plant 132(2):236–253. https://doi.org/10.1111/j.1399-3054.2007.01002.x
Verma S, Dubey RS (2003) Lead toxicity induces lipid peroxidation and alters the activities of antioxidant enzymes in growing rice plants. Plant Sci 164(4):645–655. https://doi.org/10.1016/S0168-9452(03)00022-0

Wang Y, Liu H, Xin Q (2015) Improvement of copper tolerance of Arabidopsis by transgenic expression of an allene oxide cyclase gene, GhAOC1, in upland cotton (Gossypium hirsutum L.). Crop J 3(4):343–352. https://doi.org/10.1016/j.cj.2015.02.004

Wang P et al (2018) OsWRKY28 regulates phosphate and arsenate accumulation, root system architecture and fertility in rice. Front Plant Sci 9. https://doi.org/10.3389/fpls.2018.01330

Wang F, Yu G, Liu P (2019) Transporter-mediated subcellular distribution in the metabolism and signaling of jasmonates. Front Plant Sci 10. https://doi.org/10.3389/fpls.2019.00390

Wasternack C (2007) Jasmonates: an update on biosynthesis, signal transduction and action in plant stress response, growth and development. Ann Bot:681–697. https://doi.org/10.1093/aob/mcm079

Wasternack C, Feussner I (2018) The oxylipin pathways: biochemistry and function. Annu Rev Plant Biol [Preprint]. https://doi.org/10.1146/annurev-arplant-042817-040440

Wasternack C, Strnad M (2018) Jasmonates: news on occurrence, biosynthesis, metabolism and action of an ancient group of signaling compounds. Int J Mol Sci. https://doi.org/10.3390/ijms19092539

Wingler A et al (2020) Interactions between sucrose and jasmonate signalling in the response to cold stress. BMC Plant Biol 20(1). https://doi.org/10.1186/s12870-020-02376-6

Wiszniewska A et al (2018) Evaluation of the protective role of exogenous growth regulators against Ni toxicity in woody shrub Daphne jasminea. Planta 248(6):1365–1381. https://doi.org/10.1007/s00425-018-2979-6

Xu P et al (2020) Integration of jasmonic acid and ethylene into auxin signaling in root development. Front Plant Sci. https://doi.org/10.3389/fpls.2020.00271

Xue R, Zhang B (2007) Increased endogenous methyl jasmonate altered leaf and root development in transgenic soybean plants. J Genet Genomics 34(4):339–346. https://doi.org/10.1016/S1673-8527(07)60036-8

Yadav SK (2009) Cold stress tolerance mechanisms in plants. In: Sustainable agriculture, pp 605–620. https://doi.org/10.1007/978-94-007-0394-0_27

Yadav DK, Barik BR, Pradhan G (2019) Responses of crops plant to drought and its management for crop water availability: a review. J Pharmacogn Phytochem 8(4):167–172

Yan Z, Chen J, Li X (2013) Methyl jasmonate as modulator of Cd toxicity in *Capsicum frutescens* var. *fasciculatum* seedlings. Ecotoxicol Environ Saf 98:203–209. https://doi.org/10.1016/j.ecoenv.2013.08.019

Yan Z et al (2015a) Combined toxicity of cadmium and copper in *Avicennia marina* seedlings and the regulation of exogenous jasmonic acid. Ecotoxicol Environ Saf 113:124–132. https://doi.org/10.1016/j.ecoenv.2014.11.031

Yan Z et al (2015b) Methyl jasmonate alleviates cadmium toxicity in *Solanum nigrum* by regulating metal uptake and antioxidative capacity. Biol Plant 59(2):373–381. https://doi.org/10.1007/s10535-015-0491-4

Yan C et al (2018) Injury activates Ca^{2+}/calmodulin-dependent phosphorylation of JAV1-JAZ8-WRKY51 complex for jasmonate biosynthesis. Mol Cell 70(1):136–149.e7. https://doi.org/10.1016/j.molcel.2018.03.013

Yang TH et al (2020) Jasmonate precursor biosynthetic enzymes LOX3 and LOX4 control wound-response growth restriction[OPEN]. Plant Physiol 184(2):1172–1180. https://doi.org/10.1104/pp.20.00471

Ye H et al (2009) Identification and expression profiling analysis of TIFY family genes involved in stress and phytohormone responses in rice. Plant Mol Biol 71(3):291–305. https://doi.org/10.1007/s11103-009-9524-8

You J, Chan Z (2015) Ros regulation during abiotic stress responses in crop plants. Front Plant Sci. https://doi.org/10.3389/fpls.2015.01092

Yu X et al (2019a) Effects of methyl jasmonate on growth, antioxidants, and carbon and nitrogen metabolism of *Glycyrrhiza uralensis* under salt stress. Biol Plant 63:89–96. https://doi.org/10.32615/bp.2019.011

Yu X et al (2019b) The roles of methyl jasmonate to stress in plants. Funct Plant Biol:197–212. https://doi.org/10.1071/FP18106
Yu P et al (2020) ATP hydrolysis determines cold tolerance by regulating available energy for glutathione synthesis in rice seedling plants. Rice (N Y) 13(1). https://doi.org/10.1186/s12284-020-00383-7
Zaid A, Mohammad F (2018) Methyl jasmonate and nitrogen interact to alleviate cadmium stress in Mentha arvensis by regulating physio-biochemical damages and ROS detoxification. J Plant Growth Regul 37(4):1331–1348. https://doi.org/10.1007/s00344-018-9854-3
Zhang R et al (2020a) A role for the Auxin Response Factors ARF6 and ARF8 homologs in petal spur elongation and nectary maturation in Aquilegia. New Phytol 227(5):1392–1405. https://doi.org/10.1111/nph.16633
Zhang S et al (2020b) Calcium plays a double-edged role in modulating cadmium uptake and translocation in rice. Int J Mol Sci 21(21):1–18. https://doi.org/10.3390/ijms21218058
Zhang M et al (2021) Transcriptome and metabolome reveal salt-stress responses of leaf tissues from *Dendrobium officinale*. Biomolecules 11(5). https://doi.org/10.3390/biom11050736
Zhao FJ, Wang P (2020) Arsenic and cadmium accumulation in rice and mitigation strategies. Plant Soil. https://doi.org/10.1007/s11104-019-04374-6
Zhao ML et al (2013) Induction of jasmonate signalling regulators MaMYC2s and their physical interactions with MaICE1 in methyl jasmonate-induced chilling tolerance in banana fruit. Plant Cell Environ 36(1):30–51. https://doi.org/10.1111/j.1365-3040.2012.02551.x
Zhao S et al (2016) Tomato Jasmonic acid-deficient mutant spr2 seedling response to cadmium stress. J Plant Growth Regul 35(3):603–610. https://doi.org/10.1007/s00344-015-9563-0
Zhou J et al (2018) Effects of lead stress on the growth, physiology, and cellular structure of privet seedlings. PLoS One 13(3). https://doi.org/10.1371/journal.pone.0191139
Ziegler J et al (2000) Molecular cloning of allene oxide cyclase: the enzyme establishing the stereochemistry of octadecanoids and jasmonates. J Biol Chem [Preprint]. https://doi.org/10.1074/jbc.M002133200

Chapter 10
Salicylic Acid-Mediated Regulation of Plant Biology Under Heavy Metal Stress: Physiological and Molecular Aspects

Nahla El-Sherif

Abstract Salicylic acid (SA), an essential secondary phenolic metabolite and messenger for plant defense responses, chemically known as 2-hydroxy benzoic acid which includes an aromatic ring comprising a OH moiety, is a multifaceted movable signaling plant hormone, known to be involved in many physiological processes in plants, such as germination, photosynthesis, and abiotic stress resistance such as heat, chilling, heavy metals, drought, and oxidative stress. This chapter deals with an in-depth discussion about the roles of salicylic acid in plants growing under heavy metal stress.

Keywords Abiotic stress · Heavy metals · Phenolic molecules · Plant hormones · Salicylic acid

10.1 Introduction

Due to its reputation as a plant elicitor that participates in signal transduction pathways for both inducing systemic resistance response and local defense reactions, SA has drawn particular attention. These responses protect plants from pathogen attacks and abiotic stress conditions (Klessig and Malamy 1994; Loake and Grant 2007) as well as a defense response against abiotic stress conditions (Sharma et al. 2020, 2023). Plants have two different mechanisms for the biosynthesis of salicylic acid: the phenylalanine ammonia-lyase (PAL) pathway in the cytoplasm and the isochorismate (IC) pathway in the chloroplast. This review focuses on the

N. El-Sherif (✉)
Botany Department, Faculty of Science, Ain Shams University, Cairo, Egypt

Biology Department, Faculty of Science, Taibah University,
Madinah, Kingdom of Saudi Arabia
e-mail: elsherif.nahla@sci.asu.edu.eg

A. Sharma et al. (eds.), *Hormones and Heavy Metals*, Plant Life and Environment Dynamics, https://doi.org/10.1007/978-981-96-9622-2_10

regulation of plant biology by SA under heavy metal stress from both physiological and molecular aspects.

The balance (trade-off) between plant growth and resistance induction is maintained in part by SA. In-depth research has been done on SA, a confusing signaling chemical in plants, to understand its function in response to biotic and abiotic stressors. SA signaling mechanisms interplay and are mainly related and combined with modifications of levels of reactive oxygen species (ROS) as a crucial signaling molecule in plant tissues. Acting as second messengers, in response to SA, ROS control a succession of physiological and genetic responses in adaptations to many stress factors. These adaptations include remodeling cell wall construction, processing the antioxidant defense system, and modulating photosynthesis, glutathione synthesis, and senescence. These conclusions, reached based on various experiments, explicate the extensive role of SA in phyto-toxicology (Bhardwaj et al. 2021).

Seed germination, seed growth regulation, photosynthetic activity, nutrient absorption, flower induction, thermogenesis, and especially the regulation of plant responses under biotic or abiotic stress conditions have all been effectively characterized as physiological processes in which SA may play a role. SA signaling also causes gene expression and protein synthesis to be reprogrammed. It influences cellular redox homeostasis and may alter antioxidative metabolism.

10.2 Salicylic Acid's Impact During Heavy Metal Stress

In the Periodic Table, heavy metals are mostly located in the transition elements. Elements with a specific weight of greater than 5 g cm^3 are classified as heavy metals. Additionally categorized as heavy metals are metals that have a density of at least five times that of water. According to Kiran et al. (2021), there are two types of heavy metals: non-essential (Cd, Ni, As, Hg, Pb) and essential (Mo, Mn, Cu, Ni, Fe, Zn).

According to Souri et al. (2019), heavy metals cause numerous steps in the production of chloroplasts to be compromised, as well as changes in the shape and activity of photosystems and the biosynthesis of photosynthetic pigments.

Heavy metals such as chromium (Cr), cadmium (Cd), and lead (Pb) have pressured our crop production and human health (Daud et al. 2018), which is mostly released by anthropogenic activities, i.e., leather tanning, and the paint and fertilizer industries (Nriaguand and Neiborer 1988). Heavy metals such as lead have also been shown to considerably reduce the mitotic divisions in lupin, which might be attributed to decreased microtubule formation, and thus an inhibition in cytokinesis was also detected in response to heavy metal treatment (Przymusinski and Wozny 1985). Lead was also found to accumulate in roots and leaves of privet plants, causing cell plasmolysis and swelling of the thylakoid lamellae and decreased chlorophyll content (Zhou et al. 2018).

10.2.1 Copper (Cu)

For a long time, scientists have been studying the impacts and role of SA in plant physiological processes. One of the earliest procedures in which the protective role of SA in the face of an abiotic stressor was demonstrated was heavy metal stress. Cucumber and tobacco plants treated with SA developed tolerance to copper toxicity (Strobel and Kuc 1995). The treatment of maize seeds with SA resulted in a root tissue remodeling activity to prevent maize plants from the harmful effects of excess Cu. It was found that SA pretreatment caused an ameliorating carotenoid content, lower accumulation of Cu ions with an observed lower level of H_2O_2 in roots of Cu stressed plants, and a lower MnSOD activity (Moravcová et al. 2018).

10.2.2 Lead (Pb)

Lead was found to cause a decrease in cyclin protein production or accumulation in lupine roots (Deckert and Gwóźdź 1999). This might be attributed to a decreased mitotic activity (Przymusinski and Wozny 1985). Application of SA was shown to reduce the effects of Pb^{2+} stresses in rice (Mishra and Choudhuri 1997) and ameliorated cell membrane deterioration due to electrolyte leakage, injury index, and MDA content by lipoxygenase under $PbCl_2$. SA thus reversed the effects of superoxide dismutase, peroxidase, and catalase induced by Pb (Mishra and Choudhuri 1999).

SA treatment was also reported to decrease the toxic effect of lead in wheat (Ba et al. 2019; Gillani et al. 2021) and in parsley, where SA spraying was shown to reduce chlorosis and increase photosynthetic pigments in Pb-treated plants, as compared to those only exposed to Pb. Anthocyanin content was also found to be more elevated in parsley plants treated with SA under Pb stress (Alamer and Fayez 2020). Exogenous SA resulted in an increased concentration of insoluble sugars in roots, thus decreasing the antioxidant enzyme activity of *Brassica napus* plants exposed to lead (Khozaei and Jazi 2019).

Application of SA improved growth parameters, diminished the accumulation of oxidative stress, and reduced the Pb content in *Nicotiana tabacum,* thus detoxifying Pb and alleviating its toxicity (Abdul Halim and Phang 2017). Exogenous SA was also found to exhibit a protective effect on wheat roots against the damaging effects of lead (Ba et al. 2019).

10.2.3 Mercury (Hg)

Application of mercury is toxic to plants, causing cell membrane leakage and a decrease in MDA content. Application of SA was found to have a mitigating effect on $HgCl_2$-treated plants (Mishra and Choudhuri 1999). SA has shown an ameliorating effect on lemon balm plants exposed to mercury. In response to SA, free proline levels and phenolics were elevated, accompanied by an increase in antioxidant capacity (DPPH and FRAP). SA also caused an induction in CHLG and PAL genes (Safari et al. 2019).

10.2.4 Cadmium (Cd)

Cd is one of the most common contaminants, and it can cause severe stress in plants, as well as humans, through the food chain. Plants have evolved numerous cross-linked mechanisms to decrease Cd harm, including attaching it to the cell wall, chelation with phytochelatins, and antioxidant participation. SA pretreatment of barley seedlings reduced Cd-induced oxidative damage, resulting in enhanced biomass.

The fact that exposure to Cd promotes the endogenous SA level better than any other abiotic stressor suggests that SA signaling may also be a factor of Cd tolerance.

Treatment with SA was found to alleviate Cd toxicity in barley (Metwally et al. 2003; Tamás et al. 2015), wheat (Agami and Mohamed 2013), safflower (Amjadi et al. 2021), and brown rice (Wang et al. 2021).

Guo and colleagues recently evaluated the potential techniques for using exogenous SA to protect plants from Cd toxicity, as well as the potential mechanisms that are induced by SA (Guo et al. 2019) and play a part in defense processes (Guo et al. 2019). It is critical to stress that, in addition to the treatment manner, the effects of exogenous SA are influenced by a variety of factors, including Cd and SA concentrations, plant species, and developmental stage. Guo and co-workers observed that the levels of SA in presoaking treatments or when applied as a foliar spray are generally higher than those in hydroponic treatments based on several testing systems. This suggests that SA uptake through the root system is more tightly controlled than through the leaves or seeds.

The mechanism of action of SA during Cd stress is yet unknown, and the results are frequently disputed. Despite the high number of effective exogenous SA applications, data from mutant or transgenic Arabidopsis plants suggest that a high SA level can be injurious to plants.

These findings support the idea that the SA dose should be adjusted to provide the best possible protective benefit.

The Cd defense of plants is often intricate, and it seems that SA affects multiple facets of it. SA may have a secondary effect by stimulating other defense systems, even if it is linked to other plant growth regulators. In addition to increasing Cd accumulation in the cell wall and preventing its transfer into other cell organelles, SA may have another direct protective effect.

It has also been established that SA most likely influences cell wall components in the event of biotic stressors, where the cell wall serves as the body's first line of defense.

Like many other stressors, the activation of the antioxidant system seems to be a general method for shielding plants from the oxidative damage brought on by elevated heavy metal concentrations. The elevated SA level in the root during cadmium stress was linked to a rise in the internal glutathione cycle, thereby boosting the antioxidant and metal detoxification systems, even though there was no direct correlation between the initial SA levels and wheat's ability to tolerate Cd.

SA has a role in controlling the toxicity caused by heavy metals in plants. The antioxidant defense system of potato plants exposed to Cd stress was assessed using a foliar application of 600 μM SA concentration (10 ml/plant) in Hoagland solution. Exogenous SA treatment was found to decrease the toxicity of Cd by increasing endogenous SA levels, proline, chlorophyll, and relative water content (RWC), while also causing a decrease in superoxide anion radicals (O_2^-), hydrogen peroxide (H_2O_2), and malondialdehyde (MDA) (Wang et al. 2019).

The mitigation of the toxic effect of Cd by SA was shown to occur in plants through an antistress response, which involves the redistribution of K and Mg (Drazic and Mihailovic 2005). SA was found to affect differently Cd-tolerant and Cd-sensitive genotypes. In Cd-tolerant ones, SA was found to induce Cd accumulation, whereas it inhibits the accumulation of Cd in susceptible ones. This might be attributed to differences in the efficiency of non-specific defense strategies entailing SA (Dražić and Mihailović 2009).

10.3 Oxidative Stress Under Heavy Metal Toxicity and Its Regulation by Salicylic Acid

Remarkably, most SA-regulated abiotic stresses in plants are involved in antioxidative responses, indicating that SA acts as an internal signal molecule cooperating with the ROS signal pathway.

Treatment with exogenous SA might induce phytotoxic symptoms in plants by triggering H_2O_2 accumulation, lipid peroxidation, and oxidative damage, as shown in Arabidopsis, tobacco, and mustard (Rao et al. 1997). Both exogenous SA and H_2O_2 stimulated potato micro-plant tolerance to high temperature (Dat et al. 1998), suggesting that SA may share common signaling pathways mediated by the formation of reactive oxygen species.

Because of its methyl ester form, methyl salicylate, which is very volatile, salicylic acid plays an important role as a plant growth regulator and has an impact on intra- and inter-plant communication. In heavy metal-stressed plants, SA acts in many ways to alleviate the effect of heavy metals on plants: by signaling gene activation, including genes coding for antioxidant enzymes, genes involved in both heavy metal compartmentalization and uptake. In addition, SA acts by stimulating the antioxidant apparatus responsible for moderating the antioxidant levels and controlling antioxidant enzyme activities. SA also stimulates protective compounds in plants such as polyphenols, sugars, proline, glycine betaine, and polyamines, leading to ROS scavenging and thus alleviation of oxidative stress produced by heavy metals.

SA also affects physiological processes in plants by interacting with other plant hormones (such as auxins, cytokinins, gibberellins, abscisic acid, ethylene, and brassinosteroids). All of these processes enable SA to signal HM-treated plants while also assisting in the reduction of HM stress.

Seed priming is the simplest and most cost-effective approach for limiting the negative consequences of abiotic stresses (Iqbal and Ashraf 2013). Seed priming, i.e., pre-sowing treatment of seeds, is very effective in augmenting the performance of seeds by getting better germination and seedling growth (Farooq et al. 2019; Arif et al. 2008). Integrating plant hormones like SA during presoaking and priming increases the seed activity in many crops, like rice (Basra et al. 2006). SA may directly affect the function of specific enzymes and might also switch on the genes responsible for protective processes to alleviate the effect of abiotic stress on plants (Horváth et al. 2007).

Abiotic stresses, including heavy metals, harmfully manipulate the growth, biomass production, and yield of crops (Zhu 2016), possibly by induction of variations at the biochemical and molecular level via alterations of metabolic pathways and gene expressions (Buti et al. 2019).

10.4 SA's Protective Impact on Opposing Heavy Metals

In tobacco, cucumber, and other heavy metal-sensitive plants, salicylic acid treatment has been shown to induce tolerance against copper toxicity (Strobel and Kuc 1995). In sunflower plants, on the other hand, it has been demonstrated that SA induces an adaptive response to Cu toxicity by increasing the accumulation of soluble sugars, proteins, and proline, which protects the photosynthetic pigments (El-Tayeb et al. 2006). Additionally, under Pb^{2+} or Hg^{2+} stress, SA therapy was demonstrated to have a mitigating effect on seed germination and seedling growth in two rice plants. Barley seedlings treated with SA prior to Cd exposure exhibited enhanced shoot and root fresh weight and were shielded from lipid peroxidation and membrane damage (Mishra and Choudhuri 1999). However, this protection did not result from the induction of antioxidant activities. Conversely, it was observed that in Cd-stressed seedlings, antioxidant enzyme activities increased; however, this effect was inhibited by pretreatment with SA (Metwally et al. 2003).

Concentrations of SA at 10^{-6}, 10^{-5}, and 10^{-4} M mitigated the effects of Cd on soybean (*Glycine max* L. cv SG1) seedlings that were simultaneously treated with Cd and SA. These effects included growth inhibition, desiccation, and a decrease in chlorophyll content. While SA did not prevent soybeans from absorbing Cd, it altered the distribution of Cd in plant organs (Drazic and Mihailovic 2005). The imbibition of 10^{-5} M SA increased the amount of Cd that was absorbed by the seeds in alfalfa (*Medicago sativa* L. cv. Evropa). In the roots, there was a strong negative correlation between Cd and K and Ca, whereas in the shoots, there was a positive correlation between Ca and Mg. These correlations decreased in response to SA. These findings suggest that SA may function by preserving ion homeostasis in plants (Dražić et al. 2006).

In the case of maize, although exogenous SA application (added to the hydroponic solution) reduced the uptake of Cd by the roots, SA itself caused oxidative stress to the plants, thereby aggravating the symptoms of cadmium stress (Pál et al. 2005). However, when maize seeds were presoaked in SA solution for 6 h before germination, this treatment significantly reduced the damaging effect of Cd on the photosynthetic processes (Krantev et al. 2008). In *Cassia tora* L. plants, SA enhanced Al tolerance by increasing the citrate efflux of the roots and thus inhibiting Al uptake (Yang et al. 2003). Al was also shown to increase the SA concentration of the roots. Cd treatment increased the endogenous free SA content in barley (Metwally et al. 2003) and in maize (Pál et al. 2005).

10.5 What Is a Potential Mechanism of SA to Alleviate the Negative Effects of HM on Plants?

Heavy metals, as discussed above, are related to oxidative damage. Many experiments indicated that SA is an effective preventive agent against these compounds that might induce plant oxidative damage. Treating cucumber and tobacco leaves with SA protected them from oxidative damage induced by paraquat applied 3 days after SA pretreatment (Strobel and Kuc 1995). Also, pretreatment of barley seedlings with SA protected them from the inhibitory effect of paraquat on photosynthesis and decreased the production of H_2O_2, electrolyte leakage, and lipid peroxidation (Ananieva et al. 2004), as well as induced the activity of certain antioxidant enzymes.

Earlier, it was reported that under Zn, Cu, and Mn stress, SA coupled with plant growth-promoting bacteria (PGPB) increased the activity of CAT and SOD enzymes in barley leaves (Song et al. 2014).

Furthermore, SA supplementation lowered reactive oxygen species levels by enhancing plants' antioxidant defense mechanisms and providing plant membrane stability (Faraz et al. 2019). Exogenous application of SA boosts the antioxidant system, promotes growth and yield, and reduces oxidative damage in Pb-stressed plants in *B. campestris* (Hasanuzzaman et al. 2019).

Gibberellins, auxin, cytokinins, ethylene, brassinosteroids, and SA are plant growth regulators or phytohormones that play a significant role in HM tolerance (Bhardwaj et al. 2021).

Under abiotic stress, such as high temperature, salinity, and heavy metals such as Pb, Hg, and Cd, SA modulates photosynthesis, respiration, and antioxidant defense mechanisms in different plants (Mohsenzadeh et al. 2011; Abd Ghani et al. 2015; Zengin 2015; Bhardwaj et al. 2021).

Oxidative damage in maize caused by Cr was reduced by treating the plants with SA. This enhanced the activities of antioxidant and nonantioxidant enzymes. Furthermore, when Cr-treated maize plants were exposed to SA, their carbohydrate metabolism was found to improve (Islam et al. 2016). Exogenous SA application has been shown to alleviate Cd toxicity and enhance growth and photosynthesis in mustard plants (Ahmad et al. 2011; Faraz et al. 2019). Exposure to heavy metals was found to induce ethylene production in plants, which increases ROS concentration. Salicylic acid application was also found to be associated with the induction of glutathione content, which can scavenge ROS, induced by heavy metal-induced ethylene content.

10.6 Mechanism of Action of SA in Plants on the Physiological Level

Plants are affected by a variety of physiological and biochemical pathways as a result of their exposure to various stressful circumstances. Photosynthesis is usually one of the more vulnerable systems to heavy metals. Levels of chlorophyll and photosynthesis-associated proteins were found to be elevated in SA mutants in response to exogenous SA (Hirosawa et al. 2021).

Heavy metals build up in plant organs and interfere with the production of photosynthetic pigments like carotenoids and chlorophylls. Heavy metals also disrupt the structure of the chloroplast membrane and impact electron transport, affecting light-dependent photosynthesis reactions. Stomatal and mesophyll conductance, chloroplast structures, photosystems I and II, photosynthetic enzymes, and the antioxidant defense mechanism were found to be negatively affected by heavy metals (Souri et al. 2019).

Salicylic acid was shown to increase cell division in apical meristems of wheat seedlings and thus ameliorates plant growth. It also diminishes the alteration in phytohormones (IAA and ABA) and abiotic stress conditions like salinity and drought (Sakhabutdinova et al. 2003).

10.7 Potential Mechanism of Action of SA to Reduce Heavy Metal Stress in Plants on the Molecular Level

Exogenous application of SA enhanced the expression of genes such as topoisomerase and chloroplast elongation factor (Singh et al. 2004). These genes have been associated with plant adaptation to abiotic stress conditions.

cDNA clones representing SA-induced genes were identified in cabbage, encoding a putative clone coding for a glucosyl-transferase protein, a clone with a putative metal-binding farnesylated protein, and a third clone identified as a chitinase gene with similarities to a rapeseed class IV chitinase. Class IV chitinases have deletions in the chitin-binding and catalytic domains, and the BC6-1 chitinase has an additional deletion in the catalytic domain. The BCP8-1clone is most homologous to an Arabidopsis gene that contains a tandem array of two thiJ-like sequences. These four cabbage genes were barely expressed in healthy leaves but were highly overexpressed by salicylic acid (Oh et al. 2004). In addition, a pea chloroplast translation elongation factor (tufA) was found to be regulated by salicylic acid (Singh et al. 2004).

Transcription factors such as WRKY transcription factors have also been found to be induced by SA and are involved in plant stress management and regulation (Negi and Khurana 2021). The important role of SA in protecting plants is probably played by its ability to induce expression of genes coding for not only PR proteins but also, for example, the extensin gene, a cell wall protein, in Arabidopsis plants. Data have shown the importance of the extensin gene in plant defense and development (Merkouropoulos et al. 1999; Merkouropoulos and Shirsat 2003; Roberts et al. 2013). There is evidence about SA-induced synthesis of heat shock proteins in tobacco plants (Burkhanova et al. 1999) and accumulation of wheat lectins that have been shown to protect plants against abiotic stress conditions, including salinity and heavy metals (Shakirova et al. 1997, 2013).

This suggests the involvement of SA in the realization of different antistress programs. These stress factors are of biotic as well as abiotic origin, such as salinity, heat, drought, and heavy metals. However, the way of signal regulation of plant resistance to unfavorable factors of environment induced by SA is still not clear.

Many plant proteins interact with SA. There are numerous distinct proteins with varied functions in cell regulation and potentially cell protection against stress situations, including the well-known SA receptor, NPR1, and its paralogs. NPR1 is the main protein that binds to SA. Upstream regulatory events are required for NPR1 (such as a change in cell redox status). A variety of SA-induced reactions are also unrelated to NPR1.

This was confirmed in NPR1-deficient plants, which showed that SA-induced responses, such as PR gene expression, are significantly altered.

Nevertheless, many SA-induced responses do not depend on NPR1, and many other proteins have been studied and were found to be potential SA receptors, such as SABP1-catalase, SABP2-MeSA esterase, SABP3-b carbonic anhydrase, GSTs, thioredoxins, GH3 acyl acid amido synthetase, GAPDH, alpha-ketoglutarate dehydrogenase (a Krebs cycle enzyme), thimet oligopeptidases, and TPPII exopeptidase (Pokotylo et al. 2019).

This has revealed that plant cells can perceive SA in a variety of ways, including through a variety of putative SA receptors, but not confined to salicylic acid binding proteins (SABPs). All of these proteins are potential SA receptors, and they may play a role in SA feed from numerous processors.

References

Abd Ghani I, Khan I, Ahmed I et al (2015) Amelioration of lead toxicity in Pisum sativum (L.) by foliar application of salicylic acid. J Environ Anal Toxicol 5(4). https://doi.org/10.4172/2161-0525.1000292

Abdul Halim NI, Phang IC (2017) Salicylic acid mitigates Pb stress in nicotiana tabacum. Sci Herit J 1(1):10.26480/gws.01.2017.16.19

Agami RA, Mohamed GF (2013) Exogenous treatment with indole-3-acetic acid and salicylic acid alleviates cadmium toxicity in wheat seedlings. Ecotoxicol Environ Saf 94. https://doi.org/10.1016/j.ecoenv.2013.04.013

Ahmad P, Nabi G, Ashraf M (2011) Cadmium-induced oxidative damage in mustard [Brassica juncea (L.) Czern. & Coss.] plants can be alleviated by salicylic acid. S Afr J Bot 77(1). https://doi.org/10.1016/j.sajb.2010.05.003

Alamer KH, Fayez KA (2020) Impact of salicylic acid on the growth and physiological activities of parsley plants under lead toxicity. Physiol Mol Biol Plants 26(7). https://doi.org/10.1007/s12298-020-00830-1

Amjadi Z, Namdjoyan S, Abolhasani Soorki A (2021) Exogenous melatonin and salicylic acid alleviates cadmium toxicity in safflower (Carthamus tinctorius L.) seedlings. Ecotoxicology 30(3). https://doi.org/10.1007/s10646-021-02364-y

Ananieva EA, Christov KN, Popova LP (2004) Exogenous treatment with Salicylic acid leads to increased antioxidant capacity in leaves of barley plants exposed to Paraquat. J Plant Physiol 161(3). https://doi.org/10.1078/0176-1617-01022

Arif M, Jan MT, Marwat KB, Khan MA (2008) Seed priming improves emergence and yield of soybean. Pak J Bot 40(3):1169–1177

Ba QS, Song YL, Zhang LL, Li GP, Liu Y, Chen C, Fu ZL (2019) Effects of root-applying salicylic acid on the growth and development of wheat roots under lead stress. Agric Res Arid Areas 37(1). https://doi.org/10.7606/j.issn.1000-7601.2019.01.29

Basra SMA, Farooq M, Wahid A, Khan MB (2006) Rice seed invigoration by hormonal and vitamin priming. Seed Sci Technol 34(3). https://doi.org/10.15258/sst.2006.34.3.23

Bhardwaj S, Sharma D, Kapoor D (2021) Salicylic acid signaling and ROS balance in plants. In: M Hasan (ed) Salicylic acid contribution in plant biology against a changing environment (pp. 111–126). Springer Nature

Burkhanova EA, Fedina AB, Kulaeva ON (1999) Effect of salicylic acid and (2′-5′)-oligoadenylates on protein synthesis in tobacco leaves under heat shock conditions: a comparative study. Russ J Plant Physiol 46(1):11–16

Buti M, Baldoni E, Formentin E, Milc J, Frugis G, Schiavo F l, Genga A, Francia E (2019) A meta-analysis of comparative transcriptomic data reveals a set of key genes involved in the tolerance to abiotic stresses in rice. Int J Mol Sci 20(22). https://doi.org/10.3390/ijms20225662

Dat JF, Lopez-Delgado H, Foyer CH, Scott IM (1998) Parallel changes in H_2O_2 and catalase during thermotolerance induced by salicylic acid or heat acclimation in mustard seedlings. Plant Physiol 116(4). https://doi.org/10.1104/pp.116.4.1351

Daud MK, Ali S, Abbas Z, Zaheer IE, Riaz MA, Malik A, Hussain A, Rizwan M, Zia-Ur-Rehman M, Zhu SJ (2018) Potential of duckweed (Lemna minor) for the phytoremediation of landfill leachate. J Chem 2018. https://doi.org/10.1155/2018/3951540

Deckert J, Gwóźdź EA (1999) The effect of lead on cyclin expression in lupine roots. Acta Physiol Plant 21(3):249–256. https://doi.org/10.1007/s11738-999-0039-3

Drazic G, Mihailovic N (2005) Modification of cadmium toxicity in soybean seedlings by salicylic acid. Plant Sci 168(2). https://doi.org/10.1016/j.plantsci.2004.09.019

Dražić G, Mihailović N (2009) Salicylic acid modulates accumulation of Cd in seedlings of Cd-tolerant and Cd-susceptible soybean genotypes. Arch Biol Sci 61(3). https://doi.org/10.2298/ABS0903431D

Dražić G, Mihailović N, Lojić M (2006) Cadmium accumulation in Medicago sativa seedlings treated with salicylic acid. Biol Plant 50(2). https://doi.org/10.1007/s10535-006-0013-5

El-Tayeb MA, El-Enany AE, Ahmed NL (2006) Salicylic acid-induced adaptive response to copper stress in sunflower (Helianthus annuus L). Int J Bot 2(4):372–379

Faraz A, Faizan M, Sami F, Siddiqui H, Hayat S (2019) Supplementation of salicylic acid and citric acid for alleviation of cadmium toxicity to Brassica juncea. J Plant Growth Regul. https://doi.org/10.1007/s00344-019-10007-0

Farooq M, Usman M, Nadeem F, Rehman HU, Wahid A, Basra SMA, Siddique KHM (2019) Seed priming in field crops: potential benefits, adoption and challenges. Crop Pasture Sci 70(9). https://doi.org/10.1071/CP18604

Gillani SR, Murtaza G, Mehmood A (2021) Mitigation of lead stress in *Triticum aestivum* L. seedlings by treating with salicylic acid. Pak J Bot 53(1). https://doi.org/10.30848/PJB2021-1(44)

Guo B, Liu C, Liang Y, Li N, Fu Q (2019) Salicylic acid signals plant defence against cadmium toxicity. Int J Mol Sci 20(12). NLM (Medline). https://doi.org/10.3390/ijms20122960

Hasanuzzaman M, Matin MA, Fardus J, Hasanuzzaman M, Hossain MS, Parvin K (2019) Foliar application of salicylic acid improves growth and yield attributes by upregulating the antioxidant defense system in Brassica campestris plants grown in lead-amended soils. Acta Agrobot 72(2). https://doi.org/10.5586/aa.1765

Hirosawa Y, Tada A, Matsuura T, Mori IC, Ogura Y, Hayashi T, Uehara S, Ito-Inaba Y, Inaba T (2021) Salicylic acid acts antagonistically to plastid retrograde signaling by promoting the accumulation of photosynthesis-associated proteins in Arabidopsis. Plant Cell Physiol. https://doi.org/10.1093/pcp/pcab128

Horváth E, Szalai G, Janda T (2007) Induction of abiotic stress tolerance by salicylic acid signaling. J Plant Growth Regul 26(3). https://doi.org/10.1007/s00344-007-9017-4

Iqbal M, Ashraf M (2013) Salt tolerance and regulation of gas exchange and hormonal homeostasis by auxin-priming in wheat. Pesq Agrop Brasileira 48(9):1210–1219. https://doi.org/10.1590/S0100-204X2013000900004

Islam F, Yasmeen T, Arif MS, Riaz M, Shahzad SM, Imran Q, Ali I (2016) Combined ability of chromium (Cr) tolerant plant growth promoting bacteria (PGPB) and salicylic acid (SA) in attenuation of chromium stress in maize plants. Plant Physiol Biochem 108. https://doi.org/10.1016/j.plaphy.2016.08.014

Khozaei M, Jazi SB (2019) Salicylic acid phytohormone decreased the toxicity damage of exogenous lead absorption in seedlings of an oilseed plant, Brassica napus L. grown in hydroponic conditions. J Agric Ecol Res Int. https://doi.org/10.9734/jaeri/2019/v19i130073

Klessig DF, Malamy J (1994) The salicylic acid signal in plants. Plant Mol Biol 26(5). https://doi.org/10.1007/BF00016484

Krantev A, Yordanova R, Janda T, Szalai G, Popova L (2008) Treatment with salicylic acid decreases the effect of cadmium on photosynthesis in maize plants. J Plant Physiol 165(9). https://doi.org/10.1016/j.jplph.2006.11.014

Kiran G, Sharma GN, Shrivastava B, Sudhakar Babu AMS (2021) Protective role of excoecaria agallocha L. against streptozotocin-induced diabetes and related diabetic neuropathy complication. J Glob Trends Pharm Sci 12(1):9095–9108

Loake G, Grant M (2007) Salicylic acid in plant defence-the players and protagonists. Curr Opin Plant Biol 10(5). https://doi.org/10.1016/j.pbi.2007.08.008

Merkouropoulos G, Shirsat AH (2003) The unusual Arabidopsis extensin gene atExt1 is expressed throughout plant development and is induced by a variety of biotic and abiotic stresses. Planta 217(3). https://doi.org/10.1007/s00425-003-1002-y

Merkouropoulos G, Barnett DC, Shirsat AH (1999) The Arabidopsis extensin gene is developmentally regulated, is induced by wounding, methyl jasmonate, abscisic and salicylic acid, and codes for a protein with unusual motifs. Planta 208(2). https://doi.org/10.1007/s004250050552

Metwally A, Finkemeier I, Georgi M, Dietz KJ (2003) Salicylic acid alleviates the cadmium toxicity in barley seedlings. Plant Physiol 132(1). https://doi.org/10.1104/pp.102.018457

Mishra A, Choudhuri MA (1997) Ameliorating effects of salicylic acid on lead and mercury-induced inhibition of germination and early seedling growth of two rice cultivars. Seed Sci Technol 25(2):263–270

Mishra A, Choudhuri MA (1999) Effects of salicylic acid on heavy metal-induced membrane deterioration mediated by lipoxygenase in rice. Biol Plant 42(3). https://doi.org/10.1023/A:1002469303670

Mohsenzadeh S, Shahrtash M, Mohabatkar H (2011) Interactive effects of salicylic acid and silicon on some physiological responses of cadmium-stressed maize seedlings. Iran J Sci Technol Trans A: Sci 35(1). https://doi.org/10.22099/ijsts.2011.2129

Moravcová Š, Tůma J, Dučaiová ZK, Waligórski P, Kula M, Saja D, Słomka A, Bąba W, Libik-Konieczny M (2018) Influence of salicylic acid pretreatment on seeds germination and some defence mechanisms of Zea mays plants under copper stress. Plant Physiol Biochem 122. https://doi.org/10.1016/j.plaphy.2017.11.007

Negi N, Khurana P (2021) A salicylic acid inducible mulberry WRKY transcription factor, MiWRKY53 is involved in plant defence response. Plant Cell Rep 40(11). https://doi.org/10.1007/s00299-021-02710-8

Nriaguand JO, Neiborer E (1988) Chromium in natural and human environments. Wiley-Interscience, New York, NY, USA

Oh KJ, Park YS, Lee KA, Chung YJ, Cho TJ (2004) Molecular characterization of a thiJ-like gene in Chinese cabbage. J Biochem Mol Biol 37(3). https://doi.org/10.5483/bmbrep.2004.37.3.343

Pál M, Horváth E, Janda T, Páldi E, Szalai G (2005) Cadmium stimulates the accumulation of salicylic acid and its putative precursors in maize (Zea mays) plants. Physiol Plant 125(3). https://doi.org/10.1111/j.1399-3054.2005.00545.x

Pokotylo I, Kravets V, Ruelland E (2019) Salicylic acid binding proteins (SABPs): the hidden forefront of salicylic acid signalling. Int J Mol Sci 20(18). https://doi.org/10.3390/ijms20184377

Przymusinski R, Wozny A (1985) The reactions of lupin roots on the presence of lead in the medium. Biochem Physiol Pflanz 180(4). https://doi.org/10.1016/S0015-3796(85)80007-X

Rao M v, Paliyath G, Ormrod DP, Murr DP, Watkins CB (1997) Influence of salicylic acid on H_2O_2 production, oxidative stress, and H_2O_2-metabolizing enzymes. Salicylic acid-mediated oxidative damage requires H_2O_2. Plant Physiol 115(1). https://doi.org/10.1104/pp.115.1.137

Roberts K, Merkouropoulos G, Shirsat AH (2013) Identification of promoter regions in the *Arabidopsis thaliana* at Ext1 extensin gene controlling late responses to wounding and pathogen attack. Biol Plant 57(2). https://doi.org/10.1007/s10535-012-0281-1

Safari F, Akramian M, Salehi-Arjmand H, Khadivi A (2019) Physiological and molecular mechanisms underlying salicylic acid-mitigated mercury toxicity in lemon balm (Melissa officinalis L.). Ecotoxicol Environ Saf 183. https://doi.org/10.1016/j.ecoenv.2019.109542

Sakhabutdinova AR, Fatkhutdinova DR, Bezrukova M v, Shakirova FM (2003) Salicylic acid prevents the damaging action of stress factors on wheat plants. Bulg J Plant Physiology Special Issue, 314–319

Shakirova FM, Bezrukova M, v. (1997) Induction of wheat resistance against environmental salinization by indolylacetic acid. Biol Bull 24(2)

Shakirova FM, Bezrukova M v, Yuldashev RA, Fatkhutdinova RA, Murzabaev AR (2013) Involvement of lectin in the salicylic acid-induced wheat tolerance to cadmium and the role of endogenous ABA in the regulation of its level. Dokl Biol Sci 448(1). https://doi.org/10.1134/S0012496613010158

Singh BN, Mishra RN, Agarwal PK, Goswami M, Nair S, Sopory SK, Reddy MK (2004) A pea chloroplast translation elongation factor that is regulated by abiotic factors. Biochem Biophys Res Commun 320(2). https://doi.org/10.1016/j.bbrc.2004.05.192

Song WY, Yang HC, Shao HB, Zheng AZ, Brestic M (2014) The alleviative effects of salicylic acid on the activities of catalase and superoxide dismutase in malting barley (*Hordeum uhulgare* L.) seedling leaves stressed by heavy metals. Clean (Weinh) 42(1). https://doi.org/10.1002/clen.201200310

Souri Z, Cardoso AA, Da-Silva CJ, de Oliveira LM, Dari B, Sihi D, Karimi N (2019) Heavy metals and photosynthesis: recent developments. In M Hasan (ed) Photosynthesis, productivity, and environmental stress (pp. 107–134). John Wiley & Sons Ltd, Hoboken, New Jersey, USA. https://doi.org/10.1002/9781119501800.ch7

Strobel NE, Kuc JA (1995) Chemical and biological inducers of systemic resistance to pathogens protect cucumber and tobacco plants from damage caused by paraquat and cupric chloride. Phytopathology 85(10). https://doi.org/10.1094/Phyto-85-1306

Sharma A, Sidhu GPS, Araniti F, Bali AS, Shahzad B, Tripathi DK, Brestic M, Skalicky M, Landi M (2020) The role of salicylic acid in plants exposed to heavy metals. Molecules 25(3):540. https://doi.org/10.3390/molecules25030540

Sharma A, Mohiuddin M, Gulfishan M, Shabir F, Khan SA (2023) The role of salicylic acid in plants exposed to heavy metals: a review. Int J Agric Plant Sci 5(4):37–42. https://www.agriculturejournal.in/assets/archives/2023/vol5issue4/5057-1699007703698.pdf

Tamás L, Mistrík I, Alemayehu A, Zelinová V, Bočová B, Huttová J (2015) Salicylic acid alleviates cadmium-induced stress responses through the inhibition of Cd-induced auxin-mediated reactive oxygen species production in barley root tips. J Plant Physiol 173. https://doi.org/10.1016/j.jplph.2014.08.018

Wang M, Chen S, Chen L, Wang D (2019) Responses of soil microbial communities and their network interactions to saline-alkaline stress in Cd-contaminated soils. Environ Pollut 252. https://doi.org/10.1016/j.envpol.2019.06.082

Wang F, Tan H, Zhang Y, Huang L, Bao H, Ding Y, Chen ZX, Zhu C (2021) Salicylic acid application alleviates cadmium accumulation in brown rice by modulating its shoot to grain translocation in rice. Chemosphere 263. https://doi.org/10.1016/j.chemosphere.2020.128034

Yang ZM, Wang J, Wang SH, Xu LL (2003) Salicylic acid-induced aluminum tolerance by modulation of citrate efflux from roots of Cassia tora L. Planta 217(1). https://doi.org/10.1007/s00425-003-0980-0

Zengin F (2015) Effects of exogenous salicylic acid on growth characteristics and biochemical content of wheat seeds under arsenic stress. J Environ Biol 36(1)

Zhou J, Zhang Z, Zhang Y, Wei Y, Jiang Z (2018) Effects of lead stress on the growth, physiology, and cellular structure of privet seedlings. PLoS One 13(3). https://doi.org/10.1371/journal.pone.0191139

Zhu JK (2016) Abiotic stress signaling and responses in plants. Cell 167(2). https://doi.org/10.1016/j.cell.2016.08.029

Chapter 11
Melatonin-Mediated Regulation of Plant Biology Under Heavy Metal Stress

Vasudha Maurya, Narayan Singh, Harjot Singh, Sonika Sharma, Indu Sharma, Rahul Kumar, and Ashutosh Sharma

Abstract Various forms of abiotic stresses, including salt, water (excess or scarcity), UV light, temperature, and heavy metals (HMs), cause a reduction in plant growth and productivity. Exposure to HMs is among the most detrimental forms of abiotic stress. Several anthropogenic activities have caused a considerable increase in the accumulation of HMs in almost all biological systems, leading to HMs pollution. Heavy metals severely hinder plant development and productivity in agro-ecosystems, which is a major threat to achieving the sustainable agriculture goal. The non-biodegradable nature of HMs and their accumulation may affect various biological processes in plants that include photosynthesis, the antioxidant defense system, mineral nutrition, and phytohormone production. Some HMs may have toxic effects due to their nature to impact some endogenous molecules like phytohormones, which can negatively affect a plant's ability to grow, develop, and reproduce. Exogenous application of some regulatory molecules, like melatonin, may protect the plants against the harmful effects of HMs. On the one hand, melatonin is known to immobilize HMs in cell walls and sequester them in vacuoles to directly reduce HMs-induced toxicity in plants; on the other hand, it also helps in improving photosynthesis, antioxidant defense system, nutrient uptake, and accumulation of compatible solutes to save the plants from HMs-induced toxicity indirectly. Further, it regulates the expression of key genes to confer better HM stress tolerance to plants. This chapter is, therefore, an attempt to present up-to-date information on the role of melatonin-mediated HMs stress amelioration in plants to improve crop productivity.

V. Maurya · N. Singh
Department of Biotechnology, DAV University, Jalandhar, Punjab, India

Faculty of Agricultural Sciences, DAV University, Jalandhar, Punjab, India

H. Singh · S. Sharma · R. Kumar (✉)· A. Sharma (✉)
Faculty of Agricultural Sciences, DAV University, Jalandhar, Punjab, India

I. Sharma
Department of Life Sciences, University Institute of Sciences, Sant Baba Bhag Singh University, Jalandhar, Punjab, India

A. Sharma et al. (eds.), *Hormones and Heavy Metals*, Plant Life and Environment Dynamics, https://doi.org/10.1007/978-981-96-9622-2_11

Keywords Endogenous melatonin · Reactive oxygen species · Metal chelation · Antioxidant defense · Environmental hazard · Phytohormones · Stress-responsive genes

11.1 Introduction

Abiotic stress can be increased by living things or environmental elements such as water (excess, i.e., flooding or scarcity, i.e., drought), temperature, ultraviolet (UV) radiation, presence of excess salt, and/or heavy metals (HMs) (Hasanuzzaman et al. 2012; Asgher et al. 2017). Osmotic stress, oxidative stress, ionic imbalance, and cell metabolic dyshomeostasis are all related to and appear as a consequence of abiotic stresses in plants, and impact on plant growth and productivity (Hoque et al. 2021). The agricultural productivity and growth are globally restricted by several environmental constraints and intensive plant protection interventions in agriculture like the use of pesticides (Kalefetoğlu and Ekmekci 2005; Arnao and Hernández-Ruiz 2014; Hasan et al. 2015; Ni et al. 2018; Sami et al. 2020). One of the most hazardous and dangerous abiotic stress conditions encountered by plants is exposure to HMs. Plants however, require specific quantities of certain metals (HMs), such as iron (Fe), manganese (Mn), zinc (Zn), copper (Cu), molybdenum (Mo), nickel (Ni), and cobalt (Co), but these HMs become harmful to them, when present in excess (Hossain et al. 2012).

Certain HMs are extremely damaging to plants, such as lead (Pb), cadmium (Cd), mercury (Hg), and arsenic (As). In the cell organelles like chloroplasts, mitochondria, and peroxisomes, the majority of HMs induce the enhanced production and accumulation of reactive oxygen species (ROS). This condition may lead to oxidative stress in plants and is considered a key consequence of HM toxicity (Chandrakar et al. 2016; Chen and Su 2018). A significant issue for sustainable agriculture is that HMs seriously impair plant growth and productivity. By causing membrane instability, oxidative stress, osmotic stress, ionic imbalance, cellular toxicity, and metabolic homeostasis (Ghori et al. 2019), HMs' toxicity adversely impacts the physiology and biochemistry of plants (Parmar et al. 2013). The hormone melatonin (an animal hormone that works as a plant growth regulator too) is involved in several physiological processes, including growth, development, and stress reactions (Zhao et al. 2019a, b). Melatonin (*N*-acetyl-5-methoxytryptamine) is an important molecule, which is anticipated to have a huge potential in abiotic stress management in plants. Its potential role in aiding plants in overcoming various abiotic stresses in plants, including HM stress, has been under investigation by several researchers in recent years. Keeping this in mind, this chapter was planned to discuss the harmful effects of HMs toxicity in plants, its effects at physiological and biochemical levels, and the role of melatonin in overcoming HMs toxicity in plants. An attempt was also made to review the effect of exogenous melatonin application to save crop plants from HM stress.

11.1.1 Effect of HM Stress in Plants (Hg, Cu, Ar, Zn, Cr, Pb, Cd)

Heavy metal stress can have a wide range of detrimental effects on plants. These effects may vary depending on the type of HM, its concentration, and the duration of exposure. Several natural as well as anthropogenic activities have led to higher accumulation of HMs in soil-water in agroecosystems, thereby increasing encounters of HM toxicity to crop plants. High levels of HMs in the soil can inhibit plant growth, resulting in reduced height, smaller leaves, and overall stunted development (Singh et al. 2020; Asati et al. 2016). This occurs because HMs may interfere with essential nutrient uptake and disrupt several metabolic processes (Liu and Zhang 2012; Dubey et al. 2018). Chlorosis is a common symptom of HM stress, where leaves turn yellow due to a decrease in chlorophyll production. This can impair photosynthesis and reduce the plant's ability to produce energy (Papazoglou et al. 2005). Further, HMs may damage chloroplasts, interfere with pigment synthesis, and disrupt photosynthetic electron transport chains (Rai et al. 1991; Souri et al. 2019a, b). As a result, photosynthesis is often impaired, leading to reduced plant productivity.

The second major reason for HMs' toxicity is the oxidative burst caused by the enhanced production and accumulation of reactive oxygen species (ROS) within plant cells (Berni et al. 2019). The elevated levels of ROS can cause oxidative damage to macromolecules like proteins, lipids, and DNA, leading to associated cellular dysfunction and tissue damage (Juan et al. 2021). Plants respond to HM-induced oxidative stress by activating their inherent antioxidant defense mechanisms. Moreover, HMs may compete with the uptake of some essential nutrients by the plant roots (Shiyu et al. 2020). This competition can further result in nutrient imbalances and deficiencies, affecting various physiological processes in plants under HM stress. Further, higher concentrations of HMs can accumulate in plant cells, leading to physical damage to cell membranes and organelles (Kosakivska et al. 2021). This can also disrupt cellular integrity and interface with several physiological processes. Some HMs can mimic essential nutrients and can therefore be absorbed by plant roots, leading to an altered mineral composition in plant tissues (Bhatla et al. 2018). This can negatively affect plant health and nutrient status. Plants exposed to HMs may exhibit changes in their root morphology, such as reduced root length, altered root branching, and the formation of root nodules or thickened roots as a response to metal toxicity (Ivanov et al. 2003).

Heavy metals can also affect the water balance of plants by disrupting water uptake and transport mechanisms (Rucińska-Sobkowiak 2016). This may consequently lead to water stress and reduced turgor pressure in cells. Plants respond to

HMs by altering the expression of several genes and associated biochemical pathways. They activated stress-responsive genes involved in detoxification, chelation, and defense mechanisms in *Brassica juncea* (Lai and Luo 2019). In severe cases of HM stress, the leaves of plants may develop necrotic lesions or they may die off, further reducing the plant's ability to photosynthesize (Sun and Wu 1998). Besides reducing photosynthesis, HMs can also affect the reproductive health of the plants. Heavy metal stress can lead to reduced flower production, seed set in *Potentilla anserina* (Saikkonen et al. 1998), fenugreek (Bhandari and Singh 2022), and overall reproductive success in plants. This can have long-term ecological implications for plant populations. Some HMs can accumulate in plant tissues, posing a risk to herbivores and other organisms that feed on these plants (Souri et al. 2019a, b), which may have cascading effects on ecosystems. Heavy metal accumulation in agroecosystems can result in reduced plant growth, and it can also affect soil microbial communities (Kushwaha et al. 2018). It is important to note that different plant species and even different genotypes within a species exhibit varying degrees of tolerance to HMs, indicating that the HM tolerance is genotype-dependent. Some plants have developed mechanisms to tolerate or even hyperaccumulate certain HMs as one of their survival strategies. Several researchers have studied these mechanisms to develop strategies for phytoremediation and the mitigation of HM pollution in contaminated environments to improve agricultural productivity.

11.1.2 Regulation of Stress Adaptability in Plants During HM Stress

Plants have developed various mechanisms to regulate their adaptability toward HM stress. Heavy metals are toxic to plants because they can disrupt some essential cellular processes and cause oxidative damage. To survive and adapt to HM stress, plants employ a combination of physiological, biochemical, and molecular mechanisms. Plants regulate the uptake of HMs through their roots (Sytar et al. 2013; Ghori et al. 2019). They can restrict the entry of toxic metals into the plant by limiting their uptake. This is achieved through selective ion channels and transporters (Zhao et al. 2022). Metal transporters, such as ATP-binding cassette (ABC) transporters and metallothioneins, play crucial roles in sequestering and transporting HMs within plant cells. Plants produce metal-binding molecules like phytochelatins (PCs) and metallothioneins (MTs) to chelate and detoxify HMs (Gupta et al. 2013; Jogawat et al. 2021). These molecules bind to HMs and sequester them in vacuoles, preventing their interaction with the essential cellular components (Chaudhary et al. 2018; Gupta et al. 2013). Heavy metals can induce the production of ROS in plant cells, leading to oxidative stress. Plants respond by activating antioxidant enzymes such as superoxide dismutase (SOD), catalase (CAT), and peroxidases to scavenge ROS and protect cellular components from damage (Ighodaro and Akinloye 2018). Some plants can modify their cell walls too, to reduce HM uptake. They may increase the production of pectin and lignin, which can bind with HMs and limit their movement into the plant's vascular system, thereby reducing their toxicity. To maintain cellular turgor and water uptake during HM stress, plants may accumulate compatible solutes such as proline and glycine betaine (Ali et al. 2020a, b). These compounds help to regulate the cellular osmotic potential and thus prevent water loss too. Plants may also activate some stress-responsive genes and associated signaling pathways to repair damaged DNA and mitigate the effects of HM stress (Table 11.1). This includes the upregulation of genes involved in DNA repair, stress responses, and HM/ROS detoxification (Dutta et al. 2018; Ghori et al. 2019).

Some plants can compete with HMs for obtaining their essential nutrients, like iron and calcium. By prioritizing the uptake of these nutrients, plants can reduce the absorption of toxic HMs (Rehman et al. 2021). The mycorrhizal fungi and certain bacteria, which can form symbiotic associations with plants, can help them to develop in HM tolerance. These microorganisms can enhance nutrient uptake and sequester HMs in the soil, reducing their availability to plants (Bano and Ashfaq 2013; Gamalero et al. 2009), thereby reducing HM toxicity.

Some plants are used for phytoremediation purposes for environmental clean-up, where they are intentionally grown in HM-contaminated soils to absorb and accumulate HMs. Once harvested, these plants can be disposed of safely, removing the contaminants from the environment (Leguizamo et al. 2017). Over time, plants may undergo some genetic adaptations to develop greater tolerance to specific HMs. These adaptations may involve mutations in key genes related to metal transport and detoxification. The ability of a plant to adapt to HM stress depends on its species, genetic makeup, and the specific type of HM involved (Singh et al. 2016; Emamverdian et al. 2015). Understanding these regulatory mechanisms is therefore essential for developing strategies to enhance the phytoremediation potential of such plants and mitigate the impact of HM pollution in the environment.

11.1.3 Regulation of Growth Under HM Stress

Heavy metal stress can significantly affect plant growth and development. To survive and adapt to such stresses, plants generally employ various regulatory mechanisms that influence their growth patterns (Sytar et al. 2019). Heavy metals are primarily absorbed through the roots and are transported to their aerial tissues via their vascular system. As a defense mechanism, plants often inhibit root growth when exposed to high levels of HMs. This reduces the surface area available for metal uptake (Dalvi and Bhalerao 2013; Shahid et al. 2017). Similarly, HM stress can lead to the inhibition of shoot growth, particularly in the above-ground parts of the plant. This can result from disruptions in nutrient uptake and transport processes (Gill et al. 2011). Some plants may exhibit changes in stem elongation and shoot structure to reduce metal uptake. This can include shorter internodes, altered leaf morphology, or the development of thickened cuticles to limit metal entry (Fitter and Hay 2012). Heavy metals can interfere with the uptake and utilization of

Table 11.1 Ameliorative effect of exogenous melatonin on plant physiology and their genes in the presence of heavy metal (HM) stress

S. No.	HM(s)	Melatonin conc.	Plant(s)	Effect on gene(s)	Effect on plant growth and physiology	Reference
1	As	100 μM	*Camellia sinensis*	Increased expression of anthocyanin biosynthesis genes *CsCHS* and *CsANS*	Increased anthocyanin content	Li et al. (2021)
2	V, Cr, Cd, Ni	–	*Capsicum annuum*	Increased expression of chlorophyll, proline biosynthesis genes, along with the antioxidant enzymes biosynthesis genes	Decreased ROS accumulation, increased activity of antioxidant enzymes	Altaf et al. (2023)
3	Cd	10, 25, 50, 100, and 200 μmol/L	Radish, tobacco	Increased expression of *Metallothionein 1* (*RsMT1*) gene in radish and transgenic tobacco	Increased activity of antioxidant enzymes	Xu et al. (2020)
4	Pb	10, 25, 50, 100, and 150 μM	Radish	Increased expression of *RsAPX2*, *RsPOD52*, and *RsGST*, which in turn decreased the methylation of HM transporter genes *RsABCF5*, *RsYSL7*, and *RsHMT*	Increased activity of antioxidant enzymes APX, GR, and fresh weight and decreased expression of Pb transporter genes	Tang et al. (2021)
5	Cd, Zn, and Pb	–	*Exophiala pisciphila* and *Arabidopsis*	Increased expression of melatonin biosynthesis genes *E. pisciphila* tryptophan decarboxylase (*EpTDC1*) and serotonin *N*-acetyltransferase (*EpSNAT1*) in transgenic *Arabidopsis*	Decreased in MDA and free radical content, increased SOD activity, and reduced HM accumulation	Yu et al. (2021)
6	Cd	10, 50, 200 μM	Alfalfa and *Arabidopsis*	Increased expression of the alfalfa *SNAT* gene in transgenic *Arabidopsis* plants	Upregulation of ABC transporter and *PCR2* gene in Alfalfa and *PDR8* and *HMA4* in *Arabidopsis*	Gu et al. (2017)

(continued)

Table 11.1 (continued)

S. No.	HM(s)	Melatonin conc.	Plant(s)	Effect on gene(s)	Effect on plant growth and physiology	Reference
7	V	0.1 μM	Watermelon	Increased expression of O-*methyltransferase* and *methionine S-methyltransferase*	Increased chlorophyll content, plant growth, and antioxidant enzymes gene expression	Nawaz et al. (2018)
8	Co	50 and 100 μM	*Brassica napus*	Increased expression of stress-responsive genes *HSP-90*, *methyl transfera*se (*MT*)	Decrease in the overproduction of ROS and MDA content, increased the activity of ROS-scavenging antioxidant enzymes, *PAL and PPO*	Ali et al. (2023)
9	Cd	50, 100 μM	Oat	Increased expression of *LOX*, *POX*, *Asmap1*, *NAC*, and *WRKY1*	Increased plant growth, chlorophyll and proline content, and activity of antioxidant enzymes CAT, SOD, and POD	Wang et al. (2022)
10	Cd, Pb	–	Soyabean	Decrease expression of *NCED3* and increased expression of *CYP707A1* and *CYP707A2*, *MTF-1* and *WARKY27*, as well as *gmNR* and *gmGSNOR* genes	Increased the uptake of Ca^{2+} and K^{+}, and plant growth	Imran et al. (2022)
11	Ni	–	Tomato	Increase expression of melatonin biosynthesis-related genes (*TDC, T5S, SNAT, ASMT*), and chlorophyll synthesis genes *POR, CAO,* and *CHL G*; increased expression of antioxidant enzymes encoding genes *SOD, CAT, APX, GR, GST, MDHAR,* and *DHAR*	Increased chlorophyll, phenols, flavonoids, and anthocyanin content with a decrease in ROS content	Jahan et al. (2020)

(continued)

Table 11.1 (continued)

S. No.	HM(s)	Melatonin conc.	Plant(s)	Effect on gene(s)	Effect on plant growth and physiology	Reference
12	Cd	200 μM	*Platycladus orientalis*	Increased expression of *ethanol dehydrogenase* (*ADH*) gene and antioxidant enzyme-related genes *POD, GST*, and *APX*	Increased chlorophyll content, soluble protein, soluble sugar, and proline, reduced the content of glutathione, and increased SOD and POD activities	Ou et al. (2023)
13	Cd	15 μM	Cotton	Decreased expression of Cd transporter genes (*LOC107894197, LOC107955631*, and *LOC107899273*)	Decrease in MDA and H_2O_2 content and increase in the activities of SOD, POD, CAT, and APX enzymes	Khan et al. (2023)
14	Cd	25, 50, 100, 250, 500 μM	Chinese cabbage	Decreased expression of Cd transporter gene *IRT1* and nitric oxide synthesis genes *NR, NOA1*	Increase in plant growth, chlorophyll, and carotenoid content, decrease in Cd and Fe content in seedlings	Wang et al. (2021)
15	Cd	25, 50, 100, 250 μM	*Nicotiana tabacum*	Increased expression of Cd transporter genes *IRT1, Nramp1, HMA2*, and *HMA4* and decreased expression of *IRT2* and *HMA3* genes	Increased plant growth, chlorophyll content, and antioxidant enzyme activity, while decreased in Cd accumulation and ROS production	Wang et al. (2019)

(continued)

Table 11.1 (continued)

S. No.	HM(s)	Melatonin conc.	Plant(s)	Effect on gene(s)	Effect on plant growth and physiology	Reference
16	Cd	100 μM	*Malus baccata* and *Malus micromalus*	Decreased expression of the genes involved in Cd uptake and translocation, i.e., *NRAMP3, HMA4, PCR2, HA7, NRAMP1*, and increased expression of Cd detoxification genes, i.e., *ABCC1, MHX, NAS1,* and *MT2*	Decreased Cd uptake, accumulation, translocation, and ROS production; increased growth, photosynthetic pigment, and antioxidant enzyme activities	He et al. (2020)
17	Cd	50 μM	Wheat	Decreased expression of metal transporter genes *TaNramp1, TaNramp5, TaHMA2, TaHMA3,* and *TaLCT1* and increased expression of *GSH, GR, GST, APX, MDHAR,* and the genes involved in the *ASA-GSH* cycle	Plant growth, chlorophyll, ascorbic acid, and glutathione content increased, while MDA content, Cd accumulation, and metal uptake decreased	Li et al. (2022a)
18	Ar	–	Rice	Increased expression of anthocyanin biosynthesis genes *PAL, CHS, CHI, F3H, DFR,* and *ANS* and reduced expression of transporter genes *PT-2* and *PT-8*	Increased in salicylic and abscisic acid content, and decreased in As uptake and accumulation	Jan et al. (2023)
19	Pb	50 mM	*Trigonella foenum graecum* L.	Increased expression of *SOD* and *CAT* genes	Increased plant growth, chlorophyll content, protein content, antioxidant enzymes SOD, CAT activity, and decreased MDA content, ROS synthesis, electrolyte leakage, and lipoxygenase activity	Xalxo and Keshavkant (2019)

Here, (–) indicates the unknown concentration of melatonin

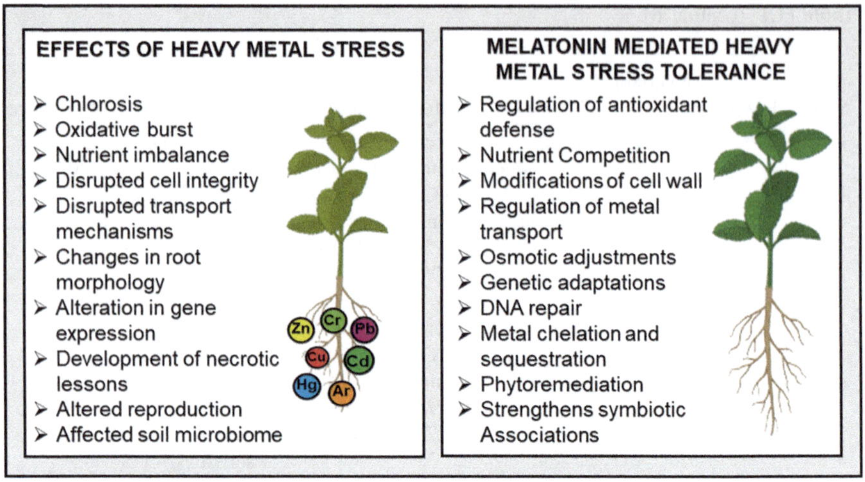

Fig. 11.1 Various biochemical/physiological changes in plants and their rhizosphere in heavy metal (HM) stress (left box) and melatonin-mediated HM stress tolerance in plants (right box)

essential nutrients, such as calcium, magnesium, and iron. This nutrient imbalance can lead to stunted growth and nutrient deficiencies in plants (Fig. 11.1).

Heavy metals can disrupt cell division and expansion processes in plant tissues, leading to reduced growth (Goyal et al. 2020). This inhibition occurs due to the toxic effects of metals on the cell cycle. Heavy metal stress can negatively impact photosynthesis, reducing the plant's ability to produce energy and grow. This occurs because HMs can damage chloroplasts, inhibit pigment synthesis, and disrupt electron transport chains (Rai et al. 2016; Souri et al. 2019a, b). To counteract the oxidative stress induced by HMs, plants activate their antioxidant defense systems. While these responses are essential for survival, they can divert energy away from the regular growth processes (Sharma et al. 2012). Phytohormones, such as auxins, cytokinins, and gibberellins, play critical roles in plant growth and development. Heavy metal stress can alter the balance of these hormones, leading to growth inhibition (Sytar et al. 2019). Plants may modify their cell walls in response to HM stress. This can include thickening the cell wall and depositing substances like lignin and pectin, which can act as barriers to metal uptake but may also reduce cell expansion. Heavy metal stress often leads to metabolic reprogramming in plants (Pandey et al. 2022). They may prioritize the production of stress-related metabolites and secondary compounds over growth-related processes. Some plants form symbiotic relationships with mycorrhizal fungi, which can help them acquire nutrients and enhance tolerance to HMs (Bano and Ashfaq 2013; Roychoudhury 2023). These relationships can influence plant growth under stress conditions. Over time, plants may undergo genetic changes or adaptations to develop greater tolerance to specific HMs. These adaptations can involve alterations in the expression of genes related to growth and stress responses (Singh et al. 2016; Dutta et al. 2018; Ghori et al. 2019). It is important to note that the response to HM stress can vary widely among plant species and even within different genotypes of the same species, as does the HM stress management. Additionally, the degree of stress, duration of exposure, and specific HM involved all play crucial roles in determining how plants regulate their growth in response to HM stress (Morkunas et al. 2018). For the appropriate management of HM stress in plants, the knowledge of the level of a plant's HM exposure is also essential.

11.1.4 Regulation of Flowering

Flowering in plants is a complex process regulated by various internal and external factors, including photoperiod, temperature, hormonal signals, and stressors such as HMs (Cho et al. 2017). Heavy metal stress can have a significant impact on the flowering process by disrupting the normal physiological and molecular mechanisms that control it. Heavy metal stress can alter the balance of plant hormones, particularly auxins, gibberellins, and abscisic acid (Kumar et al. 2022; Sharma et al. 2022). These hormonal changes can affect the flowering process. For instance, high levels of abscisic acid may inhibit flowering, while disruptions in auxin and gibberellin levels can impact the development of floral buds. Heavy metals like cadmium, lead, and mercury can compete with essential nutrients like calcium, magnesium, and iron. Nutrient imbalances can disrupt various metabolic processes in plants, affecting their overall health and, consequently, flowering (Naeem et al. 2019; Kalaivanan and Ganeshamurthy 2016). Heavy metal stress often leads to the production of ROS, such as superoxide radicals and hydrogen peroxide, in plant cells. Excessive ROS can cause oxidative damage to cellular components, including DNA, proteins, and lipids, which can interfere with the flowering process (Maleki et al. 2017; Das and Roychoudhury 2014). Heavy metals can impair photosynthesis by damaging chloroplasts and inhibiting chlorophyll synthesis. Reduced photosynthetic capacity can limit the plant's energy resources, which may be redirected away from flowering to support stress response mechanisms (Smirnoff 1993; Gill et al. 2011). Heavy metal stress can lead to changes in DNA methylation patterns and histone modifications in plants (Fig. 11.1). Epigenetic changes can alter the expression of genes related to flowering, influencing the timing and progression of flowering events. Some HMs can disrupt a plant's ability to sense photoperiod, which is crucial for flowering regulation in many species (Akhter et al. 2021). This disruption may lead to delayed or abnormal flowering responses under different day-length conditions. Plants have stress response pathways that are activated when exposed to HMs. These pathways often involve the expression of stress-responsive genes, which may compete with or inhibit the expression of genes involved in flowering (Takeno 2016). In some cases, HM-contaminated environments may deter pollinators, which are essential for the successful pollination and fertilization of flowers. This can indirectly affect plant reproduction (Xun et al. 2018). It is important to note that the specific effects of HM stress on flowering can vary depending on the plant species, the type and concentration of HMs, and the duration of exposure. Some plants may exhibit delayed flowering, while others may fail to flower altogether under severe stress conditions. To mitigate the negative impact of HM stress on flowering, it is essential to implement phytoremediation techniques or use other methods to reduce HM contamination in the soil and the surrounding environment. Additionally, genetic engineering approaches may be explored to develop plants with enhanced HM tolerance while maintaining their ability to flower and reproduce.

11.1.5 Regulation of Secondary Metabolites

The regulation of secondary metabolites in plants during HM stress is a complex process influenced by various factors, including the type and concentration of HM, the plant species, and its genetic makeup. Secondary metabolites are organic compounds that are not directly involved in the plant's growth and development but play important roles in defense against herbivores, pathogens, and environmental stressors (Aguirre-Becerra et al. 2021; Khare et al. 2020). Heavy metal stress can activate stress-responsive pathways in plants. These pathways often lead to the upregulation of genes responsible for the synthesis of secondary metabolites, especially those with antioxidant and detoxifying properties. Some secondary metabolites, such as phenolic compounds (e.g., flavonoids and polyphenols) and terpenoids (e.g., essential oils), may accumulate in response to HM stress (Chen et al. 2022; Goncharuk and Zagoskina 2023; Yadav et al. 2021). These compounds can act as antioxidants and free radical scavengers to mitigate oxidative stress induced by HMs. Heavy metal stress can influence the biosynthesis of terpenoids, which are involved in various plant functions, including defense against herbivores and pathogens. Changes in the composition and concentration of terpenoids can occur as a response to HMs (Saddique et al. 2018; Mahajan et al. 2020). Some secondary metabolites, such as organic acids and phytochelatins, have chelating properties and can bind to HMs, reducing their toxicity to plant cells. These metabolites are synthesized or upregulated in response to HM stress (Anjitha et al. 2021; Vega et al. 2022). Plants may produce phytoalexins, which are antimicrobial secondary metabolites, in response to HM stress. These compounds can help protect plants from infections that may become more prevalent under stress conditions (Tiku 2020; Akhi et al. 2021). Flavonoids and phenolic compounds are secondary metabolites known for their antioxidant properties. They can be induced in response to HM stress to scavenge reactive oxygen species (ROS) and reduce oxidative damage. Some glycosides and alkaloids can accumulate in plant tissues under HM stress. These compounds can deter herbivores and serve as a chemical defense against potential threats (Koul et al. 2021; Matsuura and Fett-Neto 2015). VOCs (Volatile organic compounds), such as terpenes, may change in composition and emission rates in response to HM stress. These compounds can influence plant-herbivore interactions and mediate indirect defense mechanisms (Lin et al. 2022; Jamieson et al. 2017). Heavy metal stress can lead to changes in gene expression, including the activation of transcription factors and regulatory genes associated with secondary metabolite biosynthesis (Jan et al. 2021). Different plant species and even different genotypes within the same species may exhibit unique responses to HM stress, resulting in variations in secondary metabolite production (Jia et al. 2020). Environmental conditions, such as light intensity, temperature, and nutrient availability, can also influence the regulation of secondary metabolites in plants under HM stress. Understanding the regulation of secondary metabolites in plants during HM stress is important not only for elucidating the plant's defense mechanisms but also for potential applications in phytoremediation and the development of stress-tolerant crop varieties (Mashabela et al. 2023). It is a complex area of research that continues to yield valuable insights into plant adaptation to environmental challenges.

11.1.6 Regulation of PGR Synthesis

Plant growth regulators (PGRs), also known as plant hormones, play a crucial role in regulating various physiological processes in plants, including growth, development, and stress responses. Heavy metal stress can impact the synthesis and regulation of PGRs in plants in several ways. The exact response may vary depending on the type and concentration of HMs, the plant species, and the duration of exposure (Gangwar et al. 2014). Heavy metals can disrupt auxin synthesis and distribution. In some cases, HM stress can lead to increased auxin levels during mild Cd stress in Braley (Demecsová et al. 2020), while in others, like arsenite, it may cause a reduction in auxin synthesis or transport (Mathur et al. 2022; Krishnamurthy and Rathinasabapathi 2013). These disruptions can affect root growth, development, and overall plant architecture. Heavy metal stress can interfere with cytokinin levels. High concentrations of HMs can lead to a decrease in cytokinin synthesis and disrupt cytokinin signaling pathways, affecting processes like cell division and shoot development (Jalmi et al. 2018). Heavy metals can influence gibberellin metabolism. Some studies have shown that HM stress can lead to a reduction in gibberellin levels, resulting in stunted growth and reduced stem elongation (Bücker-Neto et al. 2017; Chen et al. 2021). ABA plays a crucial role in plant stress responses, and its levels often increase in response to HM stress. Elevated ABA levels can lead to stomatal closure and reduced water loss, which is a common response to HM stress (Rucińska-Sobkowiak 2016). However, excessive ABA can negatively affect growth and development. Heavy metal stress can lead to an increase in ethylene production. Ethylene is involved in stress responses and can lead to leaf senescence and premature aging when produced at high levels (Thao et al. 2015). These PGRs are associated with plant defense responses. Heavy metal stress can induce the production of JA and SA, which are involved in various stress-related processes, including the activation of defense genes and the synthesis of important secondary metabolites (Isah 2019; Yu et al. 2018; Raza et al. 2021). PGRs often interact and influence one another in response to stress. The balance of different PGRs is crucial for plants to adapt to HM stress. Heavy metals can disrupt these interactions, leading to imbalances in PGR levels. Heavy metal stress can lead to epigenetic changes that affect the expression of genes involved in PGR synthesis and signaling pathways. These changes can have long-lasting effects on PGR regulation (Garcia et al. 2019; Zaid et al. 2020). Different plant species and even individual plant genotypes may exhibit varying responses to HM stress, including the regulation of PGR synthesis. Some plants may be more tolerant to HMs and maintain better PGR homeostasis under stress. To mitigate the negative effects of HM stress on PGR synthesis and overall plant health, various strategies can be employed (Singh et al. 2020; Nedjimi 2021). These may include soil remediation to reduce HM contamination, genetic engineering to enhance HM, and the use of exogenous PGR applications to restore normal growth and development in stressed plants (Zahid et al. 2023). Additionally, optimizing growth conditions and providing essential nutrients can help alleviate some of the stress-induced PGR disruptions.

11.2 Melatonin-Mediated Amelioration of HM Stress

Melatonin is an important molecule synthesized in plants, which has the capability to ameliorate a variety of abiotic stresses in plants, including HM stress (Fig. 11.1), by a variety of means, viz., antioxidant defense, metal chelation, gene regulation, and stress signaling.

11.2.1 Biosynthesis of Melatonin

The biosynthesis of melatonin in plants involves a series of enzymatic reactions starting with the amino acid tryptophan as its precursor. Plants may obtain tryptophan from the soil through their roots. Tryptophan is decarboxylated by the enzyme tryptophan decarboxylase (TDC) to produce tryptamine (Arnao and Hernández-Ruiz 2015; Mannino et al. 2021). Tryptamine is then hydroxylated by the enzyme tryptamine 5-hydroxylase (T5H) to form serotonin (5-hydroxytryptamine or 5-HT). Serotonin is then acetylated by the enzyme serotonin N-acetyltransferase (SNAT) to produce N-acetylserotonin (NAS) (Negri et al. 2021; Tan et al. 2016), which is further acetylated by the same enzyme SNAT (Fig. 11.2) to form melatonin in plants.

In plants, melatonin serves various roles, including regulation of growth, stress responses, and the plant's defense against environmental stresses (Debnath et al. 2019; Tan et al. 2014). This biosynthesis pathway in plants is similar to the one found in animals, with some differences in the enzymes involved. Melatonin in plants is believed to play a role in protecting against oxidative stress and coordinating various physiological processes in response to environmental conditions, such as day–night cycles and exposure to stressors like pathogens or abiotic stress (e.g., drought, high salinity) (Maurya et al. 2024; Sakshi et al. 2024; Khan et al. 2020; Singh et al. 2023). It is important to note that only limited information is available

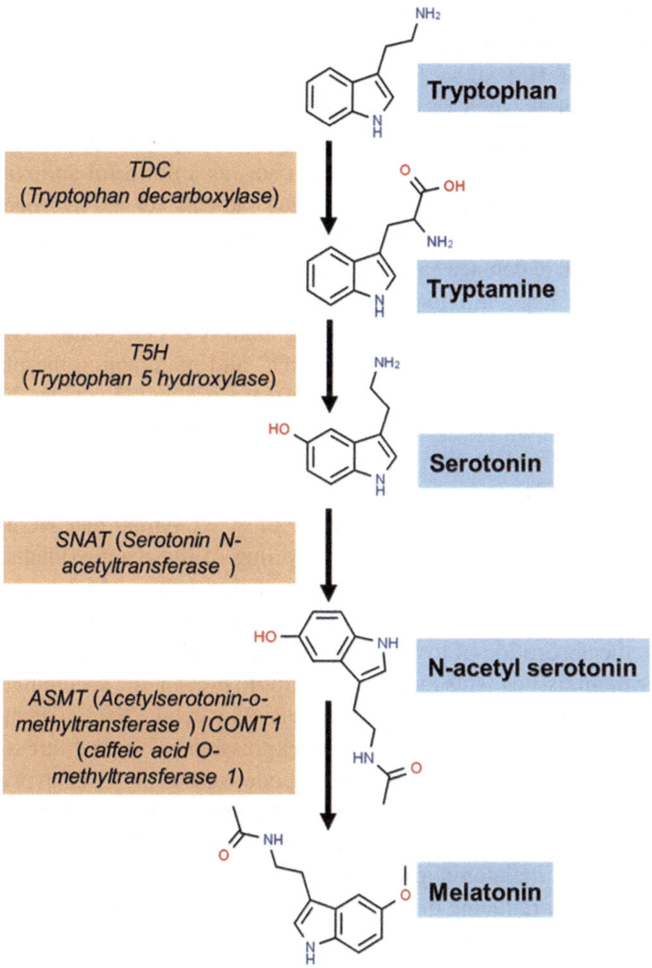

Fig. 11.2 Major steps involved in the biosynthesis of melatonin in plants from its precursor tryptophan. Pink boxes indicate the enzymes involved in each step, whereas the blue boxes represent the precursor (tryptophan), intermediate molecules synthesized, and the final product (melatonin)

on the functions of melatonin in plants and its regulatory roles therein. Since this is an area of ongoing research, more details about its roles are still being worked out.

11.2.2 Mechanism of Antioxidant Defense

The ameliorative role of melatonin in HM stress, specifically in the context of antioxidant defense mechanisms, is a well-documented and important area of research in plant physiology and stress physiology (Farouk and Al-Amri 2019). Melatonin, a hormone (both an animal hormone and a plant growth regulator) and antioxidant molecule found in various plant species, plays a significant role in mitigating the adverse effects of HM stress by regulating antioxidant systems (Rahman et al. 2021). Heavy metals, such as cadmium (Cd), lead (Pb), and mercury (Hg), can induce the overproduction of ROS in plant cells, leading to oxidative stress (Shahid et al. 2014; Ishtiyaq et al. 2018). Melatonin acts as a powerful antioxidant, effectively scavenging ROS, including superoxide radicals (O_2•-), hydrogen peroxide (H_2O_2), and hydroxyl radicals (•OH). (Kohli et al. 2017). This scavenging activity reduces the oxidative damage caused by HMs. Melatonin can upregulate the activity and expression of key antioxidant enzymes, such as superoxide dismutase (SOD), catalase (CAT), peroxidase (POD), and glutathione peroxidase (GPX). These enzymes are responsible for neutralizing ROS and maintaining the plant's redox balance (Maurya et al. 2024; Sakshi et al. 2024; Bose and Howlader 2020; Hoque et al. 2021). Melatonin's effect on these enzymes improves the plant's ability to cope with HM stress. Melatonin also influences non-enzymatic antioxidants, including glutathione (GSH) and ascorbic acid (vitamin C), which are essential for ROS detoxification and redox balance (Altaf et al. 2022; Arnao and Hernández-Ruiz 2021). By enhancing the synthesis and recycling of these antioxidants, melatonin helps protect plant cells from HM-induced oxidative damage. Melatonin can inhibit the uptake and transport of HMs in plants (Hodzic et al. 2021; Jahan et al. 2020). It can reduce the absorption of toxic metals from the soil and limit their translocation to sensitive plant tissues. This mechanism helps to prevent the accumulation of HMs in plant organs, reducing their toxic effects (Ou et al. 2023; Gu et al. 2021). Melatonin has been shown to induce the expression of several stress-responsive genes in plants (Table 11.1). These genes encode various proteins involved in stress tolerance, including antioxidant enzymes. Melatonin-mediated gene regulation enhances the plant's ability to withstand HM stress. A study by Wang et al. (2013) demonstrated that melatonin pretreatment alleviated cadmium (Cd) toxicity in cucumber plants by reducing oxidative stress and enhancing the activity of antioxidant enzymes. The research conducted by Zhang et al. (2017) revealed that melatonin application alleviated the negative effects of lead (Pb) stress in wheat seedlings by reducing ROS levels and enhancing the activities of antioxidant enzymes, including SOD, CAT, and POD. In conclusion, melatonin plays a crucial role in ameliorating HM stress in plants by regulating antioxidant defense mechanisms. Its ability to scavenge ROS, enhance antioxidant enzyme activity, modulate non-enzymatic antioxidants, and regulate gene expression makes it a promising tool for mitigating the toxic effects of HMs on plant growth and development.

11.2.3 Metal Chelation

The ameliorative role of melatonin in HM stress is not limited to its antioxidant properties but also includes its ability to chelate or bind HMs, reducing their toxicity to plants. Metal chelation is an essential mechanism through which melatonin helps

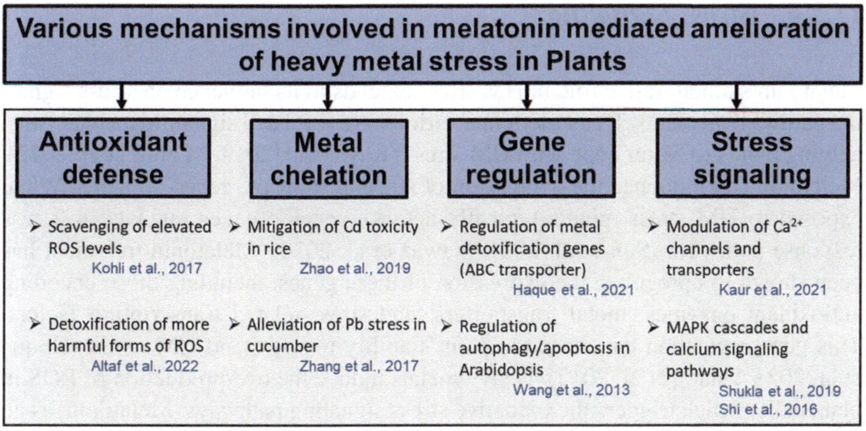

Fig. 11.3 Ameliorative role of melatonin in heavy metal (HM) stress in plants

alleviate HM stress (Fig. 11.3). Melatonin has the capability to form complexes or chelate with HMs like cadmium (Cd), lead (Pb), and copper (Cu) (Flora et al. 2008; Moustafa-Farag et al. 2020). These complexes are typically less toxic and more stable than free metal ions, making them less available for uptake by plant roots and less likely to cause cellular damage (Arnao and Hernández-Ruiz 2019). When melatonin chelates HMs in the soil, it can reduce the availability of free metal ions for plant uptake. This reduces the metal load entering plant tissues and, in turn, minimizes the toxic effects of HMs on plant growth and development (Ali et al. 2020a, b; Riaz et al. 2021; Kapoor et al. 2021). Melatonin can bind to HM ions, forming stable complexes that are less reactive and less likely to cause oxidative stress and damage to cellular components. This helps protect plant cells from HM-induced toxicity (Yang et al. 2023; Hasanein and Emamjomeh 2019). Melatonin may also influence the transport of HMs within the plant. By forming complexes with HMs, it can affect the translocation of these metals from roots to shoots, limiting their accumulation in sensitive plant organs. A study by Zhao et al. (2019a, b) investigated the role of melatonin in mitigating cadmium (Cd) toxicity in rice. The results showed that melatonin treatment led to a significant decrease in Cd accumulation in rice roots and shoots. Melatonin chelation reduced the Cd content in plant tissues, indicating its role in limiting Cd uptake and translocation. In a study by Zhang et al. (2017), melatonin was found to alleviate lead (Pb) stress in cucumber by forming complexes with Pb ions, which reduced their toxicity and limited their transport within the plant. The researchers noted that melatonin effectively chelated Pb, thereby reducing its impact on plant growth and development. In summary, melatonin's role in chelating HMs is a critical aspect of its ameliorative effects on HM stress in plants. By forming stable complexes with HMs, melatonin reduces their availability for uptake and limits their toxicity, ultimately protecting plant cells from the adverse effects of HM exposure.

11.2.4 Stress Signaling

Melatonin's ameliorative role in HM stress extends to its influence on stress signaling pathways in plants. It can modulate various stress-related signaling mechanisms, helping plants to better cope with HM stress (Kaya et al. 2019; Mathur et al. 2022). Melatonin can influence the expression of stress-responsive genes in plants. When exposed to HM stress, plants typically activate specific genes involved in stress response pathways (Sun et al. 2021; Nawaz et al. 2020). Melatonin treatment has been shown to upregulate the expression of these genes, including those encoding antioxidant enzymes, metal transporters, and stress-related transcription factors. This gene regulation enhances the plant's ability to withstand HM stress (Hoque et al. 2021; Zhang et al. 2015). Heavy metals induce the overproduction of ROS in plant cells, which triggers the oxidative stress signaling pathways. Melatonin, as an antioxidant, reduces ROS levels and can, therefore, attenuate the activation of these oxidative stress signaling cascades (Altaf et al. 2023). This, in turn, minimizes the detrimental effects of oxidative stress on plant cells. Heavy metal stress often leads to changes in plant hormone levels, particularly abscisic acid (ABA) and salicylic acid (SA). Melatonin can influence hormone signaling pathways, promoting a balance in hormone levels. For instance, it can reduce the overaccumulation of ABA, which is associated with stress-induced stomatal closure and growth inhibition (Hoque et al. 2021). This regulation of hormone signaling helps maintain normal physiological processes in plants. Calcium ions (Ca^{2+}) play a crucial role in stress signaling. Melatonin has been shown to influence calcium signaling in plants exposed to HM stress. It can modulate Ca^{2+} channels and transporters, helping to regulate the influx and efflux of calcium ions, which, in turn, impacts downstream stress signaling pathways (Kaur et al. 2021; Li et al. 2022b). In a study by Wang et al. (2013), melatonin treatment was found to enhance the occurrence of autophagy and apoptosis induced by oxidative stress in Arabidopsis seedlings under various stress conditions, including HM stress. Melatonin's role in regulating autophagy and apoptosis reflects its influence on stress signaling pathways. A study by Zhang et al. (2017) demonstrated that melatonin promoted seed germination under high salinity stress in cucumber. This effect was attributed to melatonin's modulation of stress-related signaling pathways, including antioxidant systems, ABA, and gibberellic acid (GA4) interaction, reflecting its ability to influence hormone signaling in response to stress. In conclusion, melatonin's ameliorative role in HM stress includes its ability to modulate stress signaling pathways, thereby enhancing the plant's adaptive responses to stress conditions. This regulation of gene expression, hormone signaling, and other stress-related processes contributes to improved plant tolerance to HM stress.

11.2.5 Regulation of Gene Expression

Melatonin's ameliorative role in HM stress further extends to the regulation of gene expression in plants exposed to HM toxicity. Melatonin can influence the expression of specific genes involved in stress responses, detoxification, and antioxidant defense (Romero et al. 2014). Melatonin treatment can upregulate the expression of stress-responsive genes in plants. These genes include those encoding antioxidant enzymes (e.g., superoxide dismutase, catalase, peroxidase), metal transporters, chelators (e.g., metallothioneins), and stress-related transcription factors (e.g., NAC, AP2/ERF, MYB). By enhancing the expression of these genes, melatonin primes the plant for a more effective response to HM stress (Ramegowda et al. 2020). Heavy metal stress induces oxidative stress by increasing the production of reactive oxygen species (ROS). Melatonin can stimulate the expression of genes encoding antioxidant enzymes, such as superoxide dismutase (SOD), catalase (CAT), and glutathione peroxidase (GPX). This results in higher enzyme activity, which helps plants combat ROS and oxidative damage (Tiwari et al. 2021). Some HMs, like cadmium (Cd) and lead (Pb), are detoxified in plants through metal chelation and sequestration (Fig. 11.3). Melatonin can influence the expression of metal detoxification genes, such as metallothioneins and transporters like ATP-binding cassette (ABC) transporters. These genes help in binding and sequestering HMs to reduce their toxicity (Hoque et al. 2021; Kaur et al. 2021). Heavy metal stress often perturbs the balance of plant hormones, particularly abscisic acid (ABA) and salicylic acid (SA). Melatonin can affect the expression of genes involved in hormone biosynthesis and signaling, helping to restore hormone homeostasis and maintain normal physiological processes. Melatonin can interact with various signaling pathways, including mitogen-activated protein kinase (MAPK) cascades and calcium signaling pathways, which are pivotal for stress responses (Shukla et al. 2019; Shi et al. 2016). Through these interactions, melatonin can impact gene expression by affecting transcription factors and downstream signaling components. In a study by Zhao et al. (2019a, b), melatonin treatment was shown to alleviate cadmium (Cd) toxicity in rice. This effect was attributed to melatonin's ability to upregulate the expression of stress-responsive genes, including those involved in metal detoxification and antioxidant defense. A study by Wei et al. (2018) demonstrated that melatonin increased the expression of genes involved in metal detoxification and the regulation of antioxidant enzymes in *Brassica napus* (rapeseed) exposed to HM stress. This gene expression regulation played a crucial role in reducing HM toxicity. Recently, to explore the possible molecular mechanisms of melatonin-mediated HM detoxification in cotton, a comparative transcriptomic and metabolomic analysis was conducted, which primarily indicated the possible involvement of an antioxidant defense involving both enzymatic and non-enzymatic antioxidants, accumulation of secondary metabolites, regulation of HM ion transportation (like activation of ABC transporters), and so on (Li et al. 2022c). In conclusion, this study provides new insights into the MT-mediated Cd stress response.

In summary, melatonin's ameliorative role in HM stress includes its ability to regulate gene expression, which enhances plant responses to HM toxicity. By influencing the expression of stress-responsive genes, antioxidant enzymes, metal detoxification genes, and hormone-related genes, melatonin helps plants better cope with HM stress and reduce the associated damage.

11.2.6 Conclusions and Future Prospects

Plants in general, and crop plants in particular, are susceptible to a range of stresses. Both natural and man-made activities like mining, fast urbanization, and industrialization are frequently linked to HM pollution, which disrupts the sustainability of the eco-environment and lowers worldwide plant life and productivity (Tan et al. 2007; Rubio et al. 19941994). The application of exogenous melatonin acts as a PGR that helps in reducing the toxic effects of HM stress by plant growth, photosynthesis, improvement of antioxidant defense and nutrient status, accumulation of secondary metabolites, and regulation of HM transport and their sequestration. However, preliminary information on mechanisms involved in melatonin-mediated amelioration of HM stress is available (Hoque et al. 2021; Li et al. 2022c), but a detailed analysis of different HM modes of melatonin treatment combinations is still required to be studied to reach any generalized or HM-specific mechanistic conclusion. Once the complete information is available regarding the mode of action of melatonin in melatonin-mediated alleviation of HM stress in different plant species, a suitable strategy to manage HM stress or phytoremediation of HMs may be planned and executed.

Acknowledgments The authors are thankful to the DAV University administration for providing seed money to work on melatonin and also for providing other important supporting infrastructure.

Conflict of Interest The authors declare that they do not have any conflict of competing interests.

References

Aguirre-Becerra H, Vazquez-Hernandez MC, Saenz de la OD, Alvarado-Mariana A, Guevara-Gonzalez RG, Garcia-Trejo JF, Feregrino-Perez AA (2021) Role of stress and defense in plant secondary metabolites production. In: Bioactive natural products for pharmaceutical applications. Springer, Cham, pp 151–195

Akhi MZ, Haque MM, Biswas MS (2021) Role of secondary metabolites to attenuate stress damages in plants. In: Antioxidants-benefits, sources, mechanisms of action. Intech Open, London

Akhter Z, Bi Z, Ali K, Sun C, Fiaz S, Haider FU, Bai J (2021) In response to abiotic stress, DNA methylation confers epigenetic changes in plants. Plants 10(6):1096

Ali S, Abbas Z, Rizwan M, Zaheer IE, Yavaş İ, Ünay A, Abdel-Daim MM, Bin-Jumah M, Hasanuzzaman M, Kalderis D (2020a) Application of floating aquatic plants in phytoremediation of heavy metals polluted water: a review. Sustainability 12(5):1927

Ali S, Abbas Z, Seleiman MF, Rizwan M, YavaŞ İ, Alhammad BA, Shami A, Hasanuzzaman M, Kalderis D (2020b) Glycine betaine accumulation, significance and interests for heavy metal tolerance in plants. Plants 9(7):896

Ali S, Gill RA, Ulhassan Z, Zhang N, Hussain S, Zhang K, Huang Q, Sagir M, Tahir MB, Gill MB, Mwamba TM (2023) Exogenously applied melatonin enhanced the tolerance of *Brassica napus* against cobalt toxicity by modulating antioxidant defense, osmotic adjustment, and expression of stress response genes. Ecotoxicol Environ Saf 252:114624

Altaf MA, Shahid R, Altaf MM, Kumar R, Naz S, Kumar A, Alam P, Tiwari RK, Lal MK, Ahmad P (2022) Melatonin: first-line soldier in tomato under abiotic stress current and future perspective. Plant Physiol Biochem 185:188–197

Altaf MA, Hao Y, Shu H, Mumtaz MA, Cheng S, Alyemeni MN, Ahmad P, Wang Z (2023) Melatonin enhanced the heavy metal-stress tolerance of pepper by mitigating the oxidative damage and reducing the heavy metal accumulation. J Hazard Mater 454:131468

Anjitha KS, Sameena PP, Puthur JT (2021) Functional aspects of plant secondary metabolites in metal stress tolerance and their importance in pharmacology. Plant Stress 2:100038

Arnao MB, Hernández-Ruiz J (2014) Melatonin: plant growth regulator and/or biostimulator during stress? Trends Plant Sci 19:789–797

Arnao MB, Hernández-Ruiz J (2015) Melatonin: synthesis from tryptophan and its role in higher plant. In: Amino acids in higher plants. CAB International, Wallingford, pp 390–435

Arnao BM, Hernández-Ruiz J (2019) Role of melatonin to enhance phytoremediation capacity. Appl Sci 9(24):5293

Arnao MB, Hernández-Ruiz J (2021) Regulatory role of melatonin in the redox network of plants and plant hormone relationship in stress. In: Hormones and plant response. Springer, Cham, pp 235–272

Asati A, Pichhode M, Nikhil K (2016) Effect of heavy metals on plants: an overview. Int J Appl Innov Eng Manag 5(3):56–66

Asgher M, Per TS, Anjum S, Khan MIR, Masood A, Verma S, Khan NA (2017) Contribution of glutathione in heavy metal stress tolerance in plants. In: Reactive oxygen species and antioxidant systems in plants: role and regulation under abiotic stress. Springer, Singapore, pp 297–313

Bano SA, Ashfaq D (2013) Role of mycorrhiza to reduce heavy metal stress. Nat Sci 5:16

Berni R, Luyckx M, Xu X, Legay S, Sergeant K, Hausman JF, Lutts S, Cai G, Guerriero G (2019) Reactive oxygen species and heavy metal stress in plants: impact on the cell wall and secondary metabolism. Environ Exp Bot 161:98–106

Bhandari R, Singh B (2022) Effect of heavy metals treatments on fenugreek plant. J Hazard Mater Adv 7:100114

Bhatla SC, Lal AM, Kathpalia R, Bhatla SC (2018) Plant mineral nutrition. In: Plant physiology, development and metabolism. Springer, Singapore, pp 37–81

Bose SK, Howlader P (2020) Melatonin plays multifunctional role in horticultural crops against environmental stresses: a review. Environ Exp Bot 176:104063

Bücker-Neto L, Paiva ALS, Machado RD, Arenhart RA, Margis-Pinheiro M (2017) Interactions between plant hormones and heavy metals responses. Genet Mol Biol 40:373–386

Chandrakar V, Naithani SC, Keshavkant S (2016) Arsenic-induced metabolic disturbances and their mitigation mechanisms in crop plants: a review. Biologia 71(4):367–377

Chaudhary K, Agarwal S, Khan S (2018) Role of phytochelatins (PCs), metallothioneins (MTs), and heavy metal ATPase (HMA) genes in heavy metal tolerance. In: Mycoremediation and environmental sustainability: volume 2. Springer, Cham, pp 39–60

Chen T, Su Y (2018) Uptake by rice seedlings and in-plant degradation of atrazine as influenced by the oxidative stress induced by added arsenic or phosphate deficiency. Hum Ecol Risk Assess Int J 24(6):1550–1564

Chen H, Yang R, Zhang X, Chen Y, Xia Y, Xu X (2021) Foliar application of gibberellin inhibits the cadmium uptake and xylem transport in lettuce (*Lactuca sativa* L.). Sci Hortic 288:110410

Chen D, Mubeen B, Hasnain A, Rizwan M, Adrees M, Naqvi SAH, Din GMU (2022) Role of promising secondary metabolites to confer resistance against environmental stresses in crop plants: current scenario and future perspectives. Front Plant Sci 13:881032

Cho LH, Yoon J, An G (2017) The control of flowering time by environmental factors. Plant J 90(4):708–719

Dalvi AA, Bhalerao SA (2013) Response of plants towards heavy metal toxicity: an overview of avoidance, tolerance and uptake mechanism. Ann Plant Sci 2(9):362–368

Das K, Roychoudhury A (2014) Reactive oxygen species (ROS) and response of antioxidants as ROS-scavengers during environmental stress in plants. Front Environ Sci 2:53

Debnath B, Islam W, Li M, Sun Y, Lu X, Mitra S, Hussain M, Liu S, Qiu D (2019) Melatonin mediates enhancement of stress tolerance in plants. Int J Mol Sci 20(5):1040

Demecsová L, Zelinová V, Liptáková Ľ, Tamás L (2020) Mild cadmium stress induces auxin synthesis and accumulation, while severe cadmium stress causes its rapid depletion in barley root tip. Environ Exp Bot 175:104038

Dubey S, Shri M, Gupta A, Rani V, Chakrabarty D (2018) Toxicity and detoxification of heavy metals during plant growth and metabolism. Environ Chem Lett 16:1169–1192

Dutta S, Mitra M, Agarwal P, Mahapatra K, De S, Sett U, Roy S (2018) Oxidative and genotoxic damages in plants in response to heavy metal stress and maintenance of genome stability. Plant Signal Behav 13(8):e1460048

Emamverdian A, Ding Y, Mokhberdoran F, Xie Y (2015) Heavy metal stress and some mechanisms of plant defense response. Sci World J 2015:756120

Farouk S, Al-Amri SM (2019) Ameliorative roles of melatonin and/or zeolite on chromium-induced leaf senescence in marjoram plants by activating antioxidant defense, osmolyte accumulation, and ultra-structural modification. Ind Crop Prod 142:111823

Fitter AH, Hay RK (2012) Environmental physiology of plants. Academic, San Diego

Flora SJS, Mittal M, Mehta A (2008) Heavy metal induced oxidative stress and its possible reversal by chelation therapy. Indian J Med Res 128(4):501–523

Gamalero E, Lingua G, Berta G, Glick BR (2009) Beneficial role of plant growth promoting bacteria and arbuscular mycorrhizal fungi on plant responses to heavy metal stress. Can J Microbiol 55(5):501–514

Gangwar S, Singh VP, Tripathi DK, Chauhan DK, Prasad SM, Maurya JN (2014) Plant responses to metal stress: the emerging role of plant growth hormones in toxicity alleviation. In: Emerging technologies and management of crop stress tolerance. Academic, San Diego, pp 215–248

Garcia C, Furtado de Almeida AA, Costa M, Britto D, Valle R, Royaert S, Marelli JP (2019) Abnormalities in somatic embryogenesis caused by 2, 4-D: an overview. Plant Cell Tissue Organ Cult 137:193–212

Ghori NH, Ghori T, Hayat MQ, Imadi SR, Gul A, Altay V, Ozturk M (2019) Heavy metal stress and responses in plants. Int J Environ Sci Technol 16:1807–1828

Gill SS, Khan NA, Anjum NA, Tuteja N (2011) Amelioration of cadmium stress in crop plants by nutrients management: morphological, physiological and biochemical aspects. Plant Stress 5(1):1–23

Goncharuk EA, Zagoskina NV (2023) Heavy metals, their phytotoxicity, and the role of phenolic antioxidants in plant stress responses with focus on cadmium. Molecules 28(9):3921

Goyal D, Yadav A, Prasad M, Singh TB, Shrivastav P, Ali A, Dantu PK, Mishra S (2020) Effect of heavy metals on plant growth: an overview. In: Contaminants in agriculture: sources, impacts and management. Springer, Cham, pp 79–101

Gu Q, Chen Z, Yu X, Cui W, Pan J, Zhao G, Xu S, Wang R, Shen W (2017) Melatonin confers plant tolerance against cadmium stress via the decrease of cadmium accumulation and reestablishment of microRNA-mediated redox homeostasis. Plant Sci 261:28–37

Gu Q, Wang C, Xiao Q, Chen Z, Han Y (2021) Melatonin confers plant cadmium tolerance: an update. Int J Mol Sci 22(21):11704

Gupta DK, Vandenhove H, Inouhe M (2013) Role of phytochelatins in heavy metal stress and detoxification mechanisms in plants. Springer, Berlin/Heidelberg, pp 73–94

Hasan MK, Ahammed GJ, Yin L, Shi K, Xia X, Zhou Y, Yu J, Zhou J (2015) Melatonin mitigates cadmium phytotoxicity through modulation of phytochelatins biosynthesis, vacuolar sequestration, and antioxidant potential in *Solanum lycopersicum* L. Front Plant Sci 6:601

Hasanein P, Emamjomeh A (2019) Beneficial effects of natural compounds on heavy metal–induced hepatotoxicity. In: Dietary interventions in liver disease. Academic, London, pp 345–355
Hasanuzzaman M, Hossain MA, da Silva JAT, Fujita M (2012) Plant response and tolerance to abiotic oxidative stress: antioxidant defense is a key factor. In: Crop stress and its management: perspectives and strategies. Springer Science+Business Media B.V, Dordrecht, pp 261–315
He J, Zhuang X, Zhou J, Sun L, Wan H, Li H, Lyu D (2020) Exogenous melatonin alleviates cadmium uptake and toxicity in apple rootstocks. Tree Physiol 40(6):746–761
Hodzic E, Galijasevic S, Balaban M, Rekanovic S, Makic H, Kukavica B, Mihajlovic D (2021) The protective role of melatonin under heavy metal-induced stress in *Melissa Officinalis* L. Turk J Chem 45(3):737–748
Hoque MN, Tahjib-Ul-Arif M, Hannan A, Sultana N, Akhter S, Hasanuzzaman M, Akter F, Hossain MS, Sayed MA, Hasan MT, Skalicky M (2021) Melatonin modulates plant tolerance to heavy metal stress: morphological responses to molecular mechanisms. Int J Mol Sci 22(21):11445
Hossain MA, Piyatida P, da Silva JAT, Fujita M (2012) Molecular mechanism of heavy metal toxicity and tolerance in plants: central role of glutathione in detoxification of reactive oxygen species and methylglyoxal and in heavy metal chelation. J Bot 2012:1
Ighodaro OM, Akinloye OA (2018) First line defence antioxidants-superoxide dismutase (SOD), catalase (CAT) and glutathione peroxidase (GPX): their fundamental role in the entire antioxidant defense grid. Alexandria J Med 54(4):287–293
Imran M, Khan AL, Mun BG, Bilal S, Shaffique S, Kwon EH, Kang SM, Yun BW, Lee IJ (2022) Melatonin and nitric oxide: dual players inhibiting hazardous metal toxicity in soybean plants *via* molecular and antioxidant signalling cascades. Chemosphere 308:136575
Isah T (2019) Stress and defense responses in plant secondary metabolites production. Biol Res 52:39
Ishtiyaq S, Kumar H, Varun M, Kumar B, Paul MS (2018) Heavy metal toxicity and anti-oxidative response in plants: an overview. In: Plants under metal and metalloid stress: responses, tolerance and remediation. Springer Nature, Singapore, pp 77–106
Ivanov VB, Bystrova EI, Seregin IV (2003) Comparative impacts of heavy metals on root growth as related to their specificity and selectivity. Russ J Plant Physiol 50:398–406
Jahan MS, Guo S, Baloch AR, Sun J, Shu S, Wang Y, Ahammed GJ, Kabir K, Roy R (2020) Melatonin alleviates nickel phytotoxicity by improving photosynthesis, secondary metabolism and oxidative stress tolerance in tomato seedlings. Ecotoxicol Environ Saf 197:110593
Jalmi SK, Bhagat PK, Verma D, Noryang S, Tayyeba S, Singh K, Sharma D, Sinha AK (2018) Traversing the links between heavy metal stress and plant signaling. Front Plant Sci 9:12
Jamieson MA, Burkle LA, Manson JS, Runyon JB, Trowbridge AM, Zientek J (2017) Global change effects on plant–insect interactions: the role of phytochemistry. Curr Opin Insect Sci 23:70–80
Jan R, Asaf S, Numan M, Lubna, Kim KM (2021) Plant secondary metabolite biosynthesis and transcriptional regulation in response to biotic and abiotic stress conditions. Agronomy 11(5):968
Jan R, Asif S, Asaf S, Du XX, Park JR, Nari K, Kim KM (2023) Melatonin alleviates arsenic (As) toxicity in rice plants *via* modulating antioxidant defense system and secondary metabolites and reducing oxidative stress. Environ Pollut 318:120868
Jia Y, Li X, Liu Q, Hu X, Li J, Dong R, Liu P, Liu G, Luo L, Chen Z (2020) Physiological and transcriptomic analyses reveal the roles of secondary metabolism in the adaptive responses of Stylosanthes to manganese toxicity. BMC Genomics 21:1–17
Jogawat A, Yadav B, Chhaya, Narayan OP (2021) Metal transporters in organelles and their roles in heavy metal transportation and sequestration mechanisms in plants. Physiol Plant 173(1):259–275
Juan CA, Pérez de la Lastra JM, Plou FJ, Pérez-Lebeña E (2021) The chemistry of reactive oxygen species (ROS) revisited: outlining their role in biological macromolecules (DNA, lipids and proteins) and induced pathologies. Int J Mol Sci 22(9):4642

Kalaivanan D, Ganeshamurthy AN (2016) Mechanisms of heavy metal toxicity in plants. In: Abiotic stress physiology of horticultural crops. Springer, New Delhi, pp 85–102

Kalefetoğlu T, Ekmekci Y (2005) The effects of drought on plants and tolerance mechanisms: review. Gazi Univ J Sci 18:723–740

Kapoor D, Singh S, Ramamurthy PC, Jan S, Bhardwaj S, Gill SS, Prasad R, Singh J (2021) Molecular consequences of cadmium toxicity and its regulatory networks in plants. Plant Gene 28:100342

Kaur R, Das S, Bansal S, Singh G, Sardar S, Dhar H, Ram H (2021) Heavy metal stress in rice: uptake, transport, signaling, and tolerance mechanisms. Physiol Plant 173(1):430–448

Kaya C, Okant M, Ugurlar F, Alyemeni MN, Ashraf M, Ahmad P (2019) Melatonin-mediated nitric oxide improves tolerance to cadmium toxicity by reducing oxidative stress in wheat plants. Chemosphere 225:627–638

Khan TA, Fariduddin Q, Nazir F, Saleem M (2020) Melatonin in business with abiotic stresses in plants. Physiol Mol Biol Plants 26:1931–1944

Khan A, Jie Z, Xiangjun K, Ullah N, Short AW, Diao Y, Zhou R, Xiong YC (2023) Pretreatment of melatonin rescues cotton seedlings from cadmium toxicity by regulating key physio-biochemical and molecular pathways. J Hazard Mater 445:130530

Khare S, Singh NB, Singh A, Hussain I, Niharika KM, Yadav V, Amist N (2020) Plant secondary metabolites synthesis and their regulations under biotic and abiotic constraints. J Plant Biol 63:203–216

Kohli SK, Handa N, Gautam V, Bali S, Sharma A, Khanna K, Arora S, Thukral AK, Ohri P, Karpets YV, Kolupaev YE (2017) ROS signaling in plants under heavy metal stress. In: Reactive oxygen species and antioxidant systems in plants: role and regulation under abiotic stress. Springer, Singapore, pp 185–214

Kosakivska IV, Babenko LM, Romanenko KO, Korotka IY, Potters G (2021) Molecular mechanisms of plant adaptive responses to heavy metals stress. Cell Biol Int 45(2):258–272

Koul M, Thomas L, Karmakar K (2021) Functional aspects of solanaceae trichomes in heavy metal detoxification. Nord J Bot 39(5)

Krishnamurthy A, Rathinasabapathi B (2013) Auxin and its transport play a role in plant tolerance to arsenite-induced oxidative stress in *Arabidopsis thaliana*. Plant Cell Environ 36(10):1838–1849

Kumar S, Shah SH, Vimala Y, Jatav HS, Ahmad P, Chen Y, Siddique KH (2022) Abscisic acid: metabolism, transport, crosstalk with other plant growth regulators, and its role in heavy metal stress mitigation. Front Plant Sci 13:972856

Kushwaha A, Hans N, Kumar S, Rani R (2018) A critical review on speciation, mobilization and toxicity of lead in soil-microbe-plant system and bioremediation strategies. Ecotoxicol Environ Saf 147:1035–1045

Lai JL, Luo XG (2019) High-efficiency antioxidant system, chelating system and stress-responsive genes enhance tolerance to cesium ionotoxicity in Indian mustard (*Brassica juncea* L.). Ecotoxicol Environ Saf 181:491–498

Leguizamo MAO, Gómez WDF, Sarmiento MCG (2017) Native herbaceous plant species with potential use in phytoremediation of heavy metals, spotlight on wetlands—a review. Chemosphere 168:1230–1247

Li X, Ahammed GJ, Zhang XN, Zhang L, Yan P, Zhang LP, Fu JY, Han WY (2021) Melatonin-mediated regulation of anthocyanin biosynthesis and antioxidant defense confer tolerance to arsenic stress in *Camellia sinensis* L. J Hazard Mater 403:123922

Li GZ, Wang YY, Liu J, Liu HT, Liu HP, Kang GZ (2022a) Exogenous melatonin mitigates cadmium toxicity through ascorbic acid and glutathione pathway in wheat. Ecotoxicol Environ Saf 237:113533

Li Y, Liu Y, Jin L, Peng R (2022b) Crosstalk between Ca^{2+} and other regulators assists plants in responding to abiotic stress. Plants 11(10):1351

Li L, Yan X, Li J, Wu X, Wang X (2022c) Metabolome and transcriptome association analysis revealed key factors involved in melatonin mediated cadmium-stress tolerance in cotton. Front Plant Sci 13:995205

Lin T, Zhu G, He W, Xie J, Li S, Han S, Li S, Yang C, Liu Y, Zhu T (2022) Soil cadmium stress reduced host plant odor selection and oviposition preference of leaf herbivores through the changes in leaf volatile emissions. Sci Total Environ 814:152728

Liu Q, Zhang H (2012) Molecular identification and analysis of arsenite stress-responsive miRNAs in rice. J Agric Food Chem 60(26):6524–6536

Mahajan M, Kuiry R, Pal PK (2020) Understanding the consequence of environmental stress for accumulation of secondary metabolites in medicinal and aromatic plants. J Appl Res Med Aromat Plants 18:100255

Maleki M, Ghorbanpour M, Kariman K (2017) Physiological and antioxidative responses of medicinal plants exposed to heavy metals stress. Plant Gene 11:247–254

Mannino G, Pernici C, Serio G, Gentile C, Bertea CM (2021) Melatonin and phytomelatonin: chemistry, biosynthesis, metabolism, distribution and bioactivity in plants and animals—an overview. Int J Mol Sci 22(18):9996

Mashabela MD, Masamba P, Kappo AP (2023) Applications of metabolomics for the elucidation of abiotic stress tolerance in plants: a special focus on osmotic stress and heavy metal toxicity. Plants 12(2):269

Mathur P, Tripathi DK, Baluška F, Mukherjee S (2022) Auxin-mediated molecular mechanisms of heavy metal and metalloid stress regulation in plants. Environ Exp Bot 196:104796

Matsuura HN, Fett-Neto AG (2015) Plant alkaloids: main features, toxicity, and mechanisms of action. Plant Toxins 2(7):1–15

Maurya V, Singh N, Sharma I, Sharma A, Kumar R, Sharma A, Ahammed GJ (2024) Melatonin in plants: role in plant growth development and stress response effect of melatonin in regulating salt stress responses in plants. Springer Nature, Singapore, Singapore, pp 109–139

Morkunas I, Woźniak A, Mai VC, Rucińska-Sobkowiak R, Jeandet P (2018) The role of heavy metals in plant response to biotic stress. Molecules 23(9):2320

Moustafa-Farag M, Elkelish A, Dafea M, Khan M, Arnao MB, Abdelhamid MT, El-Ezz AA, Almoneafy A, Mahmoud A, Awad M, Li L (2020) Role of melatonin in plant tolerance to soil stressors: salinity, pH and heavy metals. Molecules 25(22):5359

Naeem A, Zafar M, Khalid H, Zia-ur-Rehman M, Ahmad Z, Ayub MA, Qayyum MF (2019) Cadmium-induced imbalance in nutrient and water uptake by plants. In: Cadmium toxicity and tolerance in plants. Academic, London, pp 299–326

Nawaz MA, Jiao Y, Chen C, Shireen F, Zheng Z, Imtiaz M, Bie Z, Huang Y (2018) Melatonin pre-treatment improves vanadium stress tolerance of watermelon seedlings by reducing vanadium concentration in the leaves and regulating melatonin biosynthesis and antioxidant-related gene expression. J Plant Physiol 220:115–127

Nawaz K, Chaudhary R, Sarwar A, Ahmad B, Gul A, Hano C, Abbasi BH, Anjum S (2020) Melatonin as master regulator in plant growth, development and stress alleviator for sustainable agricultural production: current status and future perspectives. Sustainability 13(1):294

Nedjimi B (2021) Phytoremediation: a sustainable environmental technology for heavy metals decontamination. SN Appl Sci 3(3):286

Negri S, Commisso M, Avesani L, Guzzo F (2021) The case of tryptamine and serotonin in plants: a mysterious precursor for an illustrious metabolite. J Exp Bot 72(15):5336–5355

Ni J, Wang Q, Shah FA, Liu W, Wang D, Huang S, Fu S, Wu L (2018) Exogenous melatonin confers cadmium tolerance by counter balancing the hydrogen peroxide homeostasis in wheat seedlings. Molecules 23:799

Ou C, Cheng W, Wang Z, Yao X, Yang S (2023) Exogenous melatonin enhances Cd stress tolerance in *Platycladus orientalis* seedlings by improving mineral nutrient uptake and oxidative stress. Ecotoxicol Environ Saf 252:114619

Pandey AK, Zorić L, Sun T, Karanović D, Fang P, Borišev M, Wu X, Luković J, Xu P (2022) The anatomical basis of heavy metal responses in legumes and their impact on plant–rhizosphere interactions. Plants 11(19):2554

Papazoglou EG, Karantounias GA, Vemmos SN, Bouranis DL (2005) Photosynthesis and growth responses of giant reed (*Arundo donax* L.) to the heavy metals Cd and Ni. Environ Int 31(2):243–249

Parmar P, Dave B, Sudhir A, Panchal K, Subramanian RB (2013) Physiological, biochemical and molecular response of plants against heavy metals stress. Int J Curr Res 5(1):80–89

Rahman SU, Xuebin Q, Kamran M, Yasin G, Cheng H, Rehim A, Alyemeni MN (2021) Silicon elevated cadmium tolerance in wheat (*Triticum aestivum* L.) by endorsing nutrients uptake and anti-oxidative defense mechanisms in the leaves. Plant Physiol Biochem 166:148–159

Rai LC, Singh AK, Mallick N (1991) Studies on photosynthesis, the associated electron transport system and some physiological variables of *Chlorella vulgaris* under heavy metal stress. J Plant Physiol 137(4):419–424

Rai R, Agrawal M, Agrawal SB (2016) Impact of heavy metals on physiological processes of plants: with special reference to photosynthetic system. In: Plant responses to xenobiotics. Springer, Sinagpore, pp 127–140

Ramegowda V, Da Costa MVJ, Harihar S, Karaba NN, Sreeman SM (2020) Abiotic and biotic stress interactions in plants: a cross-tolerance perspective. In: Priming-mediated stress and cross-stress tolerance in crop plants. Academic, London, pp 267–302

Raza A, Charagh S, Najafi-Kakavand S, Siddiqui MH (2021) The crucial role of jasmonates in enhancing heavy metals tolerance in plants. In: Jasmonates and salicylates signaling in plants. Springer, Cham, pp 159–183

Rehman AU, Nazir S, Irshad R, Tahir K, ur Rehman K, Islam RU, Wahab Z (2021) Toxicity of heavy metals in plants and animals and their uptake by magnetic iron oxide nanoparticles. J Mol Liq 321:114455

Riaz M, Kamran M, Rizwan M, Ali S, Parveen A, Malik Z, Wang X (2021) Cadmium uptake and translocation: selenium and silicon roles in Cd detoxification for the production of low Cd crops: a critical review. Chemosphere 273:129690

Romero A, Ramos E, de Los Rios C, Egea J, Del Pino J, Reiter RJ (2014) A review of metal-catalyzed molecular damage: protection by melatonin. J Pineal Res 56(4):343–370

Roychoudhury A (ed) (2023) Biology and biotechnology of environmental stress tolerance in plants: volume 3: Sustainable approaches for enhancing environmental stress tolerance. CRC Press, Boca Raton

Rubio MI, Escrig I, Martinez-Cortina C, Lopez-Benet FJ, Sanz A (1994) Cadmium and nickel accumulation in rice plants. Effects on mineral nutrition and possible interactions of abscisic and gibberellic acids. Plant Growth Regul 14(2):151–157

Rucińska-Sobkowiak R (2016) Water relations in plants subjected to heavy metal stresses. Acta Physiol Plant 38:1–13

Saddique M, Kamran M, Shahbaz M (2018) Differential responses of plants to biotic stress and the role of metabolites. In: Plant metabolites and regulation under environmental stress. Academic, London, pp 69–87

Saikkonen K, Koivunen S, Vuorisalo T, Mutikainen P (1998) Interactive effects of pollination and heavy metals on resource allocation in *Potentilla anserina* L. Ecology 79(5):1620–1629

Sakshi VM, Singh N, Sharma S, Kumar R, Sharma A (2024) Impact of seed priming with melatonin, nano-urea and nano-zinc on growth, quality and yield of radish. Int J Adv Biochem Res 8(9):667–676. https://doi.org/10.33545/26174693

Sami A, Shah FA, Abdullah M, Zhou X, Yan Y, Zhu Z, Zhou K (2020) Melatonin mitigates cadmium and aluminium toxicity through modulation of antioxidant potential in *Brassica napus* L. Plant Biol 22:679–690

Shahid M, Pourrut B, Dumat C, Nadeem M, Aslam M, Pinelli E (2014) Heavy-metal-induced reactive oxygen species: phytotoxicity and physicochemical changes in plants. Rev Environ Contam Toxicol 232:1–44

Shahid M, Dumat C, Khalid S, Schreck E, Xiong T, Niazi NK (2017) Foliar heavy metal uptake, toxicity and detoxification in plants: a comparison of foliar and root metal uptake. J Hazard Mater 325:36–58

Sharma P, Jha AB, Dubey RS, Pessarakli M (2012) Reactive oxygen species, oxidative damage, and antioxidative defense mechanism in plants under stressful conditions. J Bot 2012:1
Sharma A, Kapoor D, Gautam S, Landi M, Kandhol N, Araniti F, Zheng B (2022) Heavy metal induced regulation of plant biology: recent insights. Physiol Plant 174(3):e13688
Shi H, Chen K, Wei Y, He C (2016) Fundamental issues of melatonin-mediated stress signaling in plants. Front Plant Sci 7:1124
Shiyu QIN, Hongen LIU, Zhaojun NIE, Rengel Z, Wei GAO, Chang LI, Peng ZHAO (2020) Toxicity of cadmium and its competition with mineral nutrients for uptake by plants: a review. Pedosphere 30(2):168–180
Shukla M, Chinchalongporn V, Govitrapong P, Reiter RJ (2019) The role of melatonin in targeting cell signaling pathways in neuro degeneration. Ann N Y Acad Sci 1443(1):75–96
Singh S, Parihar P, Singh R, Singh VP, Prasad SM (2016) Heavy metal tolerance in plants: role of transcriptomics, proteomics, metabolomics, and ionomics. Front Plant Sci 6:1143
Singh AD, Sharma P, Kohli SK, Kumar P, Singh R, Arora P, Sharma P, Kaur R, Sharma A, Bhardwaj R (2020) Role of plant growth regulators (PGRs) in mitigation of heavy metal phytotoxicity in plants. In: Faisal M, Saquib Q, Alatar AA, Al-Khedhairy AA (eds) Cellular and molecular phytotoxicity of heavy metals. Springer, Cham, pp 263–304. https://doi.org/10.1007/978-3-030-45975-8_14
Singh A, Pandey H, Pal A, Chauhan D, Pandey S, Gaikwad DJ, Sahu C, Atta K (2023) Linking the role of melatonin in plant stress acclimatization. S Afr J Bot 159:179–190
Smirnoff N (1993) Tansley review No. 52. The role of active oxygen in the response of plants to water deficit and desiccation. New Phytol 127:27–58
Souri Z, Cardoso AA, da-Silva CJ, de Oliveira LM, Dari B, Sihi D, Karimi N (2019a) Heavy metals and photosynthesis: recent developments. In: Photosynthesis, productivity and environmental stress. Wiley, Hoboken, pp 107–134
Souri MK, Hatamian M, Tesfamariam T (2019b) Plant growth stage influences heavy metal accumulation in leafy vegetables of garden cress and sweet basil. Chem Biol Technol Agric 6(1):1–7
Sun EJ, Wu FY (1998) Along-vein necrosis as indicator symptom on water spinach caused by nickel in water culture. Botanical Bulletin of Academia Sinica 39
Sun C, Liu L, Wang L, Li B, Jin C, Lin X (2021) Melatonin: a master regulator of plant development and stress responses. J Integr Plant Biol 63(1):126–145
Sytar O, Kumar A, Latowski D, Kuczynska P, Strzałka K, Prasad MNV (2013) Heavy metal-induced oxidative damage, defense reactions, and detoxification mechanisms in plants. Acta Physiol Plant 35:985–999
Sytar O, Kumari P, Yadav S, Brestic M, Rastogi A (2019) Phytohormone priming: regulator for heavy metal stress in plants. J Plant Growth Regul 38:739–752
Takeno K (2016) Stress-induced flowering: the third category of flowering response. J Exp Bot 67(17):4925–4934
Tan DX, Manchester LC, Helton P, Reiter RJ (2007) Phytoremediative capacity of plants enriched with melatonin. Plant Signal Behav 2(6):514–516
Tan DX, Zheng X, Kong J, Manchester LC, Hardeland R, Kim SJ, Xu X, Reiter RJ (2014) Fundamental issues related to the origin of melatonin and melatonin isomers during evolution: relation to their biological functions. Int J Mol Sci 15(9):15858–15890
Tan DX, Hardeland R, Back K, Manchester LC, Alatorre-Jimenez MA, Reiter RJ (2016) On the significance of an alternate pathway of melatonin synthesis via 5-methoxytryptamine: comparisons across species. J Pineal Res 61(1):27–40
Tang M, Xu L, Wang Y, Dong J, Zhang X, Wang K, Ying J, Li C, Liu L (2021) Melatonin-induced DNA demethylation of metal transporters and antioxidant genes alleviates lead stress in radish plants. Horticult Res 8
Thao NP, Khan MIR, Thu NBA, Hoang XLT, Asgher M, Khan NA, Tran LSP (2015) Role of ethylene and its cross talk with other signaling molecules in plant responses to heavy metal stress. Plant Physiol 169(1):73–84

Tiku AR (2020) Antimicrobial compounds (phytoanticipins and phytoalexins) and their role in plant defense. In: Co-evolution of secondary metabolites. Springer, Cham, pp 845–868

Tiwari RK, Lal MK, Kumar R, Chourasia KN, Naga KC, Kumar D, Das SK, Zinta G (2021) Mechanistic insights on melatonin-mediated drought stress mitigation in plants. Physiol Plant 172(2):1212–1226

Vega A, Delgado N, Handford M (2022) Increasing heavy metal tolerance by the exogenous application of organic acids. Int J Mol Sci 23(10):5438

Wang P, Sun X, Chang C, Feng F, Liang D, Cheng L, Ma F (2013) Melatonin enhances the occurrence of autophagy and apoptosis induced by oxidative stress in Arabidopsis seedlings. J Pineal Res 54(4):426–437

Wang M, Duan S, Zhou Z, Chen S, Wang D (2019) Foliar spraying of melatonin confers cadmium tolerance in *Nicotiana tabacum* L. Ecotoxicol Environ Saf 170:68–76

Wang T, Song J, Liu Z, Liu Z, Cui J (2021) Melatonin alleviates cadmium toxicity by reducing nitric oxide accumulation and IRT1 expression in Chinese cabbage seedlings. Environ Sci Pollut Res 28:15394–15405

Wang K, He J, Gao Y, Han K, Liu J, Wang Y (2022) Exogenous melatonin improved the growth and development of naked oat seedlings under cadmium stress. Environ Sci Pollut Res 29(58):88109–88118

Wei W, Li QT, Chu YN, Reiter RJ, Yu XM, Zhu DH, Yu SX (2018) Melatonin enhances plant growth and abiotic stress tolerance in soybean plants. J Exp Bot 69(7):1443–1452

Xalxo R, Keshavkant S (2019) Melatonin, glutathione and thiourea attenuates lead and acid rain-induced deleterious responses by regulating gene expression of antioxidants in *Trigonella foenum graecum* L. Chemosphere 221:1–10

Xu L, Zhang F, Tang M, Wang Y, Dong J, Ying J, Chen Y, Hu B, Li C, Liu L (2020) Melatonin confers cadmium tolerance by modulating critical heavy metal chelators and transporters in radish plants. J Pineal Res 69(1):e12659

Xun E, Zhang Y, Zhao J, Guo J (2018) Heavy metals in nectar modify behaviors of pollinators and nectar robbers: consequences for plant fitness. Environ Pollut 242:1166–1175

Yadav B, Dubey R, Gnanasekaran P, Narayan OP (2021) OMICS approaches towards understanding plant's responses to counterattack heavy metal stress: an insight into molecular mechanisms of plant defense. Plant Gene 28:100333

Yang X, Ren J, Lin X, Yang Z, Deng X, Ke Q (2023) Melatonin alleviates chromium toxicity in maize by modulation of cell wall polysaccharides biosynthesis, glutathione metabolism, and antioxidant capacity. Int J Mol Sci 24(4):3816

Yu X, Zhang W, Zhang Y, Zhang X, Lang D, Zhang X (2018) The roles of methyl jasmonate to stress in plants. Funct Plant Biol 46(3):197–212

Yu Y, Teng Z, Mou Z, Lv Y, Li T, Chen S, Zhao D, Zhao Z (2021) Melatonin confers heavy metal-induced tolerance by alleviating oxidative stress and reducing the heavy metal accumulation in *Exophiala pisciphila*, a dark septate endophyte (DSE). BMC Microbiol 21(1):1–11

Zahid G, Iftikhar S, Shimira F, Ahmad HM, Kaçar YA (2023) An overview and recent progress of plant growth regulators (PGRs) in the mitigation of abiotic stresses in fruits: A review. Sci Hortic 309:111621

Zaid A, Mohammad F, Fariduddin Q (2020) Plant growth regulators improve growth, photosynthesis, mineral nutrient and antioxidant system under cadmium stress in menthol mint (*Mentha arvensis* L.). Physiol Mol Biol Plants 26:25–39

Zhang N, Sun Q, Zhang H, Cao Y, Weeda S, Ren S, Guo YD (2015) Roles of melatonin in abiotic stress resistance in plants. J Exp Bot 66(3):647–656

Zhang HJ, Zhang N, Yang RC, Wang L, Sun QQ (2017) Melatonin promotes seed germination under high salinity by regulating antioxidant systems, ABA and GA4 interaction in cucumber (*Cucumis sativus* L.). J Pineal Res 63(4):e12438

Zhao H, Su T, Hu L, Wei H, Cui X (2019a) Effects of melatonin on alleviating cadmium toxicity in rice plants. Ecotoxicol Environ Saf 174:562–571

Zhao D, Yu Y, Shen Y, Liu Q, Zhao Z, Sharma R, Reiter RJ (2019b) Melatonin synthesis and function: evolutionary history in animals and plants. Front Endocrinol 10:249

Zhao FJ, Tang Z, Song JJ, Huang XY, Wang P (2022) Toxic metals and metalloids: uptake, transport, detoxification, phytoremediation, and crop improvement for safer food. Mol Plant 15(1):27–44

Chapter 12
Strigolactone-Mediated Regulation of Plant Biology Under Heavy Metal Stress: Physiological and Molecular Aspects

Asha Kumari, Krishna Kumar, Suresh Gali, Binny Sharma, Priyanka Sharma, Aurosha Fayaz, Ira Khan, S. Maqbool Ahmed, and Mohammad Faizan

Abstract Plant hormones known as strigolactones (SLs) are crucial to many plant species' environmental and developmental processes. It controls the symbiotic and parasitic interactions that mycorrhizal fungi, bacteria, and root parasite plants have with one another. It is well recognized that this class of endogenous phytohormones controls the morphology of the roots, secondary growth, and shoot branching. In the past few years, significant progress has been made in elucidating the function of strigolactones in plant growth, response to abiotic stressors, mesocotyl and stem elongation, secondary growth, and shoot gravitropism. Basic questions about SL biosynthesis and perception were explored with the use of several mutants from different plant species. Soil containing heavy metals (HMs) is a significant environmental issue that poses a threat to plants in a number of ways. SLs are linked to abiotic stressors like metals and nutrients, and they are essential to plants' defense mechanisms against water and nutrient shortages. With the help of metals and nutrients, SLs activate bidirectional systems that affect the structure and activities of

A. Kumari (✉)
School of Crop Science, ICAR-Indian Agricultural Research Institute, Hazaribagh, Jharkhand, India

K. Kumar · B. Sharma
Department of Plant Physiology, Institute of Agricultural Sciences, Varanasi, Uttar Pradesh, India

S. Gali
Department of Botany and Plant Physiology, Chaudhary Charan Singh Haryana Agricultural University, Hisar, Haryana, India

P. Sharma
Department of Crop Physiology, Assam Agricultural University, Jorhat, Assam, India

A. Fayaz · I. Khan · S. M. Ahmed · M. Faizan (✉)
Botany Section, School of Sciences, Maulana Azad National Urdu University, Hyderabad, Telangana, India

A. Sharma et al. (eds.), *Hormones and Heavy Metals*, Plant Life and Environment Dynamics, https://doi.org/10.1007/978-981-96-9622-2_12

roots. In response to heavy metal pollution, such as Cd stress, plant extraction, and GR24 synthetic analog stimulation, SLs may play a significant role in regulating growth, root development, volume, architecture, and ROS scavenging. The function of SLs in reducing heavy metal stress was highlighted in this chapter.

Keywords Strigolactones (SLs) · Biosynthesis · Perception · Plant growth regulator · GR24 · Heavy metal

12.1 Introduction

Plant growth regulators known as phytohormones are produced internally and are essential for many stages of the plant life cycle, including reactions to biotic and abiotic stress (EL Sabagh et al. 2015; Zheng et al. 2023; Verma et al. 2016). Auxin, gibberellin, cytokinin, ethylene, and abscisic acid (ABA) are notable phytohormones (Wang and Irving 2011; Wasternack 2014). These phytohormones control food allocation and establish source/sink transitions among other physiological, biochemical, and morphological processes (Faizan et al. 2020). Brassinosteroids (BRs), other organic molecules like jasmonic acid, salicylic acid, and strigolactones (SLs), and other substances have recently been classified as significant phytohormones (Smith and Li 2014; Faizan et al. 2024; 2023). The roots of about 80% of terrestrial plants release SLs, a tiny group of compounds produced from β-carotene. Strigolactones (SLs) were first identified as agents that stimulated germination in parasitic weeds of the genera Striga and Orobanche (Mishra et al. 2017; Yoneyama et al. 2013a, b). It was later shown that SLs function as signaling molecules for arbuscular mycorrhizal (AM) fungal host detection and hyphal branching (Kee et al. 2023; Guercio et al. 2023). In 1966, SLs were identified for the first time as cotton plant root exudates. "Strigol" was the first known naturally occurring SL to be discovered (Wakabayashi et al. 2023; Zwanenburg and Blanco-Ania 2018). Due to their diverse functions in regulating above-ground plant architecture (such as preventing bud expansion) and subterranean communication with nearby species, SLs were recognized as a novel family of phytohormones in 2008 (Yoneyama et al. 2013a, b). Currently, about 25 different plant sources—including strigol, orobanchol, sorgolactone, 20-epi-orobanchol, solanacol, and sorgomol—have yielded the isolation of these SLs (Wakabayashi et al. 2022). An enol–ether bridge unites the butenolide ring (D-ring) and the tricyclic ring (ABC-ring) in canonical SLs, while a less conserved configuration unites them in non-canonical SLs. This structural motif is characteristic of the majority of naturally occurring SLs (Waters et al. 2017; Mwakaboko and Zwanenburg 2016; Cohen et al. 2013). In root system dynamics, SLs play a variety of roles. For example, in *Solanum lycopersicum*, *Arabidopsis thaliana*, and *Pisum sativum*, they inhibit adventitious root formation, stimulate the growth of rice crown roots, and promote the elongation of primary roots and root hairs (Marzec and Melzer 2018; Sun et al. 2015). Furthermore, as noted by Urquhart et al. (2015), SLs are essential for a variety of plant growth and development processes, including seed germination, early seedling development, internode length, leaf structure, senescence, shoot gravitropism, and shape.

Strigolactones (SLs) are a class of lactones generated from carotenoid compounds that have multiple functions in the rhizosphere (Faizan et al. 2022). They also function as a novel class of hormones that control the growth and development of plants (Al-Babili and Bouwmeester 2015). Strigolactones are a type of chemical compound that is produced by plant roots. Because of their mode of action, these molecules are referred to as phytohormones or plant hormones (Umehara et al. 2015). In order to encourage symbiosis with arbuscular mycorrhizal fungi, SLs also function as signaling molecules secreted from roots into the rhizosphere. Up till now, around 30 SLs have been identified; these primarily function as germination enhancers for weeds that parasitize roots. Cook et al. (1966) noted that the first SL to be reported was (+)-strigol, which was found to be a germination stimulant from the root exudates of the false host cotton. Plants only generate trace amounts of chemically unstable SLs (pg-ng/g root FW). Since root exudates solely contain chemicals produced by the roots, they are a preferred source for SL detection. Moreover, SLs were identified in 2008 as the "branching hormone" that was lacking and was in charge of preventing the growth of lateral shoots, indicating an endogenous plant function (Umehara et al. 2008). In recent times, SLs have been connected to various aspects of plant development such as secondary thickening, adventitious roots, lateral roots, and root hair creation (Al-Babili and Bouwmeester 2015). Three distinct physiological processes have been linked to strigolactones: First, they facilitate the germination of parasitic organisms like *Striga lutea* and other plants in the genus *Striga* that develop in the roots of their host plants (Umehara et al. 2015). Second, because they form a mutualistic association with these plants and supply phosphate and other soil nutrients, strigolactones are essential for symbiotic fungi, particularly arbuscular mycorrhizal fungi, to recognize the plant (Umehara et al. 2015). Third, it has been shown that strigolactones are hormones that reduce branching in plants. These substances block the branching process in plants by preventing excess bud growth in stem terminals when they are present (Umehara et al. 2015). According to research by Akiyama et al. (2005), plant roots that emit SLs can induce hyphal branching in arbuscular mycorrhizal (AM) fungi, which can lead to the start of a mutually advantageous symbiotic connection between the plant and the fungi.

12.2 General Structure of Strigolactones

The germination stimulant strigol was the first SL to be isolated from cotton root exudates to be used against the root parasitic plant Striga (Cook et al. 1966). Furthermore, from red clover root exudates, Orobanchol was the first germination stimulant discovered for Orobanche, a different genus of root parasites (Yokota et al. 1998). The methyl butenolide D-ring in strigol is joined to the ABC-ring by an enol–ether bridge. Sorting natural SLs into two groups according to the orientation of the C-ring yields distinct results: Orobanchol-type SLs have an α-oriented C-ring, whereas strigol-type SLs have a β-oriented C-ring. Tobacco plants (*Nicotiana tabacum* CV Michinoku No. 1) generate and exude both types, including orobanchol and 5-deoxystrigol, in contrast to the majority of plant species that were studied (Xie et al. 2013). The fundamental chemical structure of strigolactones is based on a tricyclic lactone connected to a hydroxymethyl butenolide; the former is shown in the picture as the A–B–C component of the molecule, and the latter as the D part (Fig. 12.1) (Umehara et al. 2015). The D-ring is rather constant across species, despite the fact that most strigolactones exhibit changes in the ABC portion of the molecule. This prompted researchers to believe that the biological activity of the molecule depends on this portion of the molecule (Umehara et al. 2015). Numerous investigations have shown that when the molecules' C–D section is altered, the molecules' activity is diminished (Umehara et al. 2015).

A tricyclic ABC-ring system connected to a butenolide D-ring via an enol–ether bridge makes up the main structure of naturally occurring SLs that are frequently isolated from root exudates. The B–C-ring junction in the β-orientation and the butenolide linked in the R configuration at the 20 position were verified to be the absolute configuration of (+)-strigol by single crystal X-ray crystallography. It was believed for many years that all-natural SLs, such as the SL orobanchol that was discovered in red clover root exudates and that encouraged the seeds of root parasite plants in the *Orobanche* genus, included the same stereochemical arrangement (Yokota et al. 1998). Canonical strigolactones (SLs) with the ABC-ring system (such as strigol) are further divided into strigol and orobanchol types according to the C-ring's orientation: Strigol-type SLs have a β-oriented C-ring, while orobanchol-type SLs have an α-oriented C-ring. Non-canonical SLs do not have the A-, B-, or C-ring; instead, they have an enol–ether attached to the methyl butenolide D-ring, an invariable α, β-unsaturated furanone moiety that is necessary for bioactivity. In contrast, heliolactone, zealactone, and avenaol might only be found in the plant

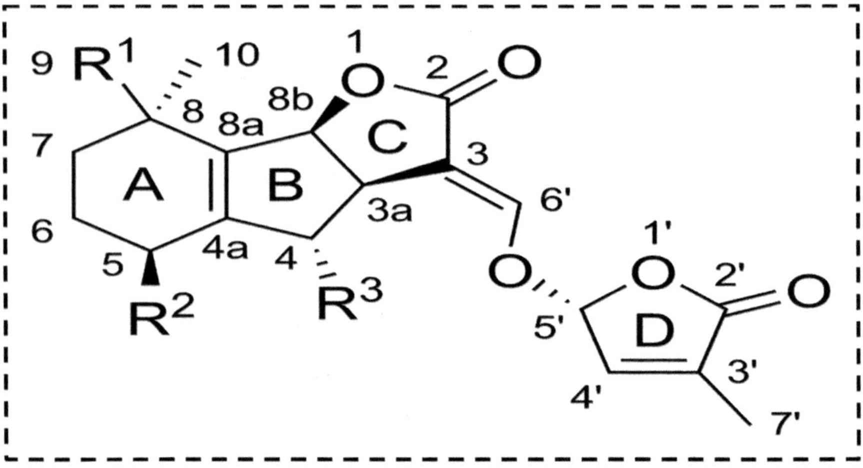

Fig. 12.1 Strigolactone structure

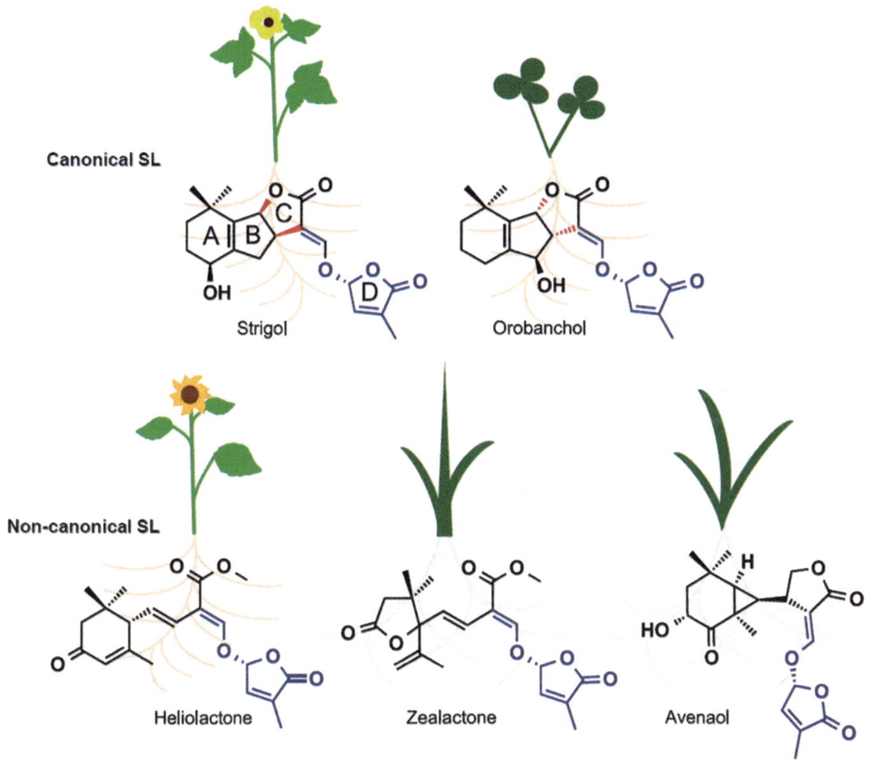

Fig. 12.2 Structures of canonical and non-canonical strigolactones. Strigolactones (SLs) having the ABC-ring system are called canonical SLs (e.g., strigol), and non-canonical SLs lack the A-, B-, or C-ring, but have an enol–ether connected to a methylbutenolide D-ring, which is essential for bioactivity. (Source: Yoneyama and Brewer 2021)

species from which they were first isolated. Strigol, and particularly orobanchol, seems to be broadly spread throughout the plant kingdom. Canonical SLs are those with the ABC-ring system; noncanonical SLs are those without the canonical A-, B-, or C-rings. From the root exudates of wild oat, sunflower, maize, and *Lotus japonicum*, respectively, avenaol, heliolactone, zealactone, and lotuslactone have been identified as germination stimulants for root parasite weeds (Fig. 12.2). These non-canonical SLs have distinct structures.

12.3 Biosynthetic Pathway of Strigolactones

While the complete process of strigolactone biosynthesis remains unknown, a number of steps have been identified, including the enzymes required to finish the chemical transformation (Alder et al. 2012). The initial stage of isomerization involves turning C40 all-trans-β-carotene into 9-cis-β-carotene and is carried out by the enzyme beta-carotene isomerase, also known as DWARF27 or D27. This enzyme requires iron as a cofactor. The chemical separation of 9-cis-beta-carotene into two distinct compounds—9-cis-aldehyde and beta-ionone—by carotenoid cleavage dioxygenase7 (CCD7) is the second stage. D17 in rice, RMS5, RMS1 in pea, DAD3, DAD1 in petunia, and MAX3 (MORE AXILLARY BRANCHING3) in Arabidopsis are the enzymes that catalyze this process. The intermediate carlactone (CL), which has the butenolide D-ring, is formed in the third step when another carotenoid cleavage oxygenase, known as CCD8, catalyzes the conversion and rearrangement of the aldehyde generated in the preceding phase into 9-cis-beta-apo-10 (Alder et al. 2012). The fundamental antecedent of all recognized SLs is carlactone. A cytochrome P450, initially discovered in Arabidopsis as MAX1 (Booker et al. 2005), was implicated in genetic and physiological research pertaining to the final stages of SL production. According to Alder et al. (2012) and Seto et al. (2014), this CL can only be created in the 20R configuration, demonstrating the extremely stereospecific character of CCD8. Lastly, carlactone changes into the various strigolactones. Numerous investigations have demonstrated that carlactone is unquestionably strigolactones' precursor (Seto and Yamaguchi 2014). In order to transform the carlactone into 5-deoxystrigol, a simple strigolactone, at least two oxygen molecules must be added during this final phase of biosynthesis. Additional oxidation should be necessary to make additional, more complicated strigolactones. Because of its function in oxidative metabolism in plants, the protein MAX1 has been suggested to catalyze the final step of the production of strigolactones (Seto and Yamaguchi 2014). *Arabidopsis thaliana* [also known as more axillary growth (max) mutants], *Oryza sativa* [dwarf (d) or high-tillering dwarf (htd) mutants], *Pisum sativum* [ramosus (rms) mutants], and *Petunia hybrida* [decreased apical dominance (dad) mutants] are among the plants that have relatively large collections of branching mutants available. Their analysis has been crucial in defining and comprehending the majority of the current assembly of SL biosynthetic (and signaling) pathways. According to Seto et al. (2014), an analysis of Arabidopsis MAX1 revealed that it is necessary for the transformation of CL into a bioactive signal. Because there are five known MAX1-type genes in rice, the situation is slightly different (Challis et al. 2013). According to Zhang et al. (2014), one of these enzymes, known as CL oxidase (CO), was discovered to convert CL directly into 4DO (ent-20-epi-5DS). Meanwhile, another MAX1-type enzyme catalyzed the hydroxylation of 4DO to form (−)-orobanchol. It is theoretically possible for a CO with a different stereospecificity to oxidize CL to 5DS, and it is likely that MAX1 enzymes in some plants, like cotton, may accomplish this. It is possible that these plants have an additional MAX1-type enzyme that hydroxylates 5DS to produce strigol. It may be feasible to switch between strigol-type and orobanchol-type SLs by alternating CO enzymes between plants. This could have significant biotechnological implications for managing crop susceptibility to parasitic weeds. The primary strigolactone in maize, zealactone, is produced by two parallel strigolactone biosynthesis routes that have been discovered recently. To create zealactone from MeCLA, the enzyme

ZmCYP706C37 catalyzes a series of sequential oxidative reactions with the probable intermediates 3-hydroxy- and 3-oxo-MeCLA. Zealactone was effectively produced from 3-hydroxy- and 3-oxo-MeCLA by ZmCYP706C37. Changes in flux across these routes can affect the equilibrium between zealactone and two additional strigolactones, zealactol and zealactonoic acid (ZA). By altering the strigolactone mix, this discovery has implications for breeding resistance to Striga in maize, thereby lessening the destructive impacts of this parasitic weed in Africa (Li et al. 2023). By promoting seed germination in *Striga hermonthica* and inhibiting shoot branching in SL biosynthesis mutants (rice, Arabidopsis), CL functions as an endogenous precursor for more specialized SLs and displays SL-like characteristics (Scaffidi et al. 2013). One of the genes that functions downstream of CCD8 in Arabidopsis is MAX1, which is responsible for converting CL into functional SLs like 5-deoxystrigol (Fig. 12.3). It encodes a cytochrome P450 belonging to the CYP711A1 lineage (Booker et al. 2005). By rearranging and changing (hydroxylating, oxidizing), MAX1 purportedly transforms CL into functional SL. This results in CL becoming carlactonic acid (CLA), which is subsequently changed into methyl carlactonoate (MeCLA) by an unidentified enzyme (Abe et al. 2014).

The latter, albeit not entirely meeting the usual requirements for SL structure, is sensed by the SL-binding moiety of the SL receptor D14 and exhibits SL-like activity, at least in Arabidopsis. The putative progenitor of rice SL, ent-20-epi-5-deoxystrigol, is produced in rice by one MAX1 paralogue converting carlactone. The conversion of ent-20-epi-5-deoxystrigol to orobanchol is subsequently catalyzed by a protein encoded by a second MAX1 homolog (Zhang et al. 2014), which at least partially explains the stereochemistry at the BC-ring junction, which contributes to the basic chemical variety of SL. Following more research, it was discovered that MAX1 and its orthologs in rice and Arabidopsis are expressed in all vascular tissues, only function in the final stages of SL synthesis, and account for a portion of the structural diversity of SL (Booker et al. 2005). Recombinant MAX1's in vitro interaction with CL in the yeast microsomal system demonstrates that MAX1 specifically converts CL stereoisomer to 9-desmethyl-9-carboxy-C2, also known as carlactonoic acid (CLA), by acting as a CL oxidase. By using LC-MS/MS to detect the endogenous presence of both CLA and MeCLA (methyl ester carlactonoate) in Arabidopsis roots, it is further confirmed that MAX1 targets CL alone. The max1 mutant phenotype in Arabidopsis was found to be rescued by exogenous application of both CLA and MeCLA, which is interesting. However, only MeCLA acts as a substrate to connect with the putative SL receptor of *Arabidopsis thaliana* DWARF14 (AtD14) (Abe et al. 2014). Zhang et al. (2014) conducted labeling experiments in rice to provide elucidative observations on the stereoselectivity of CL endogenously during interaction with MAX1. The chemically generated 13C-labeled CL was converted to 13C-20-epi-5DS and 13C-orobanchol. It was recently shown that another protein functions in Arabidopsis downstream of MAX1. The enzyme known as lateral branching oxidoreductase (LBO) belongs to the Fe(II)-dependent dioxygenase superfamily and is similar to oxidoreductase. Its expression pattern shares some similarities with that of MAX3. Branching

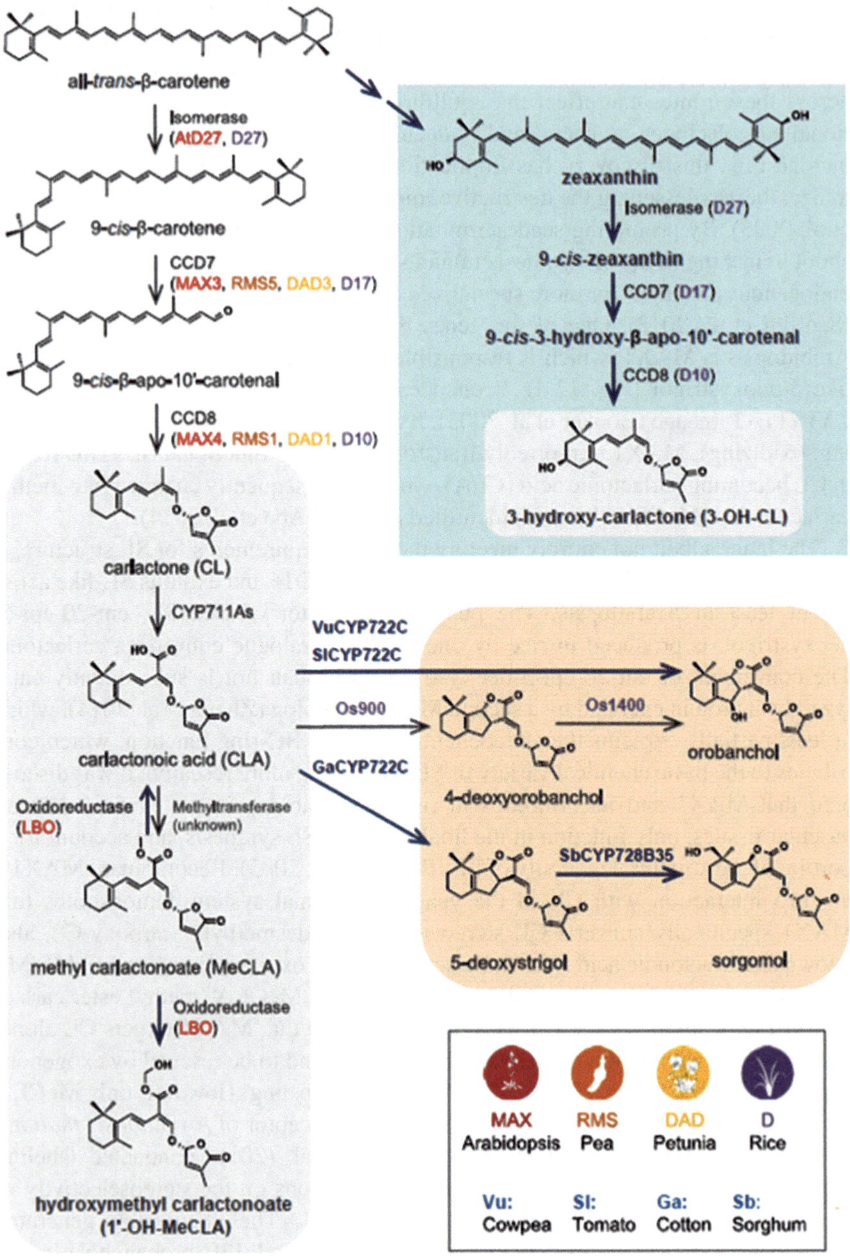

Fig. 12.3 The biosynthetic pathway of strigolactones. The genes of SL biosynthesis were identified and named MAX (MORE AXILLARY GROWTH) in Arabidopsis, RMS (RAMOSUS) in pea, DAD (DECREASED APICAL DOMINANCE) in petunia, and D (DWARF) in rice from mutants with excessive shoot branching. (Source: Yoneyama and Brewer 2021)

inhibition is thought to be caused by an unknown SL-like chemical that LBO transforms from MAX1 products like CLA or MeCLA. This substance is thought to be more potent than SL (Brewer et al. 2016). For the most recent and comprehensive findings on SL biosynthesis, see Koltai and Prandi (2016).

12.4 Role of Strigolactones in Heavy Metal Tolerance in Plants

Environmental metal contamination has increased due to an increase in anthropogenic activities such as mining, modern agriculture, sewage discharge, industrialization, and runoff from the metal-refining industry (Faizan et al. 2022, 2023, 2024a, b). Because the biomes found in the natural world may need certain elements, such as iron, zinc, and copper, in trace amounts, but may be poisonous to them in higher concentrations, heavy metals found in the ecosystem can either be harmful or beneficial to the environment (Gallego et al. 2012). On the other hand, severely poisonous metals such as arsenic (As), lead (Pb), cadmium (Cd), mercury (Hg), Cr, Al, and Be are very bad for plants and can significantly affect their development and production when their concentration rises to above acceptable levels. Based on their toxicity, abundance, and ability to affect living things, the four most dangerous heavy metals are As, Pb, Cd, and Hg, according to the Agency for dangerous Substances and Disease Registry (ATSDR 2003). Strong environmental contaminants, heavy metals, have raised serious concerns for the global ecosystem. They are made up of a collection of metals or metalloids with an atomic density of at least 4 g/cm^3. Plants take up trace amounts of heavy metals from the soil and water through their root cells and the environment. The concentration and bioavailability of HM in the soil system have a significant impact on the amount of uptake and absorption by the plants. In addition, organic matter, pH, the presence of other heavy metals or elements, temperature, redox processes, and organic acids in the rhizosphere all aid in the uptake of heavy metals. Furthermore, plant species and genotype, environmental factors, and the presence of minerals all have a direct impact on the availability and accumulation of heavy metals in the plant system (Volpe et al. 2015). The exact mechanism by which heavy metals (HMs) accumulate in plant parts is dependent upon their penetration through the roots, sequestration therein, transfer in the vascular tissues, and dilution in the plant parts throughout growth and development. Because HMs are so poisonous, they have an adverse effect on various plant species. According to Guilherme et al. (2015) and Raza et al. (2020), their toxicity to plants reduces plant growth and development, prevents germination, and modifies the physiological processes of seedlings. They prevent the germination of beans (El Rasafi et al. 2016), canola (Moosavi et al. 2012), wheat (Guilherme et al. 2015), barley (Kalai et al. 2014), peas (Majeed et al. 2019), and fenugreek (Zayneb et al. 2015). Their toxicity has an adverse influence on the physical characteristics and growth of roots (Daud et al. 2009). It has a deleterious effect on root cell DNA synthesis. Chlorosis, necrosis, reduced development, and desiccation in plants are typical signs of HMs stress. They also lower the amount of chlorophyll, impede photosynthesis, and stunt plant development. They impair the cell's redox potential and cause reactive oxygen species (ROS) to be produced, which in turn causes an oxidative burst inside the cell and destroys biomolecules (Yu et al. 2017). Additional detrimental impacts of heavy metal stress on plants include reduced photosynthesis, altered stomatal movements, altered enzymatic activity, altered metabolic activity, and altered membrane function. Furthermore, their uptake and absorption change the water balance within plant systems, impact phytochemical activity, and nourish plants with minerals (Feng et al. 2010), influencing the metabolism of carbohydrates and photosynthetic enzymes (Mobin and Khan 2007), altering antioxidant levels further (Khan et al. 2009), and ultimately influencing the growth and development of plants. The main factor influencing oxidative stress in plants is elevated HM toxicity within the plant. An increase in ROS production in plant cells, specifically H_2O_2, O_2, and OH-, is closely correlated with an increase in cadmium content. Cadmium uptake causes disruptions in the absorption and transport of mineral nutrients within plant systems, which are necessary for plant growth and development (Gill and Tuteja 2011; Parrotta et al. 2015; El Rasafi et al. 2020). In addition, it has a detrimental impact on grain output, crop growth, and yield-related characteristics. By affecting gene expression, which started specific cellular responses for different stress-related events, the toxicity of heavy metals also changed molecular mechanisms inside plant cells (Kapoor et al. 2021).

Plants use tolerance and avoidance of stress as part of their detoxification process in response to heavy metal stress. These systems either reduce the amount of heavy metal absorption or prevent the damaging effects of heavy metal stress on the plant system.

Plant hormones, also known as phytohormones, are naturally occurring chemicals that are crucial to a plant's morphogenesis, growth, uptake of nutrients, metabolic processes, and cell signaling phenomena. According to Ohri et al. (2015), phytohormones play a linked role and boost the crosstalk signaling pathway in plants. Many phytohormones, including auxin, cytokinin (CK), gibberellins (GA), abscisic acid (ABA), ethylene, brassinosteroids (BRs), and jasmonates, are important for flowering, fruiting, root development, vascular differentiation, senescence, and stem growth. Among the several types of phytohormones, strigolactones (SLs), a recently discovered phytohormone that is also known as an ecological signal and is essential to the functioning of plants, have been identified. According to Zwanenburg et al. (2016), it is a derivative of the carotenoid terpenoid complex molecule, which is composed of a conserved tricyclic lactone ring constrained by a butenolide ring (D-ring). Well-known phytohormones called SLs promote various aspects of plant growth and development, including root development, leaf senescence, and the prevention of shoot branching (Gomez-Roldan et al. 2008; Umehara et al. 2008). Moreover, SLs promote plant height, secondary development, and root hairs while suppressing the formation of lateral roots (LRs) (Kapulnik et al. 2011). Potential use of SLs in plant structures exists. The plant's processes for stress tolerance in SLs are widely recognized. They are linked to abiotic stressors like metals and minerals and are essential to plants' defense systems against water and nutrient shortages. Root architecture and functions are influenced by the bilateral mechanisms that SLs activate with metals and nutrients. Plant extraction, GR24 synthetic analog, and HM pollution, such as Cd stress, can all cause SLs to potentially play a significant role in growth regulation, root development, volume, architecture, and ROS scavenging (Marzec 2016; Wei et al. 2013). In reaction to environmental factors and interactions between flora and fauna, SLs may also be involved in the development of shoots in higher plants (Kapulnik and Koltai 2014). SL inhibits Pi uptake and modifies As accumulation, enhancing plants' ability to withstand stress. A Ca and Mg homeostasis imbalance is also brought on by SL deficiency (Mostofa et al. 2021).

12.5 Conclusion

This chapter concluded that while knowledge of SL-mediated tolerance in plants is developing, further research is required to understand the physiological interactions and downstream signaling components of plants, the molecular mechanisms and physiological effects of SLs, the efficient production of synthetic SLs, and the practical applications of SLs in field settings. Through direct or indirect effects on the actions of other hormones and/or integrands of signaling cascades of other growth regulators, SLs exert their control over physiological and morphological processes. Additionally, this chapter encourages researchers to investigate how SLs might be used to lessen the toxicity of heavy metals in plants.

References

Abe S, Sado A, Tanaka K, Kisugi T, Asami K, Ota S, Kim HI, Yoneyama K, Xie X, Ohnishi T et al (2014) Carlactone is converted to carlactonoic acid by MAX1 in *Arabidopsis* and its methyl ester can directly interact with AtD14 in vitro. Proc Natl Acad Sci USA 111:18084–18089

Agency for Toxic Substances and Disease Registry (ATSDR) (2003) Toxicological profile for Nickel U.S. Department of Health and Human Services, Public Health Service, Division of Toxicology 1600, Atlanta

Akiyama K, Matsuzaki K, Hayashi H (2005) Plant sesquiterpenes induce hyphal branching in arbuscular mycorrhizal fungi. Nature 435:824–827

Al-Babili S, Bouwmeester HJ (2015) Strigolactones, a novel carotenoid-derived plant hormone. Annu Rev Plant Biol 66:161–186

Alder A, Jamil M, Marzorati M, Bruno M, Vermathen M, Bigler P et al (2012) The path from β-carotene to carlactone, a strigolactone-like plant hormone. Science 335(6074):1348–1351

Booker J, Sieberer T, Wright W, Williamson L, Willett B, Stirnberg P, Turnbull C, Srinivasan M, Goddard P, Leyser O (2005) MAX1 encodes a cytochrome P450 family member that acts downstream of MAX3/4 to produce a carotenoid derived branch-inhibiting hormone. Dev Cell 8:443–449

Brewer PB, Yoneyama K, Filardo F, Meyers E, Scaffidi A, Frickey T, Akiyama K, Seto Y, Dun EA, Cremer JE et al (2016) Lateral branching oxidoreductase acts in the final stages of strigolactone biosynthesis in Arabidopsis. Proc Natl Acad Sci USA 113:6301–6306

Challis RJ, Hepworth J, Mouchel C, Waites R, Leyser O (2013) A role for more axillary growth1 (MAX1) in evolutionary diversity in strigolactone signaling upstream of MAX2. Plant Physiol 161:1885–1902

Cohen M, Prandi C, Occhiato EG, Tabasso S, Wininger S, Resnick N, Steinberger Y, Koltai H, Kapulnik Y (2013) Structure-function relations of strigolactone analogs: activity as plant hormones and plant interactions. Mol Plant 6:141–152

Cook CE, Whichard LP, Turner B, Wall ME, Egley GH (1966) Germination of Witchweed (*Striga lutea* Lour.): isolation and properties of a potent stimulant. Science 154(3753):1189–1190

Daud MK, Sun YQ, Dawood M, Hayat Y, Variath MT, Wu YX, Raziuddin MU, Salahuddin NU, Zhu SJ (2009) Cadmium-induced functional and ultrastructural alterations in roots of two transgenic cotton cultivars. J Hazard Mater 161:463–473

El Rasafi T, Nouri M, Bouda S, Haddioui A (2016) The effect of Cd, Zn and Fe on seed germination and early seedling growth of wheat and bean. Ekologia 35(3):213–223. https://doi.org/10.1515/eko-2016-0017

El Rasafi T, Oukarroum A, Haddioui A, Song H, Kwon EE, Bolan N, Tack FMG, Sebastian A, Prasad MNV, Rinklebe J (2020) Cadmium stress in plants: a critical review of the effects, mechanisms, and tolerance strategies. Crit Rev Environ Sci Technol 52. https://doi.org/10.1080/10643389.2020.1835435

EL Sabagh A, Islam MS, Hossain A, Iqbal MA, Mubeen M, Waleed M, Reginato M, Battaglia M, Ahmed S, Rehman A, Arif M, Athar HUR, Ratnasekera D, Danish S, Raza MA, Rajendran K, Mushtaq M, Skalicky M, Brestic M, Soufan W, Fahad S, Fahad S, Hussain S, Matloob A, Khan FA, Khaliq A, Saud S (2015) Phytohormones and plant responses to salinity stress: a review. Plant Growth Regul 75:391–404

Faizan M, Faraz A, Sami F, Siddiqui H, Yusuf M, Gruszka D, Hayat S (2020) Role of strigolactones: signalling and crosstalk with other phytohormones. Open Life Sci 15:217–228

Faizan M, Cheng SH, Tonny SH, Robab MI (2022) Specific roles of strigolactones in plant physiology and remediation of heavy metals from contaminated soil. Plant Physiol Biochem 192:186–195

Faizan M, Karabulut F, Alam P, Yusuf M, Tonny SH, Adil MF, Sehar S, Ahmed SM, Hayat S (2023) Nanobionics: a sustainable agricultural approach towards understanding plant response to heavy metals, drought, and salt stress. Nano 13:974

Faizan M, Karabulut F, Fahad S (2024) Synthetic analogs of strigolactones and their applications. In: Bashri G, Hayat S, Bajguz A (eds) Strigolactones synthesis, application and role in plants. Academic, London

Faizan M, Karabulut F, Alam P, Rajput VD, Sehar S, Hayat S (2024a) Silicon nutrition: a new perspective to improve heavy metal toxicity tolerance in plants for sustainable agriculture. SILICON 16:3191

Faizan M, Alam P, Hussain A, Karabulut F, Tonny SH, Cheng SH, Yusuf M, Adil MF, Sehar S, Alomrani SO, Albalawi T, Hayat S (2024b) Phytochelatins: key regulator against heavy metal toxicity in plants. Plant Stress 11:100355

Feng J, Shi Q, Wang X, Wei M, Yang F, Xu H (2010) Silicon supplementation ameliorated the inhibition of photosynthesis and nitrate metabolism by cadmium (Cd) toxicity in *Cucumis sativus* L. Sci Hortic 123:521–530

Gallego S, Pena L, Barcia R, Azpilicueta C, Iannone F, Rosales E, Zawoznik M, Groppa M, Benavides MP (2012) Unravelling cadmium toxicity and tolerance in plants: insight into regulatory mechanisms. Environ Exp Bot 83:33–46. https://doi.org/10.1016/j.envexpbot.2012.04.006

Gill SS, Tuteja N (2011) Cadmium stress tolerance in crop plants: probing the role of sulfur. Plant Signal Behav 6(2):215–222. https://doi.org/10.4161/psb.6.2.14880

Gomez-Roldan V, Fermas S, Brewer PB, Puech-Pagès V, Dun EA, Pillot J-P, Letisse F, Matusova R, Danoun S, Portais J-C (2008) Strigolactone inhibition of shoot branching. Nature 455:189–194

Guercio AM, Palayam M, Shabek N (2023) Strigolactones: diversity, perception, and hydrolysis. Phytochem Rev 22:339–359

Guilherme M, De FS, Oliveira HM, Silva ED (2015) Cadmium toxicity on seed germination and seedling growth of wheat Triticum aestivum. Acta Sci Biol Sci 37:499. https://doi.org/10.4025/actascibiolsci.v37i4.28148

Kalai T, Khamassi K, Teixeira da Silva JA, Gouia H, Bettaieb Ben-Kaab L (2014) Cadmium and copper stress affect seedling growth and enzymatic activities in germinating barley seeds. Arch Agron Soil Sci 60(6):765–783. https://doi.org/10.1080/03650340.2013.838001

Kapoor D, Singh S, Ramamurthy P, Jan S, Bhardwaj S, Gill S, Prasad R, Singh J (2021) Molecular consequences of cadmium toxicity and its regulatory networks in plants. Plant Gene 28:100342. https://doi.org/10.1016/j.plgene.2021.100342

Kapulnik Y, Koltai H (2014) Strigolactone involvement in root development, response to abiotic stress, and interactions with the biotic soil environment. Plant Physiol 166:560–569

Kapulnik Y, Delaux P-M, Resnick N, Mayzlish-Gati E, Wininger S, Bhattacharya C et al (2011) Strigolactones affect lateral root formation and root hair elongation in Arabidopsis. Planta 233(1):209–216. https://doi.org/10.1007/s00425-010-1310-y

Kee YJ, Ogawa S, Ichihashi Y, Shirasu K, Yoshida S (2023) Strigolactones in rhizosphere communication: multiple molecules with diverse functions. Plant Cell Physiol 64(9):955–966

Khan NA, Anjum NA, Nazar R, Iqbal N (2009) Increased activity of ATP-sulfurylase and increased contents of cysteine and glutathione reduce high cadmium-induced oxidative stress in mustard cultivar with high photosynthetic potential. Russ J Plant Physiol 56:670–677

Koltai H, Prandi C (2016) Strigolactones: past, present and future. Planta 243:1309–1309. https://doi.org/10.1007/s00425-016-2541-3

Li C, Dong L, Durairaj J, Guan J-C, Yoshimura M, Quinodoz P, Horber R, Gaus K, Li J, Setotaw YB, Qi J, De Groote H, Wang Y, Thiombiano B, Floková K (2023) Maize resistance to witchweed through changes in strigolactone biosynthesis. Science 379(6627):94–99

Majeed A, Muhammad Z, Siyar S (2019) Assessment of heavy metal induced stress responses in pea (Pisum sativum L.). Acta Ecol Sin 39(4):284–288. https://doi.org/10.1016/j.chnaes.2018.12.002

Marzec M (2016) Strigolactones as part of the plant defence system. Trends Plant Sci 21:900–903

Marzec M, Melzer M (2018) Regulation of root development and architecture by Strigolactones under optimal and nutrient deficiency conditions. Int J Mol Sci 19(7):1887

Mishra S, Upadhyay S, Shukla RK (2017) The role of strigolactones and their potential cross-talk under hostile ecological conditions in plants. Front Physiol 7:691

Mobin M, Khan NA (2007) Photosynthetic activity, pigment composition and antioxidative response of two mustard (*Brassica juncea*) cultivars differing in photosynthetic capacity subjected to cadmium stress. J Plant Physiol 164:601–610

Moosavi SA, Gharineh MH, Tavakkol Afshari R, Ebrahimi A (2012) Effects of some heavy metals on seed germination characteristics of canola (*Barassica napus*), wheat (*Triticum aestivum*) and safflower (*Carthamus tinctorious*) to evaluate phytoremediation potential of these crops. J Agric Sci 4(9):11–19

Mostofa MG, Rahman MM, Nguyen KH, Li W, Watanabe Y, Tran CD, Zhang M, Itouga M, Fujita M, Tran L-SP (2021) Strigolactones regulate arsenate uptake, vacuolar-sequestration and antioxidant defense responses to resist arsenic toxicity in rice roots. J Hazard Mater 415:125589

Mwakaboko AS, Zwanenburg B (2016) Strigolactone analogues with a D-ring modified at C-2. Eur J Org Chem 21:3495–3499

Ohri P, Bhardwaj R, Bali S, Kaur R, Jasrotia S, Khajuria A, Parihar RD (2015) The common molecular players in plant hormone crosstalk and signaling. Curr Protein Pept Sci 16(5):369–388

Parrotta L, Guerriero G, Sergeant K, Cai G, Hausman J-F (2015) Target or barrier? The cell wall of early- and later-diverging plants vs cadmium toxicity: differences in the response mechanisms. Front Plant Sci 6:133. https://doi.org/10.3389/fpls.2015.00133. PMID: 25814996; PMCID: PMC4357295

Raza A, Habib M, Kakavand SN, Zahid Z, Zahra N, Sharif R, Hasanuzzaman M (2020) Phytoremediation of cadmium: physiological, biochemical, and molecular mechanisms. Biology 9:177. https://doi.org/10.3390/biology9070177

Scaffidi A, Waters MT, Ghisalberti EL, Dixon KW, Flematti GR, Smith SM (2013) Carlactone-independent seedling morphogenesis in Arabidopsis. Plant J 76:1–9

Seto Y, Yamaguchi S (2014) Strigolactone biosynthesis and perception. Curr Opin Plant Biol 21:1–6

Seto Y, Sado A, Asami K, Hanada A, Umehara M, Akiyama K, Yamaguchi S (2014) Carlactone is an endogenous biosynthetic precursor for strigolactones. Proc Natl Acad Sci USA 111:1640–1645

Smith SM, Li J (2014) Signalling and responses to strigolactones and karrikins. Curr Opin Plant Biol 21:23–29

Sun H, Tao J, Hou M, Huang S, Chen S, Liang Z, Xie T, Wei Y, Xie X, Yoneyama K, Xu G, Zhang Y (2015) A strigolactone signal is required for adventitious root formation in rice. Ann Bot 115(7):1155–1162

Umehara M, Hanada A, Yoshida S, Akiyama K, Arite T, Takeda- Kamiya N, Magome H, Kamiya Y, Shirasu K, Yoneyama K et al (2008) Inhibition of shoot branching by new terpenoid plant hormones. Nature 455:195–200

Umehara M, Cao M, Akiyama K, Akatsu T, Seto Y, Hanada A et al (2015) Structural requirements of strigolactones for shoot branching inhibition in rice and arabidopsis. Plant Cell Physiol 56(6):1059–1072

Urquhart S, Foo E, Reid J (2015) The role of strigolactones in photomorphogenesis of pea is limited to adventitious rooting. Physiol Plant 153(3):392–402

Verma V, Ravindran P, Kumar PP (2016) Plant hormone-mediated regulation of stress responses. BMC Plant Biol 16:86

Volpe MG, Nazzaro M, Di Stasio M, Siano F, Coppola R, De Marco A (2015) Content of micronutrients, mineral and trace elements in some Mediterranean spontaneous edible herbs. Chem Cent J 9:57

Wakabayashi T, Ueno K, Sugimoto Y (2022) Structure elucidation and biosynthesis of *Orobanchol*. Front Plant Sci 13:835160

Wakabayashi T, Nakayama M, Kitano Y, Homma M, Miura K, Takikawa H, Mizutani M, Sugimoto Y (2023) Discovery of strigol synthase from cotton (*Gossypium hirsutum*): the enzyme behind the first identified germination stimulant for *Striga*. Plants People Planet 7:1–6

Wang YH, Irving HR (2011) Developing a model of plant hormone interactions. Plant Signal Behav 6(4):494–500

Wasternack C (2014) Action of jasmonates in plant stress responses and development-applied aspects. Biotechnol Adv 32:31–39

Waters MT, Gutjahr C, Bennett T, Nelson DC (2017) Strigolactone signaling and evolution. Annu Rev Plant Biol 68(1):291–322

Wei S, Wang S, Li Y, Zhu J (2013) Root system responses of hyperaccumulator Solanum nigrum L. to Cd. J Soils Sediments 13:1069–1074

Xie X, Yoneyama K, Kisugi T, Uchida K, Ito S, Akiyama K, Hayashi H, Yokota T, Nomura T, Yoneyama K (2013) Confirming stereochemical structures of strigolactones produced by rice and tobacco. Mol Plant 6:153–163

Yokota T, Sakai H, Okuno K, Yoneyama K, Takeuchi Y (1998) Alectrol and orobanchol, germination stimulants for Orobanche minor, from its host red clover. Phytochemistry 49:1967–1973

Yoneyama K, Kisugi T, Xie X, Yoneyama K (2013a) Chemistry of strigolactones: why and how do plants produce so many strigolactones? In: de Bruijin FJ (ed) Molecular microbial ecology of the rhizosphere, vol 1. Wiley, Hoboken, pp 373–379

Yoneyama K, Xie X, Yoneyama K (2013b) Strigolactones and biological activity. In: Ramawat K, Merillon JM (eds) Natural products. Springer, Berlin/Heidelberg

Yu R, Tang Y, Liu C, Du X, Miao C, Shi G (2017) Comparative transcriptomic analysis reveals the roles of ROS scavenging genes in response to cadmium in two pak choi cultivars. Sci Rep 7:9217

Zayneb C, Bassem K, Zeineb K, Grubb CD, Noureddine D, Hafedh M, Amine E (2015) Physiological responses of fenugreek seedlings and plants treated with cadmium. Environ Sci Pollut Res Int 22(14):10679–10689. https://doi.org/10.1007/s11356-015-4270-8

Zhang Y, van Dijk AD, Scaffidi A, Flematti GR, Hofmann M, Charnikhova T, Verstappen F, Hepworth J, van der Krol S, Leyser O et al (2014) Rice cytochrome P450 MAX1 homologs catalyze distinct steps in strigolactone biosynthesis. Nat Chem Biol 10:1028–1033

Zheng Y, Wang X, Cui X, Wang K, Wang Y, He Y (2023) Phytohormones regulate the abiotic stress: an overview of physiological, biochemical, and molecular responses in horticultural crops. Front Plant Sci 13:1095363

Zwanenburg B, Blanco-Ania D (2018) Strigolactones: new plant hormones in the spotlight. J Exp Bot 69(9):2205–2218

Zwanenburg B, Ćavar Zeljković S, Pospíšil T (2016) Synthesis of strigolactones, a strategic account. Pest Manag Sci 72:15–29

Chapter 13
Auxin-Induced Arsenic Regulation in Plants: New Insight

Ekhlaque A. Khan, and Amarendra N. Misra

Abstract Arsenic (As) is a metalloid that is mostly found in soil and water. The presence of As over permitted levels becomes poisonous and harmful to living creatures, making it a major global concern. Because the plants are sessile and acquire As as well as critical minerals and water through their roots, they are more vulnerable to arsenic poisoning and the resulting stress. Arsenic and phosphorus share many physical and chemical properties, and they frequently compete to generate physiological anomalies in biological systems that lead to further stress. A fall in yield and a loss in plant biomass are early markers of arsenic's proclivity to generate toxicity in plants. This is accompanied by significant physiological changes, including an immediate oxidative surge, followed by critical biomolecule oxidation. The phytohormone auxin is essential for plant growth and regulates a variety of developmental processes. The vast array of processes plants use to manage auxin levels, the transit of auxin through the plant, the evolving view of auxin-signaling systems, and various interactions between auxin and arsenic stress are all discussed in this chapter. Interactions between plant hormones and environmental signals are critical for maintaining root development flexibility in the face of constantly changing environmental conditions. This review outlines the nature of arsenic's influence on plants, the physio-biochemical systems that have developed to deal with As stress, and the mitigation techniques that can be used to minimize the detrimental effects of As through auxin supply.

Keywords Auxin · Arsenic · Heavy metal · Stress biology · Toxicity

E. A. Khan (✉)
Department of Biotechnology, Chaudhary Bansi Lal University, Bhiwani, Haryana, India

A. N. Misra
Office of the Vice Chancellor, Jagdguru Kripalu University, Banara, Cuttack, Odisha, India

A. Sharma et al. (eds.), *Hormones and Heavy Metals*, Plant Life and Environment Dynamics, https://doi.org/10.1007/978-981-96-9622-2_13

13.1 Introduction

Plant growth regulators are mainly involved in maintaining plant growth, development, and their interaction with the environmental factors and stimulated responses. Auxin is the key regulator in maintaining plant growth in all parts of plants, predominantly in the root and shoot. Indole-3-acetic acid (IAA) is the naturally occurring auxin form in plants. This small molecule has its local as well as distant effects by synthesis, accumulation, and transport in different cells, tissues, and organs of a plant by some mechanisms (Friml 2022). Auxin is transported from its site of synthesis to a new location by some transporters like MDR/PGP (MULTIDRUG RESISTANCE/P-GLYCOPROTEIN) FAMILY, PIN Protein Family, and AUX/LAX (AUXIN PERMEASE/LIKE AUX) FAMILY. Auxin plays a vital role in regulating the growth and development of plants and is involved in various important processes such as the initiation of roots, elongation of cells, division of cells, differentiation, tropic responses, and apical dominance, in addition to a multitude of other developmental mechanisms. Auxin plays a crucial part in the development, upkeep, and production of primary as well as axillary meristems.

Various types of abiotic stress are present in nature, affecting plants. On the basis of factors causing stress conditions, it may be drought stress, salt stress, chilling stress, osmotic stress, oxidative stress, heavy metal stress, etc., that hampers the activity of plants.

Heavy metals are generally present in the environment, and among these, some potentially toxic elements (PTEs) cause pollution in all the natural resources and are harmful to the living factors of the ecosystem. PTEs present in soil and water are taken up by plants and cause toxicity in the plant system. These toxic elements deteriorate the morphological, physical, biochemical, and molecular processes in plants. PTEs toxicity in the environment arises from some natural processes like volcanic eruptions, pedogenesis, and sedimentation and some anthropogenic activities like industrial and domestic leftovers, mining, smelting, excess use of agrochemicals, and defense activities of armed forces (Palansooriya et al. 2020).

Toxic elements causing contamination of ecosystem are arsenic (As), zinc (Zn), lead (Pb), selenium (Se), cadmium (Cd), vanadium (V), cobalt (Co), silver (Ag), manganese (Mn), copper (Cu), mercury (Hg), antimony (Sb), chromium (Cr), barium (Ba), molybdenum (Mo), titanium (Ti), nickel (Ni), and tin (Sn) (Li et al. 2015). Arsenic is a heavy metal present in different forms in the ecosystem, namely inorganic [arsenate; As(V)], organic [arsenite; As(III)], and methylated derivatives monomethylarsonic acid and dimethyl arsinic acid). Among these, As(V) and As(III) are the most familiar types detected in soil and water, with inorganic As being more abundant than organic As. Aerobic soils have a higher concentration of As(V), while under a submerged environment, As(III) predominates (Abedin et al. 2002). Arsenic phyto-uptake is dependent on its overall concentration as well as its speciation in the rhizosphere (Martínez-Sánchez et al. 2011). Different types of As are absorbed by plant roots through various mechanisms. Arsenic(V) is chemically similar to phosphate (Pi); therefore, Pi transporters (PHTs) are principally responsible for its transport into most cells. Inorganic phosphate and As(V) share the same transporters since they compete with one another for uptake in a variety of plant species (Jiang et al. 2014). Plasma membrane-associated PHT1 transporters primarily engage in Pi absorption and its remobilization in plant systems.

13.2 Auxin Biosynthesis Pathways

In plants, auxin is synthesized in all the parts: roots, stems, and leaves. IAA is synthesized by different tryptophan-dependent and tryptophan-independent pathways. Multiple Trp-dependent pathways are named after the intermediate molecule formed during synthesis of IAA, like the indole-3-acetamide (IAM) pathway, indole-3-pyruvic acid (IPA) pathway, tryptamine pathway, and indole-3-acetaldoxime (IAOx) pathway.

Indole-3-pyruvic acid (IPA) pathway is generally observed in microbes and in plants. In this pathway, tryptophan is oxidatively transaminated to IPA by Trp aminotransferase. After this, indole-3-acetaldehyde (IAAld) is formed from IPA by IPA decarboxylase, and finally, IAAld is converted to IAA by IAAld-specific aldehyde oxidase protein (AAO1) reported in the IAA-overproducing superroot1 (sur1) mutant (Omarov et al. 1999).

The second pathway of IAA biosynthesis in microbes and plants is the indole-3-acetamide (IAM) pathway. Trp is converted to IAM by the enzyme Trp monooxygenase (IaaM) and then IAM to IAA by an IAM hydrolase (IaaH) (Patten and Glick 1996) in *Agrobacterium tumifaciens* and *Pseudomonas syringae*.

Another pathway for the biosynthesis of IAA is the tryptamine (TAM) pathway. In this mechanism, Trp is converted to IAA by various intermediates. As studied in *Catharanthus roseus,* firstly Trp is converted into TAM by Trp decarboxylase (Facchini et al. 2000). Normally, the enzyme Trp decarboxylase is absent in the Arabidopsis plant, but after the discovery of an IAA-accumulating mutant, yucca phenotypes (Zhao et al. 2001), it became evident that the TAM pathway occurs in some plant genera. Mutant yucca shows resistance to harmful Trp analogues, showing that the accumulating IAA is Trp-based (Zhao et al. 2001). The IAA-accumulating mutant arises due to overexpression of an enzyme, flavin monooxygenase (FMO), causing oxidation of tryptamine to N-hydroxyl-tryptamine in vitro (Zhao et al. 2001). In the tryptamine pathway, YUCCA acts as a rate-limiting enzyme, but genetic variants disrupt this statement. The ten YUCCA gene family has ten enzymes in Arabidopsis, and mutations in YUCCA and YUCCA2 show no morphological variants (Zhao et al. 2001). Dehydrogenation of YUCCA produces N-hydroxyl-tryptamine, which is converted into IAOx, then hydrolyzed to IAAld, and finally transformed into IAA.

Indole-3-acetaldoxime (IAOx) pathways of auxin biosynthesis are found in plants producing indolic glucosinolate secondary metabolites (Abdel-Massih et al. 2023). In Arabidopsis, P450 oxygenases CYP79B2 and CYP79B3 lead to conversion of Trp to IAOx in vitro conditions, where it further converts to indole-3-acetonitrile and finally to IAA (Zhao et al. 2002).

13.3 Auxin Transport Across the Plant

Auxin is synthesized from its precursor tryptophan as well as Trp-independent manner in all the plant parts. Even some IAA is also stored in conjugate form with amino acids and sugar. IAA is transported from its synthesis site to different plant parts by two mechanisms: (i) non-directional transport at a speed of 5–10 cm/h; (ii) polar transport at a speed of 5–10 mm/h. Phloem-based transport of auxin takes place in addition to polar transport in the early seedling growth stage (Ljung et al. 2005). Inside plant roots, auxin transport is bidirectional, and due to this, an auxin gradient develops, causing gravitropism (Kramer and Bennett 2006). Inside the shoot system, auxin transport by a polar mechanism from epidermis to vascular bundle in leaves (Swarup and Bennett 2003).

For the distribution of auxin within and between tissues, several auxin import and export proteins are present. Two methods were used to identify auxin transport genes in Arabidopsis mutants: first, abnormal tropisms and stimulus-response coupling were examined using forward and reverse genetics; second, a changed sensitivity to synthetic auxin analogues or inhibitors was noted (Parry et al. 2001).

13.3.1 AUX/LAX (AUXIN PERMEASE/LIKE AUX) FAMILY

In Arabidopsis mutants, acropetal and basipetal auxin transport is controlled by AUX1, which shows expression in root apex portions, generally in protophloem, columella, lateral root cap, and epidermal cells. Apart from AUX1, three closely related LAX1, LAX2, and LAX3 genes also show expression in root and shoot vascular tissues (Swarup et al. 2001). In plants, the AAAP (amino acid and auxin permease) family subgroup is formed from closely related AUX and LAX IAA importers. AUX/LAX shows a very close similarity with the amino acid permease found in most organisms. In yeast, the amino acid permease shows 15% similarity with AUX1, and due to this, IAA has a strong impact on the yap-1 gene.

Structurally AUX/LAX consists of a molecular weight of 55 kDa from its 460–490 amino acid chain, which is not prone to glycosylation, showing that O-linked sites for glycosylation are absent here, and only N-linked sites are found at four sites at membrane-embedded or intra-membrane locations as depicted by SDS/PAGE (Swarup et al. 2004). Apart from this, the AUX/LAX protein sequence shows 60% similarity in sequence, while a 100 amino acid sequence shows dissimilarity in *Momordica charantia*.

13.3.2 PIN Protein Family

The PIN protein family is the second auxin efflux carrier that helps in the polar transport of auxin in roots and vascular tissue. Generally, multiple members of the PIN protein family are reported in different plants, but the main transporters of the PIN family are PIN1 and PIN2. Even a clone of PIN1, known as AGR (AGRAVITROPIC)/EIR(ETHYLENE INSENSITIVE ROOT), is prepared from a mutant. Location of both of these proteins is fixed like PIN1 in the stele region, while PIN2 is in the lateral root cap and outer cortical parts (Paponov et al. 2005).

The PIN protein family is further divided into subfamilies showing 35% of sequence similarity (Paponov et al. 2005). The PIN protein family is made of 650 amino acids with a molecular weight of 65–70 kDa. SDS/PAGE analysis of PIN2 protein also shows the absence of O-linked glycosylation sites and only one N-linked site in an intracellular position.

13.3.3 MDR/PGP (MULTIDRUG RESISTANCE/ P-GLYCOPROTEIN) FAMILY

ATP-binding cassette (ABC) transporter is the third family of auxin carrier found in plants. It is further divided into numerous clades and subfamilies, and out of all, MDR and MRP proteins have sequence homology with mammalian drug pumps (Sánchez-Fernández et al. 2001). Both ABC protein and auxin transport are resistant to 5-nitro-2-(3-phenylpropylamino) benzoic acid (NPPB), an anion channel inhibitor. The Arabidopsis homologue, namely Atmdr1, is identified from a gene showing overexpression after treatment with NPPB in *Brassica,* encoding for proteins AtMDR11/AtPGP19 (Sánchez-Fernández et al. 2001). AtPGP1 has two TM domains, followed by a nucleotide binding domain, an auxin transporter with 1200–1300 amino acids (Tusnády et al. 2006).

13.4 Role of Auxin in Plants

13.4.1 Lateral Root (LR) Patterning and Development

In plant development, LR priming is the process where a group of pericycle cells is arranged to form a pre-branching site (PBS), which can divide to form an LR (Torres-Martínez et al. 2022). In the primary root growth axis, reiterative patterning is present, i.e., these PBSs are arranged at equal distance and equal time every 6 h (Moreno-Risueno et al. 2010). Generally, establishment of PBSs was found in the apical meristem and root elongation zone controlled by the activity of a large number of auxin-responsive genes (Moreno-Risueno et al. 2010). Several models were proposed to show the effect of auxin in LR priming. The first model is the root clock model, showing that cells will autonomously control the cyclic gene expression (Moreno-Risueno et al. 2010), and auxin develops a negative loop involving ARF7 and IAA18 in the root zones (Perianez-Rodriguez et al. 2021). The second model for LR development is the efflux and growth model, which shows the integrated effect of auxin concentration and cell growth in LR development. First, enhanced auxin levels will lead to PBS initiation, and then, auxin signaling leads to the development of PBSs (Teixeira et al. 2022).

13.4.2 Control Fast and Slow Responses During Gravitropism

Auxin shows proximodistal root meristem patterning and controls the gradient in the quiescent center. In this process, several proteins like PLETHORA (PLT) 1 and 2 are the main driving agents, along with auxin, to maintain this auxin gradient and its localization during gravitropism. PLT2 expression is activated by auxin, and this is a slow vector for the redistribution of auxin. Instead of this, by receiving a gravity stimulus, auxin relocation occurs transiently within 6 h at a fast rate (Mähönen et al. 2014). This proves that auxin signaling in the root meristem occurs at two speeds, fast and slow, and these correspond to one another.

13.4.3 Regulation of Organogenesis in the Shoot Apex

On the shoot apex, new organ formation occurs at spatial positions and in a plastochron time gap (average time gap between the initiation of two successive leaves). PIN efflux carriers transported auxin to SAM, where it can initiate organogenesis (Vernoux et al. 2021). It is observed that cells having higher concentrations of auxin do not always show a higher transcriptional rate (Galvan-Ampudia et al. 2020). During studies, it is observed that genetic and epigenetic modifications in the NAPs are controlled by auxin signaling, which controls the transcriptional response in a time gap between the initiation of successive leaf primordial, i.e., plastochron, and mediates the organ development in spatial position in a fixed interval of time (Galvan-Ampudia et al. 2020).

13.4.4 Regulation of Female Gametophyte Development in Arabidopsis

In Arabidopsis embryo sac 7-celled, 8-nucleus arrangement of cells is present before the fertilization event. After fertilization, gametic and non-gametic cells divide in different planes to maintain their segregation properly. At the time of gametophyte development, it is found that an auxin reporter DR5-GFP is found in the embryo sac with its maxima at the micropylar end (Pagnussat et al. 2009). Even at this time of development, the auxin efflux carriers (PIN proteins) are not working, and YUC genes for IAA biosynthesis are expressed in the embryo sac, which proves local synthesis of IAA. Even faulty expression of YUC1 leads to a change in the cell fate of gametic and non-gametic cells (Pagnussat et al. 2009).

13.5 Abiotic Stress in Plants

Plant growth, development, yield, and seed quality are affected by various abiotic stresses, such as water stress, cold, frost, heat stress, waterlogging, salinity stress, and toxic metals. Due to excessive exploitation of water resources, its availability will be less in the future, and the severity of abiotic stress will increase. Plant hormones help plants adapt to abiotic stress. Some of the abiotic stresses occurring in plants are discussed below.

13.5.1 Cold Stress

Both chilling and freezing conditions in plants at low temperatures stop the action of cellular enzymes and retard the growth and development of the plant body. In cold stress, the fluidity of the cell membrane changes due to modifications in protein-lipid composition (Wang et al. 2006). Some plants can adapt themselves to this extreme condition of cold and show acclimation. Decreased membrane fluidity leads to the activation of several calcium-permeable channels and the expression of downstream genes. Several signal transduction pathways exist by which these cold stresses are transduced, like components of reactive oxygen species (ROS), protein kinase, protein phosphatase, abscisic acid (ABA), and calcium ion. ABA plays a main role in mitigating the response to abiotic stress in plants.

13.5.2 Drought Stress

In water-constrained conditions, all the mechanisms of plants involving water, like stomatal movement, respiration, and photosynthesis, are affected. Plant produces various responses under drought stress, and signal molecules are produced. Growth of plant shoots under drought will be reduced due to decreased cell expansion, increased defoliation, and abnormal mitosis (Yang et al. 2021). In cotton, drought stress inhibits the development of seedlings, enhances the elongation and thinning of roots, shortens lifespan, promotes the elongation of root hairs, and accelerates death (Xiao et al. 2020). In plants, LEA (late embryogenesis abundant protein) plays a main role in resistance against abiotic stress.

13.5.3 Oxidative Stress

During the entire abiotic stress, like drought, cold, salinity, and heavy metals, in plants, reactive oxygen species (ROS) are formed. These ROS cause damage in the plant cell due to their high penetration ability. To control the negative effect of ROS, some regulators like glutathione (GSH) and thioredoxin (TRX) are present (Hasanuzzaman et al. 2020). Auxin receptor and transporters help reduce the level of H_2O_2 accumulation, increase the antioxidant activity of enzymes, and elevate chlorophyll and ABA levels in Arabidopsis mutants (Iglesias et al. 2010).

13.5.4 Salinity Stress

Salinity stress is a major factor affecting our crops. Approximately 45 million hectares of agricultural land are under stress all over the world, and by 2050, more than 50% of agricultural land could be converted into salt-affected land (Munns and Tester 2008). Mostly semi-arid and arid regions are mostly affected by salt stress. So, due to aridity and high temperature conditions, the salt stress became more severe in these areas due to the cumulative effect of these factors (Huang et al. 2019). Salt stress induces the formation of ROS, which causes damage to proteins, lipids, and nucleic acid biomolecules, and changes redox homeostasis of the plant (Waqas et al. 2019). Recent studies show the crucial role of mitochondria in plant defense against stress caused by salt (Paiva et al. 2019).

13.5.5 Heavy Metal Stress

Plants intake micronutrients as well as macronutrients from the soil solution for their proper development and growth. Due to industrialization, contamination of metals with agricultural land with metals results in a decline in crop yield (Proshad et al. 2018). This exposure of heavy metals to the ecosystem is a threat to both our environment and agriculture. The use of excessive chemical fertilizers, synthetic nutrients, and improper irrigation methods is the main cause of metal toxicity to the crops (Minhas et al. 2017). Fe, Cu, Ni, Zn, Co, and Se are heavy metals; they act as essential elements required for plants for their proper functions and become toxic for plants when deposited in soil in excess amounts (Khan et al. 2018). On the other hand, arsenate (As), cesium (Cs), lead (Pb), and cadmium (Cd), like non-essential elements, can adversely affect the productivity of crops even when deposited in soil in very low amounts (Khalid et al. 2018).

13.6 Arsenic Stress in Plants

Plants absorb micronutrients as well as macronutrients from the soil to grow properly. Some heavy metals are also required in plants in trace amounts. These heavy metals were absorbed by the plant system along with other elements. Arsenic is found in soil as well as water in organic form as arsenite, as well as inorganic form as arsenate, which are most common. In plants, phosphate transporters are present in roots for uptake of Pi, but arsenic (V) has a similar structure to phosphate, so it enters rapidly through these transporters (Li et al. 2016). Aquaporin subfamily proteins Nodulin 26-like intrinsic proteins (NIPs) help in the transport of As(III) and silicon (Mitani-Ueno et al. 2011).

Arsenic induces multiple harmful effects in plants at morphological, biochemical, and physiological levels (Fig. 13.1). Plant roots at the site of As exposure are small and stop to proliferate; after this, As passes to the shoot, and it also stops

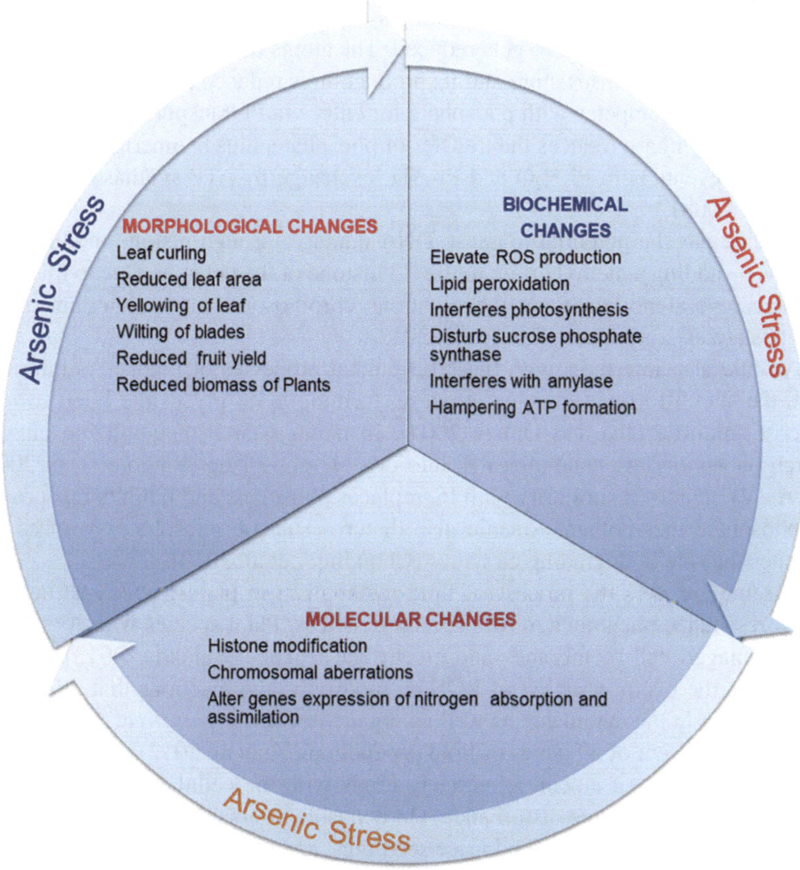

Fig. 13.1 Arsenic-induced multiple harmful effects in plants

division and growth. Due to this, both nutrient transfer and plant biomass decrease. The toxicity of As also affects the grain production and yield in plants. Affected grains show low germination potential, reduced leaf area, yellowing of leaves, and wilting of blades. This enhances leaf defoliation and reduces fruit yield (Srivastava et al. 2022). At higher concentrations, As inhibits metabolic and biochemical processes in sunflowers, which can also cause death of the plant (Muhammad et al. 2013). In soybean plants, high tissue injury and reduced biomass of root and shoot are reported by As exposure (Reichman 2014).

Arsenic also enhances the production of reactive oxygen species (ROS) in plants. In general, mitochondria produce ROS, but As toxicity on succinic dehydrogenase enhances additional ROS formation (Thakur et al. 2021). Also, NAD(P)H oxidase, a membrane-bound enzyme, is used for ROS formation (Vermot et al. 2021). Increased ROS and oxidative stress lead to lipid peroxidation in membranes, causing cellular content leakage (Ahmad et al. 2020).

Arsenic also reduces photosynthesis in plants by interfering with photosystem II, chlorophyll, and carotenoids (Gajić et al. 2020). During As absorption, PSII activity decreases, hampering the light-harvesting system and other processes. Along with this, chlorophyll amount also gets reduced. The atoms of As occupy the position of Mg in the chlorophyll structure and its production is reduced (Zhang et al. 2021).

Arsenic (V) competes with phosphate for entry via Pi transporters. So, As at the site of ATP synthase reduces the transfer of phosphate, thus hampering ATP formation. As prevents both Pi-H_2O and Pi-ATP exchange by ATP synthase (Hong and Pedersen 2008).

Arsenic has the potential to cause DNA damage through histone modification, such as by adding a methyl group to the H3 histone of the octamer. Due to this DNA damage, several modifications like mutations, chromosomal aberrations, and abnormalities are also produced (Faita et al. 2013).

Arsenic also interferes with the biochemical processes of plants. As interferes with the starch-forming enzyme amylase, forming a 1,4-glycosidic bond to form glucose, maltose (Jha and Dubey 2004). In plants, starch metabolizing enzyme starch phosphorylase catabolizes Pi into organic phosphate (Rathore et al. 2009), but As(V), due to its similarity with Pi, replaces phosphate and inhibits this process of phosphate metabolism. Arsenic also disturbs sucrose phosphate synthase and elevates the role of degrading enzymes (Choudhury et al. 2010).

Arsenic hampers the process of lipid metabolism in plants, which is the main stress resistance phenomenon against abiotic stress. Plant species with As toxicity have damaged cell membranes and electrolyte leakage (Talukdar 2013). Arsenic stimulates the overproduction of ROS by modifying the reactions that affect lipid peroxidation in the cytoplasm as well as organelles. In *Oryza sativa*, As(III) modified the expression of 59 genes of lipid production (Yu et al. 2012).

Inorganic form of arsenic effectively binds with the cellular proteins, causing cell death and membrane disturbance. On exposure to arsenic, protein content will be reduced (Zhang et al. 2021). In rice seedlings, arsenic reduces protein-degrading enzyme and RNA-degrading enzyme activity, so accumulation of RNA and protein starts (Mishra and Dubey 2006).

13.7 Molecular Mechanisms and Regulation of Auxin-Induced Arsenic Regulation in Plants

In plants, auxin is the main growth hormone regulating root growth, division, and differentiation of cells in different root zones. Auxin works in collaboration with other plant growth hormones to regulate plant growth. In roots ASA1 (Anthranilate Synthase A1) and ASB1 (Anthranilate Synthase B1) genes are involved in auxin biosynthesis. In the soil, arsenic is found in inorganic as well as organic form and transported to the plant cell via Pi transporters due to its similarity with phosphate. In the root cells, As toxicity leads to inhibition of root growth (Yoon et al. 2015). In Arabidopsis, it was reported that As(V) stimulated the production of auxin at the early stage of root response. Within 12 h of As(V) exposure, the endogenous level of auxin increases by stimulating the expression of DR5:GUS in the root tips (Tu et al. 2021). As(V) also upregulates the expression of several auxin biosynthetic genes like ASA1 and ASB1, as analyzed in their transcription study. Due to increased expression of these auxin biosynthetic genes and local auxin biosynthesis, the level of auxin increases in As(V) affected plants (Zhao et al. 2010), which enhances the root growth. Generally, a high level of auxin by the activity of ASA1 and ASB1 is reported in the root cap, proving that the root cap is the main site in direct contact with arsenic during transport from water or soil to the plant.

In Arabidopsis, root growth inhibition is observed in asa1 asb1 double mutant seedlings, but after exposure to As(V), auxin synthesis is elevated. This shows that ASA1 and ASB1 are important genes in plants to maintain auxin concentration at a higher level during arsenic stress.

To find out the mechanism of auxin synthesis gene upregulation in Arabidopsis, a transcription level analysis of the *ASA1* gene and its promoter was carried out using a yeast one-hybrid (Y1H) screening (Tu et al. 2021). The *ASA1* gene promoter controls HISTIDINE 3(HIS3) reporter genes. Seven transcription factors like ARR1 and ARR10 were found by coding of 35 cDNA clones, and ARR1 is the regulator of ASA1. The ASA1 promoter has ARR1-binding sites (ABSs) by which its regulator ARR1 binds and controls its expression. The repression of ASA1 with its regulator ARR1 is also reported in the leaves of *Nicotiana benthamiana,* and the activity of the luciferase (LUC) gene was observed under the ASA1 promoter (Tu et al. 2021).

In roots, the opposite action of auxin and cytokinin hormones regulates the root apical meristem (RAM) development and differentiation, and cell elongation in the zone of elongation (ZE) (Dello Ioio et al. 2008). During As(V) stress, the size of both RAM and ZE reduced. Within 6 h of As toxicity, the level of auxin increases in the terminal zone (TZ) of the root by expression of DII-VENUS auxin reporter (Tu et al. 2021).

13.8 Conclusion

During the last 50 years, our understanding of auxin signaling in cell growth has gone through at least three stages. The emphasis in the premolecular era was on physiological and biochemical studies of auxin action, such as the discovery of apoplast pH as a critical mechanism for auxin promotion of growth, known as the acid growth theory. The development of modern molecular biology stimulated interest in the regulation of gene expression in the nucleus by auxin. The discipline is currently entering a new phase that emphasizes extranuclear signaling. In the future, it is expected that the integration of intranuclear and extranuclear routes will dominate our understanding of the processes underpinning a specific auxin response, providing fresh insights into auxin's adaptability.

Arsenic, a metalloid, is a major environmental pollutant that causes serious health problems in living things, including humans, making it a local and global concern. Arsenic exists as both an inorganic and an organic contaminant with varying oxidation states, making it highly reactive and mobile, determining its bioavailability and toxicity. Other parameters that influence As′s physio-biochemical qualities include soil pH, redox potential, solubility, and microbial activity. These parameters influence As sensing, uptake, transport, and distribution in plant cells and tissues, as well as its accumulation, detoxification, and homeostasis. Arsenic causes unfavorable effects in plants at the morpho-physio-biochemical levels, resulting in a decrease in expansion, development, and productivity. Researchers are currently working to develop eco-friendly solutions, such as phytoremediation, as well as boosting plants with exogenous auxin supplementation to help alleviate the harmful effects of As. A new season of lucrative auxin research awaits.

References

Abdel-Massih RM et al (2023) Glucosinolates, a natural chemical arsenal: more to tell than the myrosinase story. Front Microbiol [Preprint]. https://doi.org/10.3389/fmicb.2023.1130208

Abedin MJ, Feldmann J, Meharg AA (2002) Uptake kinetics of arsenic species in rice plants. Plant Physiol [Preprint]. https://doi.org/10.1104/pp.010733

Ahmad P et al (2020) Zinc oxide nanoparticles application alleviates arsenic (As) toxicity in soybean plants by restricting the uptake of as and modulating key biochemical attributes, antioxidant enzymes, ascorbate-glutathione cycle and glyoxalase system. Plants [Preprint]. https://doi.org/10.3390/plants9070825

Choudhury B, Mitra S, Biswas AK (2010) Regulation of sugar metabolism in rice (Oryza sativa L.) seedlings under arsenate toxicity and its improvement by phosphate. Physiol Mol Biol Plants [Preprint]. https://doi.org/10.1007/s12298-010-0008-8

Dello Ioio R et al (2008) A genetic framework for the control of cell division and differentiation in the root meristem. Science [Preprint]. https://doi.org/10.1126/science.1164147

Facchini PJ, Huber-Allanach KL, Tari LW (2000) Plant aromatic L-amino acid decarboxylases: evolution, biochemistry, regulation, and metabolic engineering applications. Phytochemistry [Preprint]. https://doi.org/10.1016/S0031-9422(00)00050-9

Faita F et al (2013) Arsenic-induced genotoxicity and genetic susceptibility to arsenic-related pathologies. Int J Environ Res Public Health [Preprint]. https://doi.org/10.3390/ijerph10041527
Friml J (2022) Fourteen stations of auxin. Cold Spring Harb Perspect Biol [Preprint]. https://doi.org/10.1101/cshperspect.a039859
Gajić G et al (2020) Phytoremediation potential, photosynthetic and antioxidant response to arsenic-induced stress of dactylis glomerata L. Sown on fly ash deposits. Plants [Preprint]. https://doi.org/10.3390/plants9050657
Galvan-Ampudia CS et al (2020) Temporal integration of auxin information for the regulation of patterning. eLife [Preprint]. https://doi.org/10.7554/eLife.55832
Hasanuzzaman M et al (2020) Reactive oxygen species and antioxidant defense in plants under abiotic stress: revisiting the crucial role of a universal defense regulator. Antioxidants [Preprint]. https://doi.org/10.3390/antiox9080681
Hong S, Pedersen PL (2008) ATP synthase and the actions of inhibitors utilized to study its roles in human health, disease, and other scientific areas. Microbiol Mol Biol Rev [Preprint]. https://doi.org/10.1128/mmbr.00016-08
Huang M et al (2019) Increasing aridity affects soil archaeal communities by mediating soil niches in semi-arid regions. Sci Total Environ [Preprint]. https://doi.org/10.1016/j.scitotenv.2018.07.305
Iglesias MJ et al (2010) Auxin signaling participates in the adaptative response against oxidative stress and salinity by interacting with redox metabolism in Arabidopsis. Plant Mol Biol [Preprint]. https://doi.org/10.1007/s11103-010-9667-7
Jha AB, Dubey RS (2004) Carbohydrate metabolism in growing rice seedlings under arsenic toxicity. J Plant Physiol [Preprint]. https://doi.org/10.1016/j.jplph.2004.01.004
Jiang W et al (2014) Evaluation of potential effects of soil available phosphorus on soil arsenic availability and paddy rice inorganic arsenic content. Environ Pollut [Preprint]. https://doi.org/10.1016/j.envpol.2014.02.014
Khalid S et al (2018) A review of environmental contamination and health risk assessment of wastewater use for crop irrigation with a focus on low and high-income countries. Int J Environ Res Public Health [Preprint]. https://doi.org/10.3390/ijerph15050895
Khan ZI et al (2018) Determination of toxic metals in fruits of Abelmoschus esculentus grown in contaminated soils with different irrigation sources by spectroscopic method. Int J Environ Res [Preprint]. https://doi.org/10.1007/s41742-018-0110-2
Kramer EM, Bennett MJ (2006) Auxin transport: a field in flux. Trends Plant Sci [Preprint]. https://doi.org/10.1016/j.tplants.2006.06.002
Li N et al (2015) Concentration and transportation of heavy metals in vegetables and risk assessment of human exposure to bioaccessible heavy metals in soil near a waste-incinerator site, South China. Sci Total Environ [Preprint]. https://doi.org/10.1016/j.scitotenv.2015.03.081
Li N, Wang J, Song WY (2016) Arsenic uptake and translocation in plants. Plant Cell Physiol [Preprint]. https://doi.org/10.1093/pcp/pcv143
Ljung K et al (2005) Sites and regulation of auxin biosynthesis in arabidopsis roots. Plant Cell [Preprint]. https://doi.org/10.1105/tpc.104.029272
Mähönen AP et al (2014) PLETHORA gradient formation mechanism separates auxin responses. Nature [Preprint]. https://doi.org/10.1038/nature13663
Martínez-Sánchez MJ et al (2011) Evaluation of arsenic in soils and plant uptake using various chemical extraction methods in soils affected by old mining activities. Geoderma [Preprint]. https://doi.org/10.1016/j.geoderma.2010.11.001
Minhas PS, Rane J, Pasala RK (2017) Abiotic stress management for resilient agriculture. https://doi.org/10.1007/978-981-10-5744-1
Mishra S, Dubey RS (2006) Inhibition of ribonuclease and protease activities in arsenic exposed rice seedlings: role of proline as enzyme protectant. J Plant Physiol [Preprint]. https://doi.org/10.1016/j.jplph.2005.08.003
Mitani-Ueno N et al (2011) The aromatic/arginine selectivity filter of NIP aquaporins plays a critical role in substrate selectivity for silicon, boron, and arsenic. J Exp Bot [Preprint]. https://doi.org/10.1093/jxb/err158

Moreno-Risueno MA et al (2010) Oscillating gene expression determines competence for periodic Arabidopsis root branching. Science [Preprint]. https://doi.org/10.1126/science.1191937
Muhammad AI et al (2013) Toxicity of arsenic (As) on seed germination of sunflower (Helianthus annuus L.). Int J Phys Sci [Preprint]. https://doi.org/10.5897/ijps2013.3894
Munns R, Tester M (2008) Mechanisms of salinity tolerance', Annu Rev Plant Biol [Preprint]. https://doi.org/10.1146/annurev.arplant.59.032607.092911
Omarov RT et al (1999) Aldehyde oxidase in roots, leaves and seeds of barley (Hordeum vulgare L.). J Exp Bot [Preprint]. https://doi.org/10.1093/jxb/50.330.63
Pagnussat GC et al (2009) Auxin-dependent patterning and gamete specification in the Arabidopsis female gametophyte. Science [Preprint]. https://doi.org/10.1126/science.1167324
Paiva ALS et al (2019) Mitochondrial glutathione peroxidase (OsGPX3) has a crucial role in rice protection against salt stress. Environ Exp Bot [Preprint]. https://doi.org/10.1016/j.envexpbot.2018.10.027
Palansooriya KN et al (2020) Soil amendments for immobilization of potentially toxic elements in contaminated soils: a critical review. Environ Int [Preprint]. https://doi.org/10.1016/j.envint.2019.105046
Paponov IA et al (2005) The PIN auxin efflux facilitators: evolutionary and functional perspectives. Trends Plant Sci [Preprint]. https://doi.org/10.1016/j.tplants.2005.02.009
Parry G et al (2001) Novel auxin transport inhibitors phenocopy the auxin influx carrier mutation aux1. Plant J [Preprint]. https://doi.org/10.1046/j.1365-313X.2001.00970.x
Patten CL, Glick BR (1996) Bacterial biosynthesis of indole-3-acetic acid. Can J Microbiol [Preprint]. https://doi.org/10.1139/m96-032
Perianez-Rodriguez J et al (2021) An auxin-regulable oscillatory circuit drives the root clock in arabidopsis. Sci Adv [Preprint]. https://doi.org/10.1126/sciadv.abd4722
Proshad R et al (2018) Heavy metal toxicity in agricultural soil due to rapid industrialization in Bangladesh: a review. Int J Adv Geosci [Preprint]. https://doi.org/10.14419/ijag.v6i1.9174
Rathore RS et al (2009) Starch phosphorylase: role in starch metabolism and biotechnological applications. Crit Rev Biotechnol [Preprint]. https://doi.org/10.1080/07388550902926063
Reichman SM (2014) Probing the plant growth-promoting and heavy metal tolerance characteristics of Bradyrhizobium japonicum CB1809. Eur J Soil Biol [Preprint]. https://doi.org/10.1016/j.ejsobi.2014.04.001
Sánchez-Fernández R et al (2001) The Arabidopsis thaliana ABC protein superfamily, a complete inventory. J Biol Chem [Preprint]. https://doi.org/10.1074/jbc.M103104200
Srivastava PK et al (2022) Arsenic in plants: uptake, consequences and remediation techniques. https://doi.org/10.1002/9781119791461
Swarup R, Bennett M (2003) Auxin transport: the fountain of life in plants?. Dev Cell [Preprint]. https://doi.org/10.1016/S1534-5807(03)00370-8
Swarup R et al (2001) Localization of the auxin permease AUX1 suggests two functionally distinct hormone transport pathways operate in the Arabidopsis root apex. Genes Dev [Preprint]. https://doi.org/10.1101/gad.210501
Swarup R et al (2004) Structure-function analysis of the presumptive Arabidopsis auxin permease AUX1. Plant Cell [Preprint]. https://doi.org/10.1105/tpc.104.024737
Talukdar D (2013) Arsenic-induced oxidative stress in the common bean legume, Phaseolus vulgaris L. seedlings and its amelioration by exogenous nitric oxide. Physiol Mol Biol Plants [Preprint]. https://doi.org/10.1007/s12298-012-0140-8
Teixeira JS, van den Berg T, ten Tusscher K (2022) Complementary roles for auxin and auxin signalling revealed by reverse engineering lateral root stable prebranch site formation. Development (Cambridge) [Preprint]. https://doi.org/10.1242/dev.200927
Thakur M et al (2021) Molecular mechanism of arsenic-induced neurotoxicity including neuronal dysfunctions. Int J Mol Sci [Preprint]. https://doi.org/10.3390/ijms221810077
Torres-Martínez HH, Napsucialy-Mendivil S, Dubrovsky JG (2022) Cellular and molecular bases of lateral root initiation and morphogenesis. Curr Opin Plant Biol [Preprint]. https://doi.org/10.1016/j.pbi.2021.102115

Tu T et al (2021) Coordinated cytokinin signaling and auxin biosynthesis mediates arsenate-induced root growth inhibition. Plant Physiol [Preprint]. https://doi.org/10.1093/PLPHYS/KIAA072
Tusnády GE et al (2006) Membrane topology of human ABC proteins. FEBS Letters [Preprint]. https://doi.org/10.1016/j.febslet.2005.11.040
Vermot A et al (2021) Nadph oxidases (Nox): An overview from discovery, molecular mechanisms to physiology and pathology. Antioxidants [Preprint]. https://doi.org/10.3390/antiox10060890
Vernoux T, Besnard F, Godin C (2021) What shoots can teach about theories of plant form. Nat Plants [Preprint]. https://doi.org/10.1038/s41477-021-00930-0
Wang X et al (2006) Profiling lipid changes in plant response to low temperatures. Physiol Plant [Preprint]. https://doi.org/10.1111/j.1399-3054.2006.00622.x
Waqas MA et al (2019) Potential mechanisms of abiotic stress tolerance in crop plants induced by Thiourea. Front Plant Sci [Preprint]. https://doi.org/10.3389/fpls.2019.01336
Xiao S et al (2020) Fine root and root hair morphology of cotton under drought stress revealed with RhizoPot. J Agron Crop Sci [Preprint]. https://doi.org/10.1111/jac.12429
Yang X et al (2021) Response mechanism of plants to drought stress. Horticulturae [Preprint]. https://doi.org/10.3390/horticulturae7030050
Yoon Y, Lee WM, An YJ (2015) Phytotoxicity of arsenic compounds on crop plant seedlings. Environ Sci Pollut Res [Preprint]. https://doi.org/10.1007/s11356-015-4317-x
Yu LJ et al (2012) Comparative transcriptome analysis of transporters, phytohormone and lipid metabolism pathways in response to arsenic stress in rice (Oryza sativa). New Phytologist [Preprint]. https://doi.org/10.1111/j.1469-8137.2012.04154.x
Zhang J et al (2021) Arsenic transport and interaction with plant metabolism: clues for improving agricultural productivity and food safety. Environ Pollut [Preprint]. https://doi.org/10.1016/j.envpol.2021.117987
Zhao Y et al (2001). A role for flavin monooxygenase-like enzymes in auxin biosynthesis. Science [Preprint]. https://doi.org/10.1126/science.291.5502.306
Zhao Y et al (2002) Trp-dependent auxin biosynthesis in Arabidopsis: involvement of cytochrome P450s CYP79B2 and CYP79B3. Genes Dev [Preprint]. https://doi.org/10.1101/gad.1035402
Zhao FJ, McGrath SP, Meharg AA (2010) Arsenic as a food chain contaminant: mechanisms of plant uptake and metabolism and mitigation strategies. Annu Rev Plant Biol [Preprint]. https://doi.org/10.1146/annurev-arplant-042809-112152

Chapter 14
Phytohormones and Heavy Metal Stress Combination in Plants: An Overview

Shivam Jasrotia

Abstract Heavy metal (HM) stress is one of the main abiotic stresses that affect biochemical, physiological, and molecular processes in plants, hence reducing the plant's production potential and yield. Globally, agricultural crops are at risk due to heavy metal stress. Because HM toxicity disrupts cellular ionic equilibrium, cell membrane integrity, metabolic balance, and enzymatic activities, it has a negative impact on the development and physiological processes of plants and their yield. Scientists are employing helpful techniques to guarantee food security in response to abiotic stresses and the growing global population. It has long been thought that using organic compounds might increase resistance to abiotic stresses. Under heavy metal stress, plants naturally generate defence mechanisms against their adversities by using phytohormones. Recently, there has been a lot of interest in controlling the harmful consequences of heavy metal stress by the exogenous administration of phytohormones. Such signalling transduction compounds, i.e. phytohormones, help plants grow and develop while reducing the toxicity of heavy metals.

Keywords Abiotic stress · Ecophysiology · Hormones · Metal stress · Plant stress

14.1 Introduction

14.1.1 Overview of Phytohormones

Small endogenous signalling molecules called auxin, gibberellins, cytokinin, abscisic acid (ABA), ethylene, brassinosteroids, salicylic acid (SA), and jasmonic acid (JA) are examples of plant growth regulators also known as phytohormones. By activating a number of antioxidative enzymes, phytohormones play a significant role in primary root development, leaf senescence, upregulation of photosynthesis, cell signalling, boosting plant biomass, and heavy metal detoxification. The

---

S. Jasrotia (✉)
Department of Biosciences, Chandigarh University, Mohali, Punjab, India

A. Sharma et al. (eds.), *Hormones and Heavy Metals*, Plant Life and Environment Dynamics, https://doi.org/10.1007/978-981-96-9622-2_14

hormones ABA, SA, JA, and ethylene are engaged in modulating plant defence responses against biotic and abiotic constraints, while auxin, gibberellin, and cytokinin are known to play key roles in plant developmental responses (Khan et al. 2020; Sytar et al. 2019).

14.1.2 Plants and HM (Heavy Metal) Toxicity

Since plants are sessile organisms, they are powerless to defend against the consequences of climate change. Agriculture soils around the world have been found to contain heavy metal toxicity where the contamination ranges from a minor to a significant level due to lead (Pb), chromium (Cr), cadmium (Cd), zinc (Zn), copper (Cu), nickel (Ni), and arsenic (As) (Rahman et al. 2022; Rasheed et al. 2021; Wang et al. 2021a). These pollutants may originate from the following sources: continuously using phosphatic fertilisers for a long time; sewage sludge application; smelting dust; hospital and industrial waste; and inadequate irrigation practices on agricultural land (Yasir et al. 2021; Hussain et al. 2021). Plants' major reaction to elevated concentrations of heavy metals is an overproduction of reactive oxygen species (ROS), which is detrimental to the development and survival of the plant (Rahman et al. 2021b; Rahman et al. 2021a). Lipid peroxidation, which can result in biomembrane damage, is thought to be the most dangerous adverse effect of heavy metal exposure, aside from the overproduction of ROS. Malondialdehyde (MDA) overexpression has been found to be a reliable marker of oxidative stress among the breakdown products of polyunsaturated fatty acids in membranes (Riaz et al. 2021). Plants that are exposed to heavy metals in polluted areas where heavy metal levels rise to substantial levels often tend to contaminate the food chain (Yan et al. 2021). Heavy metals are absorbed into human and animal organs through the contaminated food chain, which increases the susceptibility of these organisms to a variety of ailments, from skin irritation to cancer (Chen et al. 2020).

Numerous regulatory and signalling molecules in plants that react quickly to both internal and external stimuli are responsible for regulating HM toxicities, according to a number of studies (Nguyen et al. 2021; Wei et al. 2020). These signalling molecules, also known as phytohormones, consist of chemical substances or groups of compounds such as salicylic acid, ethylene, auxin, polyamines, jasmonates, cytokinins, abscisic acid, and brassinosteroids (Bücker-Neto et al. 2017; Lymperopoulos et al. 2018). In recent years, phytohormones' eco-friendly methods of supporting plants' resistance to heavy metal toxicity have grown significantly in significance (Sytar et al. 2021). Figure 14.1 explains heavy metal toxicity in plants.

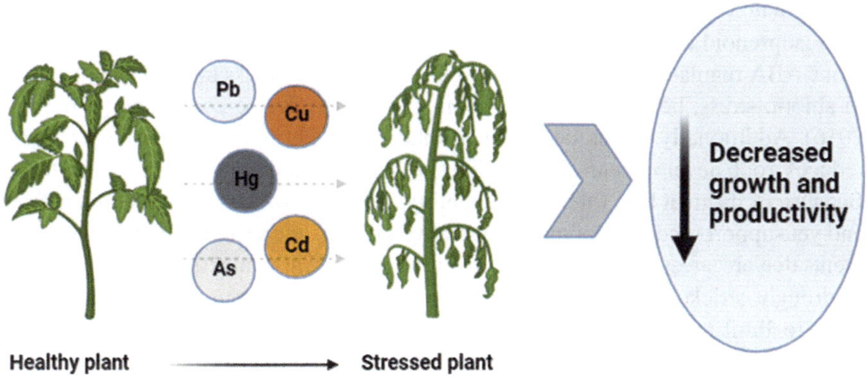

Fig. 14.1 Toxic effects of heavy metals on plants

14.2 Phytohormones and Plants' Growth and Development

14.2.1 Auxin

Auxin is a phytohormone that has several activities in plants, including plant cell division, cell enlargement, stimulation of vascular tissue differentiation, root initiation, tropic response, and delay in leaf senescence (Davies 2010). According to studies by Syu et al. (2014), Gui et al. (2015), Xiong et al. (2021), and other authors, auxin is important for controlling the development of plants and stress caused by nanoparticles. Indole-3-acetic acid (IAA) is the primary auxin that is biosynthesised through the reactions involving tryptophan aminotransferase from the Arabidopsis-1 (TAA1) family and the YUC (YUCCA) (flavin monooxygenases family). IAA is generated in the apical meristem zone of plant buds, shoots, and root tips. Auxin's action in plants is proportional to its concentration. (Madhaiyan et al. 2007).

14.2.2 Abscisic Acid

Abscisic acid is a significant phytohormone that participates in a number of developmental processes, including the dormancy of seeds and buds, control over the opening and shutting of stomata, regulation of the growth of different plant components, and reaction to environmental stress (Hu et al. 2020). In developmental processes like seed dormancy and germination, ABA and gibberellic acid (GA) have an antagonistic interaction because ABA promotes the production of storage molecules and inhibits premature seed germination (Spence et al. 2015). According to Vishwakarma et al. (2017a), ABA is a well-known stress phytohormone that is crucial to plant adaptability and abiotic stress tolerance. Inhibiting lateral root initiation, differentiating vascular tissues in roots, regulating cell division, photomorphogenic cell differentiation in expanding leaves and shoots, and inhibiting leaf senescence (Hai et al. 2020) are just a few of the critical processes of plant growth and development that are mediated by cytokinins, a class of plant hormone that are adenine derivatives with isoprenoid or aromatic side chains (Haberer and Kieber 2002). The phytohormone ABA regulates the expression of genes that confer stress tolerance in response to abiotic stress, hence augmenting plant growth in adverse conditions (Sah et al. 2016). Additionally, it has been shown that in drought-prone environments, ABA reduces root development and regulates water conductivity (Cutler et al. 2010). Because of its great heat tolerance capability, ABA can withstand high temperatures and yet support seed development, vegetative growth, and seed sprouting in environments that are stressful to the environment (Vishwakarma et al. 2017b). Furthermore, in drought-stricken plants, ABA is crucial for stomatal closure, which allows the plant to limit transpiration and prevent excessive water loss from the plant body when CO_2 absorption from the environment is not required (Assmann 2003). Using carotenoid as a precursor molecule, ABA was biosynthesised in plant plastids via the MEP route (Lievens et al. 2017; Spence et al. 2015). It involves a number of enzymes that mediate the zeaxanthin epoxidation, carotenoid breakdown into ABA molecules, and isomerisation (Ton et al. 2009). Through transcriptional reprogramming of plant cell metabolism and by promoting the production of catalase, which scavenges H_2O_2, ABA also affects plant defensive responses (Bari and Jones 2009).

14.2.3 Gibberellins

The tetracycline diterpenoid growth promoter hormones known as gibberellic acids (GAs) are a larger class of compounds that may play a role in the growth and development of plants. These biomolecules are known to control the harmful effects of stressors by preserving hormonal levels in addition to promoting growth (Jamla et al. 2021). The community of plant scientists has come to value GAs greatly because of their many positive effects. Many important physiological and biochemical characteristics of plants, including seed germination and growth, stem growth, root elongation, and the phases of fruit and reproductive development, are regulated by phytohormone GA (Saini et al. 2021; Singh et al. 2021). Gibberellins are also one of the most significant phytohormones that promote plant development by degrading the negative growth regulator protein DELLA (Bari and Jones 2009). GA is recognised to be a potent accelerator of shoot development but has minimal effect on root growth. Plants create a variety of phytohormones to battle various environmental stress situations, increasing their ability to flourish under such settings. Gibberellin is a phytohormone that is released in response to stress. GA3 levels were found to be higher in wheat and rice, which improved growth in saline circumstances (Kaya et al. 2009). GAs play a role in stem development, seed germination via generating enzymes during seed germination, and fruit setting and growth (Davies 2010). GAs were discovered to limit the production of free radicals that promote lipid peroxidation (Bari and Jones 2009). The most obvious effect of GAs on plant growth was the lengthening of internodes in shoots, growth and strengthened apical dominance, and boosted leaf growth. An increase in GAs has relatively little effect on a plant, as seen by increased dry weight. This demonstrated GA's secondary influence on leaf growth. This secondary impact was mostly caused by an increase in carbon assimilation (Brian 1959). There have been very few studies on the control of GAs in root development. Externally administered GAs had no influence on root development in a variety of plants. GAs interact antagonistically with ABA in various developmental stages, such as germination, growth, and blooming (Razem et al. 2006).

14.2.4 Ethylene

An essential plant growth hormone, ethylene, is associated with a number of plants' physiological processes, including senescence, leaf abscission, fruit ripening, and "triple signalling" during the seedling stage. Throughout the life cycle of a plant, ethylene, a gaseous hormone, functions as a signalling molecule and is engaged in a number of biochemical, physiological and molecular processes (Iqbal et al. 2017). The ACS enzyme 1-aminocyclopropane-1-carboxylic acid synthase and 1-aminocyclopropane-1-carboxylic acid oxidase (ACO) catalyse the biosynthesis of ethylene from methionine using S-adenosyl L methionine and 1-aminocyclopropane-1-carboxylic acid (ACC). From germination to the integration of environmental changes to fruit ripening, ethylene performs a crucial regulatory role. Considering its significance, ethylene's function in plant growth along with its biosynthesis and signalling pathways, have been the subject of ongoing research since the time it has been discovered as a growth factor for plant development (Neljubow 1901). A review of the vast amount of information on plant ethylene signalling that resulted from that research may be found in recent studies (Bakshi et al. 2015; Binder 2020; Pattyn et al. 2021). Nowadays, ethylene is also being acknowledged more and more as a crucial stress signalling mediator, with links to both mitochondrial and plastid signalling.

14.2.5 Cytokinin

One well-known phytohormone that is found in the environment both naturally and artificially is cytokinin. Naturally occurring in plants, cytokinin is essential for cell division, differentiation, and other biological processes, including apical dominance, leaf growth and senescence, nutrient mobilisation, chloroplast differentiation, and shoot meristem activation (Barciszewski et al. 2000). Smaller apical meristems and stunted shoots are the results of cytokine deficiency. Furthermore, it has been discovered that cytokinin plays a crucial role in plant morphogenesis and meristem activity. Through signalling pathways, cytokinin is crucial for the transmission of signals in response to many environmental challenges in plants, including salt, drought, and temperature (Ha et al. 2012). The phytohormone cytokinin

(CK) affects a number of physiological and biochemical characteristics of plants, such as cell division and morphogenesis, in addition to several agriculturally significant plant systems, including the regulation of environmental shocks (Emery and Kisiala 2020).

14.2.6 Brassinosteroids

According to Ahanger et al. (2018), brassinosteroids are a class of phytohormones that control a number of biological processes in plants, including photosynthesis, antioxidant defence mechanisms, osmolyte accumulation, nitrogen metabolism, and plant–water interaction, both under normal and stressful conditions. Previous studies have shown that foliar application of BRs is frequently used to increase crop yield and resilience to stress in many different plant species, such as *Triticum aestivum* (Filek et al. 2017), *Solanum lycopersicum*, *Adamant Vigna unguiculata*, *Oryza sativa* (Tadaiesky et al. 2021; Alam et al. 2020), *Solanum lycopersicum* (Jan et al. 2020), (Santos et al. 2018), and *Glycine max* (Chen et al. 2021a).

14.2.7 Salicylic Acid

According to Koo et al. (2020), SA is a defence-related plant hormone that is produced from the shikimate pathway and is essential for the regulation of plant immunity. In addition, SA has a key role in controlling abiotic stress tolerance, fruit output, seed germination, DNA damage and repair, and plant disease resistance. It is a crucial signalling molecule that triggers the activation of genes linked to pathogenesis (PR) and aids in the development of systemic acquired resistance (SAR) in plants.

14.2.8 Jasmonic Acid

A signalling chemical called jasmonic acid (JA) is involved in many different developmental processes, including fruit ripening, senescence, plant growth, pollen fertility, and stress tolerance (Ghorbel et al. 2021). Under stress, JA controls the antioxidant defence system's activity (superoxide anion radical, peroxidase, NADPH-oxidase), the build-up of amino acids such as isoleucine and methionine, soluble carbohydrates, and the opening and shutting of stomata (Wang et al. 2020). The term "jasmonate" refers to both jasmonic acid (JA) and a wide range of its precursors and derivatives. They control a wide range of functions, including defence against stress and the growth and development of plants (Wasternack and Hause

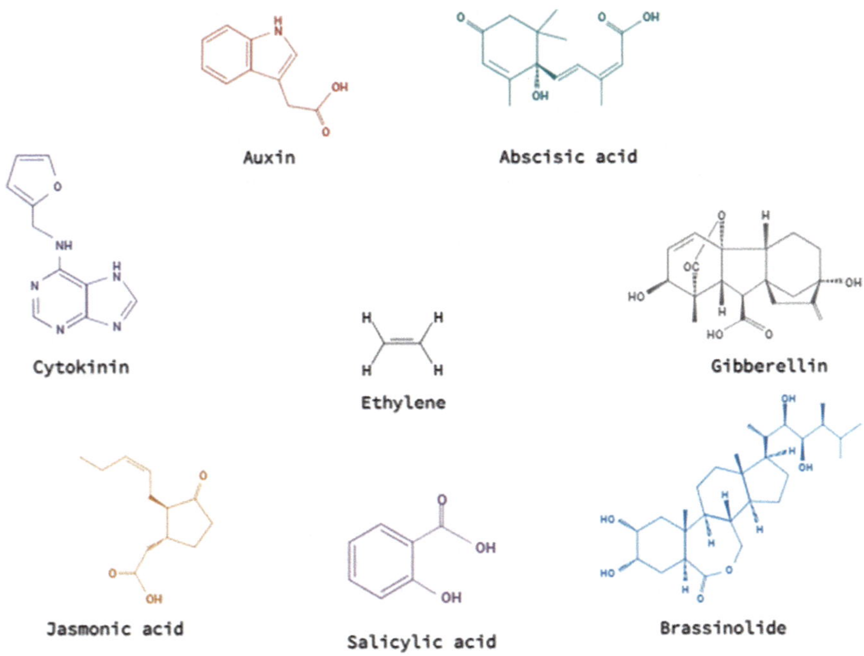

Fig. 14.2 Phytohormones

2013; Koo 2018). The basic structure of different hormones is mentioned in Fig. 14.2.

14.3 Participation of Phytohormones Against Heavy Metal Stress in Plants

14.3.1 Auxins and Heavy Metals

The incredibly complex HMs regulating pathways are highlighted by plant growth flexibility and its sensitivity to a variety of environmental incitement, including the improvement of stress management. In plants, among such environmental conditions, one is the auxin homeostasis and heavy metal toxicity interaction (Wang et al. 2021b). This indole-3-acetic acid, or IAA, is a chemical compound that is identical to a hormone and gained importance as a result of its acknowledged mediating function throughout a number of plant growth and development characteristics. Several results showed that the auxin endogenous status in shoots was differentially altered and root tissues, exhibiting both antagonistic and cooperative relationships between the level of auxin and HM stress (Yuan and Huang 2016). Even though heavy metals negatively impact auxin production, it has been found that auxin applied topically can aid in restoring auxins' intracellular state inside the plant. For example, foliar auxin (IAA) treatment increased the total biomass of plants (root + stem) in *Helianthus annuus* plants growing in Pb-spiked soil (Liphadzi et al. 2006). Furthermore, the foliar use of IAA improved *Brassica juncea* during oxidative damage mediated by As by reducing ROS and lipid peroxidation (MDA contents) and by improving membrane integrity and stiffness (Srivastava et al. 2013). Comparably, adding different amounts of auxin precursor (L-TRP) to the roots of *Oryza sativa* exposed to soil laced with Cd resulted in a considerable increase in root elongation and growth compared to control plants (which did not receive either Cd or auxin treatment) (Farooq et al. 2015).

Moreover, it has also been shown that auxin interacts synergistically with ethylene, leading to the growth of root hair creation and root elongation processes, where the opposed activity reveals formation of lateral roots and hypocotyl elongation in plants (Gloria and Castellanos 2012). Gustafson proved for the first time in 1937 (Gustafson 1937) that the incorporation of auxin-like components into tomato's stigma, as well as in many other plant species, led to parthenocarpic fruit formation. The amount of auxin and gibberellins increased continuously during anthesis, or most likely after pollination and fertilisation, leading to the production of seedless fruit. Auxin can stimulate the formation of gibberellins and raise auxin levels in plant ovaries (de Jong et al. 2009). Antagonistic action of auxin route was observed with salicylic acid pathway during plant defence, whereas a constructive interaction was observed in auxin and JA (jasmonic acid) signalling pathways (Kemal and Manners 2009). Resistance to necrotrophic infections is shared by the auxin and jasmonic acid signalling pathways. TIR1 and COI1 (F-box proteins for auxin and JA) play roles in conferring resistance to a variety of necrotrophic diseases (Mockaitis and Estelle 2008).

14.3.2 Abscisic Acid and Heavy Metals

Many plants exposed to heavy metals (HMs) have increased endogenous ABA concentrations; these include *Helianthus annuus* L., *Lactuca sativa*, *Populus canescens*, *Tritium aestivum*, *Oryza sativa*, *Cucumis sativus*, and *Solanum tuberosum*. These findings indicate that ABA is important in mitigating the negative effects of HM toxicities (Kim et al. 2014; Lou et al. 2021; Shi et al. 2019; Stroiński et al. 2010; Tang et al. 2020; Vasyuk et al. 2019; Zengin and Kirbag 2007). According to earlier research, ABA is essential for controlling the water status, stomatal conductance, and stomatal functioning of plants during heavy metal stress (Bücker-Neto et al. 2017). When ABA was applied to grapevine and lettuce plants that were stressed with Cd and Zn, Zn and Cd bioaccumulation was greatly reduced while root and shoot biomass rose overall (Tao et al. 2021). The zinc/iron transporter protein (ZIP) family, which is in charge of cross-cellular HM translocation and depositions during Zn cytotoxicity, was also upregulated by the ABA administration. Moreover, Zn and the ABA application enhanced a number of HM decontaminated genes, including yellow strip-like (YSL) plant-cadmium resistance 2 (PCR2) and natural resistance-associated macrophage (NRAMP3) in grapevine plants (Song et al. 2019). It has recently been determined that nourishing *Solanum lycopersicum* with ABA and 6-benzylaminopurine (BAP) at the same time lessened the negative effects of cobalt (Co) stress (Kamran et al. 2021).

14.3.3 Gibberellic Acid and Heavy Metals

In addition to its growth-promoting functions, GA may have a role in controlling the toxicities of heavy metals (HMs) by triggering a variety of defence mechanisms (Gong et al. 2021; Javed et al. 2021). By enhancing antioxidant enzymes and proline activities, a moderate concentration of GA in spinach plants was shown to be responsible for correcting Cu-induced oxidative damage, whereas a greater concentration of GA was found to be responsible for amplifying the harmful effects of Cu-induced oxidative burst (Gong et al. 2021). Furthermore, GA exogenous administration in *Pisum sativum* strengthened the antioxidative defence mechanism under Cu-mediated oxidative damage, therefore initiating the Cu phytoremediation processes from Cu-spiked soil (Javed et al. 2021). It can be summarised from the prior research that concentration-dependent exogenous GA therapy helps plants avoid, withstand, or adapt to HM stress.

14.3.4 Ethylene and Heavy Metals

Several studies have demonstrated that ethylene modulates a number of abiotic stressors, including the stress caused by heavy metals (Asgher et al. 2018; Betti et al. 2021). This is achieved by up-regulating ethylene synthesis. However, further mechanical research is required to fully understand the ethylene-mediated heavy metal regulation routes in plants, as they are currently unknown.

14.3.5 Cytokinin and Heavy Metals

According to Zhou et al. (2019), a number of studies have shown that CK may have a function in controlling HM stress in plants by affecting HM biosorption and altering the HM toxicity adaptation process. For example, by reversing HM-mediated oxidation in the mitotic index, CK administration enhanced cell division in algal plants, resulting in plant growth and development (Piotrowska-Niczyporuk et al. 2012). Nonetheless, it is uncertain if exogenous CK administration can reduce HM phytotoxicity, despite this being acknowledged in earlier research. Veselov et al. (2003) found that the production of CK oxidase enzymes involved in CK biosynthesis caused a considerable deterioration in CK in wheat seedlings under Cd stress.

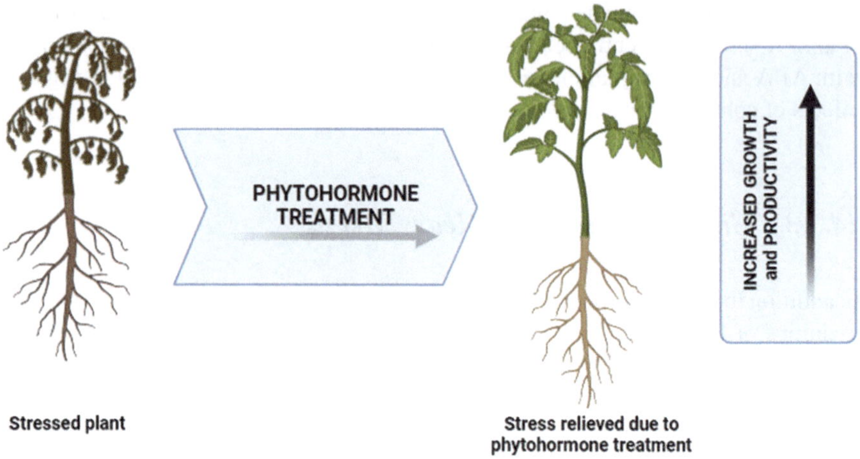

Fig. 14.3 Stress ameliorative potential of phytohormones

14.3.6 Brassinosteroids and Heavy Metals

Brassinosteroids (BRs), another name for phytohormones, are hormones that can control ion absorption in plant cells to drastically lower the build-up of heavy metals. Moreover, BRs can benefit plants with different biotic and abiotic stressors, such as pathogenic organisms, salt, heavy metals, and drought assault, as well as variations in temperature (Vardhini et al. 2010; Chen et al. 2021b).

Figure 14.3 explains the roles of plant hormones under heavy metal stress.

14.4 Conclusion

The harmful effects of heavy metals (HMs) on plants have drawn a lot of attention from all around the world because they pose a direct danger to the food supply chain. Plants that are exposed to surroundings polluted with heavy metals sometimes accumulate high concentrations of these metals, which also contaminates the food chain. While a number of steps have been taken to lessen the negative effects of heavy metal stress on plants, but application of phytohormones is considered to be the best solution because of their eco-friendly and eco-safe management of the same.

References

Ahanger MA, Ashraf M, Bajguz A, Ahmad P (2018) Brassinosteroids regulate growth in plants under stressful environments and crosstalk with other potential phytohormones. J Plant Growth Regul 37(4):1007–1024

Alam P, Kaur Kohli S, Al Balawi T, Altalayan FH, Alam P, Ashraf M, Bhardwaj R, Ahmad P (2020) Foliar application of 24-Epibrassinolide improves growth, ascorbate-glutathione cycle, and glyoxalase system in brown mustard (Brassica juncea (L.) Czern.) under cadmium toxicity. Plants 9(11):1487

Asgher M, Khan MIR, Anjum NA, Verma S, Vyas D, Per TS, Masood A, Khan NA (2018) Ethylene and polyamines in counteracting heavy metal phytotoxicity: a crosstalk perspective. J Plant Growth Regul 37(4):1050–1065

Assmann SM (2003) OPEN STOMATA1 opens the door to ABA signaling in Arabidopsis guard cells. Trends Plant Sci 8(4):151–153. https://doi.org/10.1016/S1360-1385(03)00052-9

Bakshi A, Shemansky JM, Chang CR, Binder BM (2015) History of research on the plant hormone ethylene. J Plant Growth Regul 34:809–827

Barciszewski J, Siboska G, Clark BF, Rattan SI (2000) Cytokinin formation by oxidative metabolism. J Plant Physiol 157(5):587–588. https://doi.org/10.1016/S0176-1617(00)80116-9

Bari R, Jones JD (2009) Role of plant hormones in plant defence responses. Plant Mol Biol 69(4):473–488. https://doi.org/10.1007/s11103-008-9435-0

Betti C, Della Rovere F, Piacentini D, Fattorini L, Falasca G, Altamura MM (2021) Jasmonates, ethylene and brassinosteroids control adventitious and lateral rooting as stress avoidance responses to heavy metals and metalloids. Biomolecules 11(1):77

Binder BM (2020) Ethylene signaling in plants. J Biol Chem 295:7710–7725

Brian PW (1959) Effects of gibberellins on plant growth and development. Biol Rev 34(1):37–77. https://doi.org/10.1111/j.1469-185x.1959.tb01301.x

Bücker-Neto L, Paiva ALS, Machado RD, Arenhart RA, Margis-Pinheiro M (2017) Interactions between plant hormones and heavy metals responses. Genet Mol Biol 40:373–386

Chen X-X, Liu Y-M, Zhao Q-Y, Cao W-Q, Chen X-P, Zou C-Q (2020) Health risk assessment associated with heavy metal accumulation in wheat after long-term phosphorus fertilizer application. Environ Pollut 262:114348

Chen H, Yang R, Zhang X, Chen Y, Xia Y, Xu X (2021a) Foliar application of gibberellin inhibits the cadmium uptake and xylem transport in lettuce (*Lactuca sativa* L.). Sci Hortic 288:110410

Chen L, Yang H, Fang Y, Guo W, Chen H, Zhang X, Dai W, Chen S, Hao Q, Yuan S (2021b) Overexpression of GmMYB14 improves high-density yield and drought tolerance of soybean through regulating plant architecture mediated by the brassinosteroid pathway. Plant Biotechnol J 19(4):702–716

Cutler SR, Rodriguez PL, Finkelstein RR, Abrams SR (2010) Abscisic acid: emergence of a core signaling network. Annu Rev Plant Biol 61:651–679. https://doi.org/10.1146/annurev-arplant-042809-112122

Davies PJ (2010) The plant hormones: their nature, occurrence, and functions. In: Plant hormones. Springer, Dordrecht, pp 1–15. https://doi.org/10.1007/978-1-4020-2686-7_1

de Jong M et al (2009) The role of auxin and gibberellin in tomato fruit set. J Exp Bot 60(5):1523–1532. https://doi.org/10.1093/jxb/erp094

Emery RJN, Kisiala A (2020) The roles of cytokinins in plants and their response to environmental stimuli. Plants 9(9):1158. https://doi.org/10.3390/plants9091158

Farooq H, Asghar HN, Khan MY, Saleem M, Zahir ZA (2015) Auxin-mediated growth of rice in cadmium-contaminated soil. Turk J Agric For 39(2):272–276

Filek M, Rudolphi-Skórska E, Sieprawska A, Kvasnica M, Janeczko A (2017) Regulation of the membrane structure by brassinosteroids and progesterone in winter wheat seedlings exposed to low temperature. Steroids 128:37–45

Ghorbel M, Brini F, Sharma A, Landi M (2021) Role of jasmonic acid in plants: the molecular point of view. Plant Cell Rep 40(8):1471–1494

Gloria AM, Castellanos J (2012) Desafíos y bendiciones: a multiperspective examination of the educational experiences and coping responses of first-generation college Latina students. J Hisp High Educ 11(1):82–99

Gong Q, Li Z-H, Wang L, Zhou J-Y, Kang Q, Niu D-D (2021) Gibberellic acid application on biomass, oxidative stress response, and photosynthesis in spinach (*Spinacia oleracea* L.) seedlings under copper stress. Environ Sci Pollut Res 28(38):53594–53604

Gui X, Deng Y, Rui Y et al (2015) Response difference of transgenic and conventional rice (*Oryza sativa*) to nanoparticles (γFe2O3). Environ Sci Pollut Res 22(22):17716–17723

Gustafson FG (1937) Parthenocarpy induced by pollen extracts. Am J Bot 24(2):102–107. https://doi.org/10.1002/j.1537-2197.1937.tb09075.x

Ha S, Vankova R, Shinozaki KY, Shinozaki K, Tran LSP (2012) Cytokinins: metabolism and function in plant adaptation to environmental stresses. Trends Plant Sci 17(3):172–179. https://doi.org/10.1016/j.tplants.2011.12.005

Haberer G, Kieber JJ (2002) Cytokinins. New insights into a classic phytohormone. Plant Physiol 128(2):354–362

Hai NN, Chuong NN, Tu NHC, Kisiala A, Hoang XLT, Thao NP (2020) Role and regulation of cytokinins in plant response to drought stress. Plants 9(4):422

Hu B, Deng F, Chen G, Chen X, Gao W, Long L, Xia J, Chen Z-H (2020) Evolution of abscisic acid signaling for stress responses to toxic metals and metalloids. Front Plant Sci 11:909

Hussain B, Ashraf MN, Abbas A, Li J, Farooq M (2021) Cadmium stress in paddy fields: effects of soil conditions and remediation strategies. Sci Total Environ 754:142188

Iqbal N, Khan NA, Ferrante A, Trivellini A, Francini A, Khan MIR (2017) Ethylene role in plant growth, development and senescence: interaction with other phytohormones. Front Plant Sci 8:475

Jamla M, Khare T, Joshi S, Patil S, Penna S, Kumar V (2021) Omics approaches for understanding heavy metal responses and tolerance in plants. Curr Plant Biol 27:100213

Jan S, Noman A, Kaya C, Ashraf M, Alyemeni MN, Ahmad P (2020) 24- Epibrassinolide alleviates the injurious effects of Cr (VI) toxicity in tomato plants: insights into growth, physio-biochemical attributes, antioxidant activity and regulation of Ascorbate–glutathione and Glyoxalase cycles. J Plant Growth Regul 39(4):1587–1604

Javed T, Ali MM, Shabbir R, Anwar R, Afzal I, Mauro RP (2021) Alleviation of copper-induced stress in pea (Pisum sativum L.) through foliar application of gibberellic acid. Biology 10(2):120

Kamran M, Danish M, Saleem MH, Malik Z, Parveen A, Abbasi GH, Jamil M, Ali S, Afzal S, Riaz M (2021) Application of abscisic acid and 6-benzylaminopurine modulated morpho-physiological and antioxidative defense responses of tomato (Solanum lycopersicum L.) by minimizing cobalt uptake. Chemosphere 263:128169

Kaya C, Tuna AL, Yokaş I (2009) The role of plant hormones in plants under salinity stress. In: Salinity water stress. Springer, Dordrecht, pp 45–50. https://doi.org/10.1007/978-1-4020-9065-3_5

Kemal K, Manners JM (2009) Linking development to defense: auxin in plant–pathogen interactions. Trends Plant Sci 14(7):373–382

Khan N, Bano A, Ali S, Babar M (2020) Crosstalk amongst phytohormones from planta and PGPR under biotic and abiotic stresses. Plant Growth Regul 90(2):189–203

Kim Y-H, Khan AL, Kim D-H, Lee S-Y, Kim K-M, Waqas M, Jung H-Y, Shin J-H, Kim J-G, Lee I-J (2014) Silicon mitigates heavy metal stress by regulating P-type heavy metal ATPases, Oryza sativa low silicon genes, and endogenous phytohormones. BMC Plant Biol 14(1):1–13

Koo AJ (2018) Metabolism of the plant hormone jasmonate: a sentinel for tissue damage and master regulator of stress response. Phytochem Rev 17:51–80

Koo YM, Heo AY, Choi HW (2020) Salicylic acid as a safe plant protector and growth regulator. Plant Pathol J 36(1):1

Lievens L, Pollier J, Goossens A, Beyaert R, Staal J (2017) Abscisic acid as pathogen effector and immune regulator. Front Plant Sci 8:587. https://doi.org/10.3389/fpls.2017.00587

Liphadzi M, Kirkham M, Paulsen G (2006) Auxin-enhanced root growth for phytoremediation of sewage-sludge amended soil. Environ Technol 27(6):695–704

Lou X, Zhang X, Zhang Y, Tang M (2021) The synergy of arbuscular mycorrhizal fungi and exogenous abscisic acid benefits Robinia pseudoacacia L. growth through altering the distribution of Zn and endogenous abscisic acid. J Fungi 7(8):671

Lymperopoulos P, Msanne J, Rabara R (2018) Phytochrome and phytohormones: working in tandem for plant growth and development. Front Plant Sci 9:1037

Madhaiyan M, Poonguzhali S, Sa T (2007) Characterization of 1-aminocyclopropane-1-carboxylate (ACC) deaminase containing Methylobacterium oryzae and interactions with auxins and ACC regulation of ethylene in canola (Brassica campestris). Planta 226(4):867–876. https://doi.org/10.1007/s00425-007-0532-0

Mockaitis K, Estelle M (2008) Auxin receptors and plant development: a new signaling paradigm. Annu Rev Cell Dev Biol 24:55–80. https://doi.org/10.1146/annurev.cellbio.23.090506.123214

Neljubow D (1901) Ueber die horizontale Nutation der Stengel von *Pisum sativum* und einiger anderen Pflanzen. Beihefte zum botanischen Zentralblatt 10:128–139

Nguyen TQ, Sesin V, Kisiala A, Emery RN (2021) Phytohormonal roles in plant responses to heavy metal stress: implications for using macrophytes in phytoremediation of aquatic ecosystems. Environ Toxicol Chem 40(1):7–22

Pattyn J, Vaughan-Hirsch J, Van de Poel B (2021) The regulation of ethylene biosynthesis: a complex multilevel control circuitry. New Phytol 229:770–782

Piotrowska-Niczyporuk A, Bajguz A, Zambrzycka E, Godlewska-Żyłkiewicz B (2012) Phytohormones as regulators of heavy metal biosorption and toxicity in green alga Chlorella vulgaris (Chlorophyceae). Plant Physiol Biochem 52:52–65

Rahman SU, Xuebin Q, Kamran M, Yasin G, Cheng H, Rehim A, Riaz L, Rizwan M, Ali S, Alsahli AA (2021a) Silicon elevated cadmium tolerance in wheat (Triticum aestivum L.) by endorsing nutrients uptake and antioxidative defense mechanisms in the leaves. Plant Physiol Biochem 166:148–159

Rahman SU, Xuebin Q, Yasin G, Cheng H, Mehmood F, Zain M, Shehzad M, Ahmad MI, Riaz L, Rahim A (2021b) Role of silicon on root morphological characters of wheat (Triticum aestivum L.) plants grown under Cd-contaminated nutrient solution. Acta Physiol Plant 43(4):1–13

Rahman SU, Yasin G, Nawaz MF, Cheng H, Azhar MF, Riaz L, Li Y (2022) Evaluation of heavy metal phytoremediation potential of six tree species of Faisalabad city of Pakistan during summer and winter seasons. J Environ Manag 320:115801

Rasheed A, Hassan MU, Fahad S, Aamer M, Batool M, Ilyas M, Shang F, Wu Z, Li H (2021) Heavy metals stress and plants defense responses. In: Sustainable soil and land management and climate change. CRC Press, Boca Raton, pp 57–82

Razem FA, Baron K, Hill RD (2006) Turning on gibberellin and abscisic acid signaling. Curr Opin Plant Biol 9(5):454–459. https://doi.org/10.1016/j.pbi.2006.07.007

Riaz M, Kamran M, Fang Y, Wang Q, Cao H, Yang G, Deng L, Wang Y, Zhou Y, Anastopoulos I (2021) Arbuscular mycorrhizal fungi-induced mitigation of heavy metal phytotoxicity in metal contaminated soils: a critical review. J Hazard Mater 402:123919

Sah SK, Reddy KR, Li J (2016) Abscisic acid and abiotic stress tolerance in crop plants. Front Plant Sci 7:571. https://doi.org/10.3389/fpls.2016.00571

Saini S, Kaur N, Pati PK (2021) Phytohormones: key players in the modulation of heavy metal stress tolerance in plants. Ecotoxicol Environ Saf 223:112578

Santos L, Batista B, Lobato A (2018) Brassinosteroids mitigate cadmium toxicity in cowpea plants. Photosynthetica 56(2):591–605

Shi W-G, Liu W, Yu W, Zhang Y, Ding S, Li H, Mrak T, Kraigher H, Luo Z-B (2019) Abscisic acid enhances lead translocation from the roots to the leaves and alleviates its toxicity in Populus× canescens. J Hazard Mater 362:275–285

Singh NB, Singh A, Khare S, Yadav V, Bano C, Yadav RK (2021) Mitigating strategies of gibberellins in various environmental cues and their crosstalk with other hormonal pathways in plants: a review. Plant Mol Biol Report 39(1):34–49

Song C, Yan Y, Rosado A, Zhang Z, Castellarin SD (2019) ABA alleviates uptake and accumulation of zinc in grapevine (Vitis vinifera l.) by inducing expression of ZIP and detoxification-related genes. Front Plant Sci 10:872

Spence CA, Lakshmanan V, Donofrio N, Bais HP (2015) Crucial roles of abscisic acid biogenesis in virulence of rice blast fungus *Magnaporthe oryzae*. Front Plant Sci 6:1082. https://doi.org/10.3389/fpls.2015.01082

Srivastava S, Srivastava AK, Suprasanna P, D'souza S (2013) Identification and profiling of arsenic stress-induced microRNAs in Brassica juncea. J Exp Bot 64(1):303–315

Stroiński A, Chadzinikolau T, Giżewska K, Zielezińska M (2010) ABA or cadmium induced phytochelatin synthesis in potato tubers. Biol Plant 54(1):117–120

Sytar O, Kumari P, Yadav S, Brestic M, Rastogi A (2019) Phytohormone priming: regulator for heavy metal stress in plants. J Plant Growth Regul 38(2):739–752

Sytar O, Ghosh S, Malinska H, Zivcak M, Brestic M (2021) Physiological and molecular mechanisms of metal accumulation in hyperaccumulator plants. Physiol Plant 173(1):148–166

Syu YY, Hung JH, Chen JC, Chuang HW (2014) Impacts of size and shape of silver nanoparticles on Arabidopsis plant growth and gene expression. Plant Physiol Biochem 83:57–64

Tadaiesky LB, da Silva BR, Batista BL, Lobato, A.K.d. S. (2021) Brassinosteroids trigger tolerance to iron toxicity in rice. Physiol Plant 171(3):371–387

Tang Y, Wang L, Xie Y, Yu X, Li H, Lin L, Liao MA, Wang Z, Sun G, Wang X (2020) Effects of exogenous abscisic acid on the growth and cadmium accumulation of lettuce under cadmium-stress conditions. Int J Environ Anal Chem 100(6):720–731

Tao Q, Jupa R, Dong Q, Yang X, Liu Y, Li B, Yuan S, Yin J, Xu Q, Li T (2021) Abscisic acid-mediated modifications in water transport continuum are involved in cadmium hyperaccumulation in Sedum alfredii. Chemosphere 268:129339

Ton J, Flors V, Mani BM (2009) The multifaceted role of ABA in disease resistance. Trends Plant Sci 14(6):310–317. https://doi.org/10.1016/j.tplants.2009.03.006

Vardhini BV, Anuradha S, Sujatha E, Rao SSR (2010) Role of brassinosteroids in alleviating various abiotic and biotic stresses—a review. Plant Stress 4(1):55–61

Vasyuk V, Voytenko L, Shcherbatiuk M, Kosakivska I (2019) Effect of exogenous abscisic acid on seed germination and growth of winter wheat seedlings under zinc stress. J Stress Physiol Biochem 15(2):68–78

Veselov D, Kudoyarova G, Symonyan M, Veselov S (2003) Effect of cadmium on ion uptake, transpiration and cytokinin content in wheat seedlings. Bulg J Plant Physiol 29(3–4):353–359

Vishwakarma K, Upadhyay N, Kumar N, Yadav G, Singh J, Mishra RK, Kumar V, Verma R, Upadhyay RG, Pandey M, Sharma S (2017a) Abscisic acid signaling and abiotic stress tolerance in plants: a review on current knowledge and future prospects. Front Plant Sci 8:161. https://doi.org/10.3389/fpls.2017.00161

Vishwakarma K, Upadhyay N, Kumar N et al (2017b) Abscisic acid signaling and abiotic stress tolerance in plants: a review on current knowledge and future prospects. Front Plant Sci 8:161

Wang J, Song L, Gong X, Xu J, Li M (2020) Functions of jasmonic acid in plant regulation and response to abiotic stress. Int J Mol Sci 21(4):1446

Wang H, Cui S, Wu D, Yang X, Wang H, Wang Z (2021a) Effects of kinetin on arsenic speciation and antioxidative enzymes in fronds of the arsenic hyperaccumulator Pteris cretica var. nervosa and non-hyperaccumulator Pteris ensiformis. Environ Exp Bot 191:104622

Wang Q, Huang D, Niu D, Deng J, Ma F, Liu C (2021b) Overexpression of auxin response gene MdIAA24 enhanced cadmium tolerance in apple (Malus domestica). Ecotoxicol Environ Saf 225:112734

Wasternack C, Hause B (2013) Jasmonates: biosynthesis, perception, signal transduction and action in plant stress response, growth and development. An update to the 2007 review in Annals of Botany. Ann Bot 111:1021–1058

Wei L, Zhang M, Wei S, Zhang J, Wang C, Liao W (2020) Roles of nitric oxide in heavy metal stress in plants: cross-talk with phytohormones and protein S-nitrosylation. Environ Pollut 259:113943

Xiong T, Zhang S, Kang Z, Zhang T, Li S (2021) Dose-dependent physiological and transcriptomic responses of lettuce (*Lactuca sativa* L.) to copper oxide nanoparticles insights into the phytotoxicity mechanisms. Int J Mol Sci 22(7):3688
Yan C, Jie L, Pingping J, Guo Y, Xusheng J, Zhifan D, Ling L, Bing Z (2021) Mitigation effect of Mn on Cd toxicity in a Hyperaccumulator Celosia argentea Linn. Asian J Ecotoxicol 1:223–232
Yasir MW, Siddique MBA, Shabbir Z, Ullah H, Riaz L, Shah AA (2021) Biotreatment potential of co-contaminants hexavalent chromium and polychlorinated biphenyls in industrial wastewater: individual and simultaneous prospects. Sci Total Environ 779:146345
Yuan HM, Huang X (2016) Inhibition of root meristem growth by cadmium involves nitric oxide-mediated repression of auxin accumulation and signalling in Arabidopsis. Plant Cell Environ 39(1):120–135
Zengin FK, Kirbag S (2007) Effects of copper on chlorophyll, proline, protein and abscisic acid level of sunflower (Helianthus annuus L.) seedlings. J Environ Biol 28(3):561
Zhou M, Ghnaya T, Dailly H, Cui G, Vanpee B, Han R, Lutts S (2019) The cytokinin trans-zeatine riboside increased resistance to heavy metals in the halophyte plant species Kosteletzkya pentacarpos in the absence but not in the presence of NaCl. Chemosphere 233:954–965

Chapter 15
Signaling Molecules and Phytohormonal-Induced Heavy Metal Stress Regulation in Plants

Savita Bhardwaj, Dhriti Sharma, Sadaf Jan, Ambika Choudhary, Dhriti Kapoor, Puja Panda, and Usha Thakur

Abstract Due to the increasing world population, food demand is increasing. However, metal toxicity poses a serious threat to agricultural productivity as well as to human health by entering the food chain. Various techniques are in use in the present day to alleviate the negative effects of metal stresses in plant species. Plants have a variety of defense mechanisms to fend off the damaging effects of heavy metal toxicity. These systems include the enzymatic and non-enzymatic antioxidants, metal transporters, and the synthesis of phytohormone and signaling molecules. Exogenous application of signaling molecules and phytohormones has also been demonstrated to be a successful strategy for counteracting the harmful effects of heavy metals, in addition to their endogenous synthesis. Heavy metal stress in plants is mitigated by these signaling molecules and phytohormones through augmenting the photosynthetic and gaseous exchange attributes, antioxidative defense mechanisms, and defense-associated genes such as glutathione biosynthesis genes, osmoregulation, and heavy metal transporter genes. Moreover, the crosstalk between different phytohormones will open new doors in comprehending the intricacies involved in heavy metal stress tolerance for enhancing crop yield.

Keywords Heavy metal · Phytohormones · Signaling molecules · Food · Antioxidant

---

S. Bhardwaj · U. Thakur
Department of Botany, MCM DAV College, Kangra, Himachal Pradesh, India

D. Sharma
Department of Botany, School of Bioengineering and Biosciences, Lovely Professional University, Phagwara, Punjab, India

S. Jan · P. Panda
Department of Biotechnology, School of Bioengineering and Biosciences, Lovely Professional University, Phagwara, Punjab, India

A. Choudhary · D. Kapoor (✉)
School of Biological and Environmental Sciences, Shoolini University, Solan, Himachal Pradesh, India

A. Sharma et al. (eds.), *Hormones and Heavy Metals*, Plant Life and Environment Dynamics, https://doi.org/10.1007/978-981-96-9622-2_15

15.1 Introduction

Lead (Pb), cadmium (Cd), chromium (Cr), copper (Cu), zinc (Zn), arsenic (As), and nickel (Ni) have been discovered to induce heavy metal toxicity in agricultural soils around the world, with levels of pollution ranging from negligible to significant (Bhardwaj et al. 2022). These contaminants may originate from the following sources: long-term use of phosphatic fertilizers; sewage sludge application; smelting dust; industrial and hospital waste; and inadequate irrigation practices on agricultural fields (Hussain et al. 2021; Yasir et al. 2021). Reactive oxygen species (ROS) overproduction is a significant reaction of plants to high concentrations of heavy metals, and it damages the growth and survival of the plant (Ur Rahman et al. 2021). Oxidative stress is caused by these metals in plants by the overaccumulation of ROS (Kohli et al. 2017). The activity of the antioxidant system, the electron transport chain, and the metabolism of important biomolecules are disturbed by oxidative stress (Azeez et al. 2021). Lipid peroxidation, which can lead to biomembrane damage, is thought to be the most dangerous adverse effect of heavy metal exposure, aside from the overproduction of ROS. Overexpression of malondialdehyde (MDA), one of the breakdown products of polyunsaturated fatty acids in membranes, has been proven to be an indicator of oxidative stress (Riaz et al. 2021).

The use of plant hormones is a promising strategy for attaining sustainable agriculture to boost crop yield worldwide, which is essential for meeting global food demand (Sytar et al. 2019). Phytohormones are organic chemical compounds that exhibit advantageous effects in plants when present at minute concentrations (Jamla et al. 2021). Several phytohormones, such as auxins, gibberellic acid, abscisic acid, jasmonic acid, ethylene, and brassinosteroids, have been shown to exert a positive impact in improving crop performance and productivity under stressed conditions (Sabagh et al. 2022). Phytohormones are beneficial in regulating signaling mechanisms to modulate various developmental processes in plants (Rhaman et al. 2020). For the improvement of crop productivity under stress environments, it is essential to understand the mode of action of PGRs for physiological, biochemical, and molecular processes brought on by PGRs treatments Sabagh et al. 2021). Exogenous application of several phytohormones has been successfully utilized to ameliorate the adverse effects of heavy metals in plants (Rostami and Azhdarpoor 2019). Overall, phytohormones are essential modulators of heavy metal stress tolerance in plants. These hormones help in strengthening the antioxidant defense machinery in response to heavy metal toxicity. Signaling molecules, including phytohormones and other gasotransmitters, including nitric oxide and hydrogen sulfide, play an important role in alleviating metal toxicity in plants by regulating various growth, as well as physiological and biochemical parameters.

15.2 Role of Phytohormones in Heavy Metal Stress Tolerance in Plants

Phytohormones are plant hormones that act as chemical carriers at a very low concentration. These hormones play a very crucial role in heavy metal (HM) stress tolerance in plants. Heavy Metals intrude various plants' biological processes, leading to noxious impacts such as retarded nutrient supply, ion imbalance, and decreased photosynthetic activity (Morkunas et al. 2018; Ghori et al. 2019). HM accumulation in plants has witnessed a remarkable expansion in the present day due to accelerated economic development and current farming strategies (Roy and McDonald 2015). Phytohormones play a significant role in order to alleviate the negative effects of heavy metal toxicity in plants.

15.2.1 Auxin

Auxins are plant hormones that play an indispensable role in plant growth and development (Korver et al. 2018). Various abiotic constraints are regulated by auxins via adapting their formation, distribution, assimilation, and degeneration progressions (Saini et al. 2017). Plant-root progress is also maintained by auxins under both stressful and non-stressed environments (Bücker-Neto et al. 2017). As roots are highly exposed plant parts that are in constant contiguity with HMs, comprehending the interrelation between auxin and metal toxicity is important (Ronzan et al. 2018). Several studies have shown that stressed conditions change the internal content of auxin distinctively in underground and aboveground plant parts. In this sense, metal stress and auxin content exhibit both negative and beneficial correlation as revealed by such investigations (Srivastava et al. 2013; Yuan and Huang 2016). Modulation of auxin-associated genes alters the site and accretion of auxin to cope with stressed conditions. Cd stress upregulated the endogenous auxin content and functioning of auxin synthesizing gene, viz., *YUCCA6,* in roots of *Arabidopsis thaliana* (Fattorini et al. 2017). In contrast, down-regulation in the expression of the *YUCCA6* gene and internal IAA content was noticed in the arsenic (As) exposed cultivars. Additionally, As and Cd treatment decreased the levels of *PIN 1* and *AUX1* in roots of *Arabidopsis*, reducing the auxin movement to disrupt the auxin balance (Fattorini et al. 2017). The Cd application also increased the gretchen-hagen-3 (GH3) functioning. In *Oryza sativa* roots, Cd supplementation resulted in major changes in *MAPK* signaling, ultimately causing decreased root growth caused by disturbance of auxin homeostasis (Zhao et al. 2013). Apart from this, auxin is known to mitigate the negative impacts of heavy metals by interacting with ROS detoxification machinery (Krishnamurthy and Rathinasabapathi 2013). Besides, an association between cadmium and auxin has been studied in Popular (Elobeid et al. 2012). From this study, it was discerned that cadmium improved the *GH3* functioning, thereby mediating auxin linkage and deprivation processes. This further increased the functioning of lignin and peroxidase and lowered the IAA content to induce the protection mechanisms against heavy metal stress (Elobeid et al. 2012).

15.2.2 Gibberellic Acid

Gibberellic acid (GA) regulates various key elements of plant growth and development, including seed germination, stem elongation, flowering time, flower and fruit development (Hernández-García et al. 2021). They have also been involved in increasing the HM stress tolerance in plants by managing various plant processes (Emam Verdian et al. 2020; Javed et al. 2021). In *Vicia faba* L., GA treatment aided in invigorating the lead and cadmium-induced reduced mitotic index. The level of chromosomal abnormalities mediated by heavy metal stress was reduced in exogenous GA treated plants (Mansour and Kamel 2005). In lupin plants, GA supplementation during Cd stress improved the amylase activity (Sharaf et al. 2009). In chromium-exposed *Pisum sativum*, the minimal concentrations of GA have positively induced various plant processes which are associated with stress tolerance, though higher concentrations have a negative effect on such processes. Supplementation of 10 μM of GA increased the seed germination rate and overall growth of chromium-exposed pea seedlings. Moreover, in pea plants, GA application regulated the enzyme activities that are associated with nitrogen assimilation and ROS detoxification processes (Gangwar et al. 2011). In cadmium-exposed *Arabidopsis,* the GA application decreased the nitric oxide levels and inhibited expression of *iron-regulated transporter 1* (*IRT1*), which is involved in the uptake of cadmium (Zhu et al. 2012). In barley, GA_3 supplementation mitigated the harmful effects on various processes, including seed germination, which resulted from cadmium and molybdenum exposure (Amri et al. 2016).

The role of exogenous GA in alleviating heavy metal stress is widely known through literature, yet few reports about the interplay between endogenous GA content, signaling, and heavy metal stress resistance in plants are available. In the *Oryza sativa* transcriptomic study, two *gibberellin-2-oxidase* (*OsGA2ox*) genes, viz. *GA2ox3* and *GA2ox9* showed increased expression under As stress, indicating an association between GA and HM stress endurance (Huang et al. 2012). Thus, it is evident that heavy metal stress can regulate the GA biosynthesis as well as signaling in plants, though further research endeavors in this direction are demanded to ascertain a strong link between GA and metal toxicity.

15.2.3 Cytokinin

Cytokinin (CK) regulates different plant processes under stressed conditions (Cortleven et al. 2019). CK can improve metal stress tolerance by regulating the uptake and transport of heavy metals in plants (Zhou et al. 2019). Piotrowska-Niczyporuk et al. (2012) reported that photosynthetic functioning and antioxidant activity were increased by CK to significantly mitigate the metal stress. Lately, the application of CK in combination with ABA escalated the tolerance of *Solanum lycopersicum* to Co stress by regulating its uptake, transport, and accumulation (Kamran et al. 2021). In *Triticum*, CK catabolism is facilitated by Cd toxicity by mediating the functioning of Cd oxidase in *T. aestivum* (Veselov et al. 2003). Additionally, transcriptional activity and plant development were lowered in wheat cultivars under Cd stress (Veselov et al. 2003). Cd exposure to *Arabidopsis* resulted in increased cytokinin signaling within the root apex (Bruno et al. 2017). In Arabidopsis, the CK levels were increased in shoots due to reduced expression of CK catabolic gene *cytokinin oxidase* (*CKX*) and enhanced levels of cytokinin biosynthesis gene *isopentyl transferase* (*IPT*) in the presence of Cd stress (Vitti et al. 2013). Also, enhancing the CK content in tobacco plants via transformation of the *IPT* gene regulates biosynthesis of CK and increases tolerance toward Zn stress. The treatment of INCYDE, an obstructer of cytokinin catabolism, promotes the growth of heavy metal-stressed plants (Gemrotová et al. 2013). Further, INCYDE also enhances the levels of different types of antioxidants in plants. The exact role of endogenous CK levels in the modulation of heavy metal stress resistance is still unclear and needs further clarification in the future.

15.2.4 Ethylene

The crucial association of ethylene in a plant's heavy metal stress response has been a subject of concern (Keunen et al. 2016). Various reports propose that the regulation of the ROS system is a vital aspect modulated by ethylene under metal toxicity (Khan et al. 2019) by regulating the functioning of NADPH oxidases (Montero Palmero et al. 2014). Mercury exposure to the *ein2-5* (ethylene insensitive mutant) of *Arabidopsis* resulted in elevated release of H_2O_2 (Montero Palmero et al. 2014). As treated rice roots exhibited improved transcript levels of ethylene biosynthesis genes (Huang et al. 2012). In *Arabidopsis*, Cd application regulated the expression of *ACS2, ACS6, ACO2,* and *ACO4* genes (Schellingen et al. 2014). Conversely, *Arabidopsis* biosynthesis mutants devoid of *ACS2* and *ACS6* did not demonstrate elevated ethylene production during Cd stress (Schellingen et al. 2014). The literature clearly indicates the association of ethylene dynamics with a plant's heavy metal sensitivity versus tolerance.

15.2.5 Jasmonic Acid

The jasmonate phytohormone acts as a signaling biomolecule and modulates various biochemical processes within plants, such as responses against abiotic/biotic stresses (Siddiqi and Husen 2019). It has been observed that methyl jasmonates reduce the impact of heavy metal stress within plants chiefly by improving the antioxidant defense system (Farooq et al. 2016). Methyl jasmonate supplementation improves the enzymatic activities, viz., POD, SOD, CAT, and APX, which are involved in ROS detoxification. In *Brassica napus*, treatment with methyl jasmonate during As stress considerably decreased the MDA and As levels in leaves and roots, thereby improving chlorophyll fluorescence and biomass production (Farooq et al. 2016). Likewise, in faba beans, the jasmonic acid (JA) significantly reduced Cd-induced stress by increasing osmolyte content and antioxidant enzyme activities, which are imperative in lowering the oxidative stress via hindering the accretion of Cd, H_2O_2, and MDA (Ahmad et al. 2017). In the rapeseed plant, treatment of JA decreased the MDA levels and uptake of Cd in leaves, thus enhancing heavy metal stress tolerance in plants (Ali et al. 2018). Methyl jasmonate also has an impact on thiol metabolism in heavy metal-exposed plants (Per et al. 2016). In Cd-exposed *Brassica juncea,* the methyl jasmonate supplementation led to higher levels of GSH and sulfur (S)-assimilation and improved activity of ATP-S (Per et al. 2016). Additionally, JA application regulated the expression of those genes which are involved in GSH biosynthesis, viz., *glutathione synthetase* (*GS*) and *gamma-glutamylcysteine synthetase* (*gamma-GCS*) under heavy metal stress in *Arabidopsis* (Xiang and Oliver 1998). Thus, through literature, it is concluded that jasmonates have a positive regulatory role toward heavy metal stress tolerance in plants. Hence, more research directed toward hidden molecular mechanisms associated with JA-mediated heavy metal stress tolerance is of utmost importance.

15.2.6 Melatonin

Heavy metal-induced stress has been reported to enhance the endogenous measures of melatonin (MT) to enhance the metal tolerance potential in plants like rice, tobacco, tomato, soybean, and barley (Arnao and Hernández-Ruiz 2009, 2013; Byeon et al. 2016; Li et al. 2016; Lee and Back 2017). Considering this, external supplementation of melatonin is now being adopted to confer on the plants the ability to combat metal toxicity. Its mechanism of action involves either a direct role in scavenging ROS and counteracting metal-generated oxidative stress or an indirect role whereby it improves the photosynthetic potential and metabolite content better together with stimulation of the antioxidant defense system and genes regulating stress resistance in plants (Zhang et al. 2016; Al-Huqail et al. 2020). Homeostasis among different nutrients, osmolytes, and metabolites is also maintained by melatonin treatment in stressed plants.

For instance, melatonin mitigated negative after-effects of Al stress in Chinese cabbage (Brassica rapa ssp. pekinensis) and soybean (*Glycine max*) by performing osmotic adjustments, scavenging ROS, protecting photosynthetic machinery, increasing exudation of malate and citrate from roots along with elevating the expression levels of antioxidant enzymes (SOS, POD, and CAT) (Zhang et al. 2017). In case of rosemary (*Salvia rosmarinus*) plants subjected to As stress, foliar treatment of MT piled up osmoregulatory compounds and enhanced the quantities of non-proteinaceous thiols plus phytochelatins for providing tolerance to As. (Farouk and Al-Amri 2019). Cd stress-exposed tobacco (*Nicotiana tabacum*) plants, apple rootstalks, and seedlings of alfalfa (*Medicago sativa)* and cucumber (*Cucumis sativus*) exhibited reversal of adverse impacts of Cd on growth when supplemented with MT by drastically reducing the Cd absorption and its accumulation, enhancing the leaf area and photosynthetic efficiency while upkeeping ROS equilibrium and antioxidant defense mechanisms (Gu et al. 2017; Wang et al. 2019a; Wu et al. 2019).

In apple plants, treatment of MT modulated the transcript levels of genes regulating Cd absorption and translocation (He et al. 2020). In melon (*Cucumis melo*) and seedlings of cucumber, negative impacts on growth induced due to Cu stress are ameliorated by chelating the Cu ions and binding them to cell walls together with elevating the expression of genes targeting redox equilibrium and biosynthesis of jasmonic acid (Cao et al. 2019; Hu et al. 2020). Furthermore, melatonin treatment has been reported to alleviate:

(a) Fe deficiency induced stress in pepper (*Capsicum annuum*) by enhancing measures of active Fe, intracellular NO plus H_2S, and antioxidant enzymes while drastically reducing the contents of oxidative stress indicators (Kaya et al. 2020).
(b) Fe excess stress in cucumber by escalating the levels of secondary metabolites, antioxidant enzymes, photosynthetic pigments, and rates of CO_2 assimilation while declining the measures of ROS and acquisition of Fe by the plant (Ahammed et al. 2020).
(c) Ni stress in seedlings of tomato (*Lycopersicum esculentum*) by modulating the transcript levels of genes regulating melatonin biosynthesis together with those of chlorophyll biosynthesis such as chlorophyll synthase (CHL G), cytochrome P450 oxidoreductase (POR), and chlorophyll oxygenase (CAO) for alleviating adverse impacts of Ni on photosynthetic efficiency (Jahan et al. 2020).
(d) Cr stress in canola (*Brassica napus*) by improving the growth attributes with that of efficiency of PS II and redox rate of photosynthetic quotient (Ayyaz et al. 2020).
(e) Pb stress in tobacco (*Nicotiana tabacum*) suspension cells by reduction in cell death and promotion of cell proliferation, prevention of DNA fragmentation while lowering the levels of H_2O_2 (Kobylińska and Posmyk 2016).

15.2.7 Brassinosteriods

Metal toxicity restrains cell elongation, leading to underdevelopment in plants. This slowdown of plant growth could be attributed to interruptions in various cellular processes as metal toxicity upsets the nutrient balance in plants (Schopfer 1996). Brassinosteroids (BRs), on the other hand, are extensively acknowledged for their potential of being growth enhancers (Davies 2010). For instance, biomass of roots and shoots was improved and growth damage was mitigated by 24-Epibrassinolide (24-EBL) in Zn-stressed *Medicago sativa* and Cr-stressed *Lycopersicum esculentum* (Jan et al. 2020; Ren et al. 2023). Further, BRs promote growth by stimulating cell elongation, progression of the cell cycle, together with expressing the genes responsible for encoding xyloglucans of the cell wall and its expansion in the plants exposed to metal stress (González-García et al. 2011; Zhang et al. 2009).

BRs are also disclosed to modulate the membrane permeability, consequently leading to more elongation plus expansion of the plant cell (Fariduddin et al. 2009). They further perform a key role in controlling multifarious physiological responses of plants, including growth that can be calibrated by percentage germination, rate of flowering, cell necrosis, pollen tube lengthening, and photo-morphogenetic development (Bajguz 2010). Furthermore, it was suggested that cascades of BR signaling modulate loosening or stiffening of components of cell walls to ensure their properties with regard to stressful conditions (Rao and Dixon 2017). BRs also maintain the optimal functioning of the electron transport system and rectify the disorganization of photosystem II (PS II) and its RC (reaction center) in winter rape (*Brassica napus* var. *napus*) and *Vitis vinifera* plants under Cd toxicity (Janeczko et al. 2005; Li et al. 2022). Even the net rate of photosynthesis is made better by the supplementation of BRs in Cd-stressed Indian mustard (*Brassica juncea*) and Radish (*Raphanus sativus*) (Anuradha and SSR 2009).

Reduction of metal uptake has been largely linked to the augmentation of BRs in plants exposed to metal contamination (Kohli et al. 2013; Sharma and Bhardwaj 2007). This decline in metal uptake could be due to BR-mediated stimulation of antioxidative defense mechanisms, abatement of oxidative stress, and enhancement in growth (Sharma and Bhardwaj 2007; Yadav et al. 2016). Besides reducing lipid peroxidation, metal stress-induced damage to membrane integrity and oxidation of proteins can also be prevented by EBL supplementation (Ramakrishna and Rao 2015). Additionally, EBL treatment elevated the protein contents in tall fescue (*Festuca arundinacea*) and *Lycopersicum esculentum* under Pb and Ni stress, respectively (Nazir et al. 2019; Zhong et al. 2020). Similarly, secondary metabolites also upscale their production in metal-stressed plants upon receiving EBL application as observed in Cr and Cd-stressed *Lycopersicum esculentum* and Brassica napus plants, respectively (Sanjari et al. 2019; Jan et al. 2020). It was suggested by Xu et al. (2015) that there exists a direct link in-between the antioxidative potential of plants and enhanced levels of these secondary metabolites.

Extensive studies have been made on the ability of BRs to induce the cascading of antioxidative defense for inhibiting ROS production and lipid degradation (Soares et al. 2016). Activities of antioxidants (ascorbate, glutathione, and tocopherol) of non-enzymatic character, with those of enzymatic ones (APOX, CAT, GR, GST, POD, SOD, etc.), are modulated by BRs addition to the plants exposed to heavy metal toxicity. This can be substantiated by multifarious reports on the protective influence of BRs whereby the antioxidative defense line of action gets upregulated in plants. For instance, a noticeable boost has been noted in the enzymatic performance of APX, CAT, POD, and SOD in EBL-treated Fe- and Pb-stressed rice plants (Guedes et al. 2021; Tadaiesky et al. 2021).

Furthermore, application of BRs modulates the expression levels of various stress-responsive genes as documented in Mn-stressed thale cress (*Arabidopsis thaliana*), where EBL treatment upregulated the expression levels of genes, namely Cu/Zn SD3 (superoxide dismutase), CAT2 (Catalase), and two FSD2&3 (iron dismutase) (Surgun-Acar and Zemheri-Navruz 2022).

15.2.8 Salicylic Acid

Ions of heavy metals, when entering plants, adversely affect growth and developmental processes due to their inhibitory action on various vital physiological processes like photosynthesis, nutrient absorption, transpiration, and respiration. However, treatment of such plants with salicylic acid (SA) could efficiently improve the tolerance and acclimatization potential of such plants to heavy metals (Ur Rahman et al. 2023). Salicylic acid is reported to enhance stress tolerance by the restoration of metal-induced negative impacts on growth and photosynthetic traits as seen in common bean (Khalil et al. 2021; Saini et al. 2021). In Cd stress subjected wheat and tomato plants, supplementation of SA made the fixation of CO_2 better with improvements in measures of chlorophyll and enzymatic action of RuBP carboxylase and PEP carboxylase (Moussa and El-Gamal 2010; Guo et al. 2018). Parameters of gaseous exchange along with the activity of carbonic anhydrase exhibited an enhancement in Ni and Mn-stressed *Brassica juncea* plants upon receiving SA treatment (Ventrella et al. 2009; Parashar et al. 2014).

A similar metal stress alleviatory role by SA application has been reported in Pb-stressed barley plants by improving measures of photosynthetic pigments via possibly blocking the Ca channels responsible for transport of Pb inside the roots (Arshad et al. 2017). This consistently protective nature of SA toward the photosynthetic machinery under heavy metal stress could possibly be on account of enhanced scavenging of ROS or by stimulation of the antioxidant defense system (Rivas-San Vicente and Plasencia 2011). This involves the production of more proline, as in common olive (*Olea europaea*) and date palm (*Phoenix dactylifera*), inclusive of other non-enzymatic antioxidant metabolites along with enzymatic antioxidants (CAT, SOD, and POD) to combat heavy metal-induced toxicity (Zouari et al. 2016; Saini et al. 2021). Documentation of the role of SA in conferring Cd tolerance to plants like water lily (*Nymphaea tetragona*) and duckweed (*Lemna minor*) has been successfully achieved through upregulation of antioxidant enzymes, including APX and SOD, along with different peroxidases (Gu et al. 2018; Lu et al. 2018). Applying SA to Cd-stress-exposed tomato plants drastically reduces the accumulation and distribution of Cd in them (Jia et al. 2021).

Exogenous treatment with SA mitigates the oxidative stress by drastically reducing lipid peroxidation, membrane disruption, and electrolyte leakage in Mn-stressed Indian mustard (*Brassica juncea*) and Cd-stressed melon plants (Parashar et al. 2014; Zhang et al. 2015). Moreover, this is made possible through stimulation of different pathways involving hormones of stress signaling, together with overexpression of specific stress-regulating genes and enzymes like *OsWRKY45* as recorded in rice plants (Chao et al. 2010; Ur Rahman et al. 2023). The gene topoisomerase, TLC1 (a retrotransposon family with long terminal repeats), and elongation factor (chloroplast) are effectively upregulated by SA supplementation, which enables the plants to adapt to stressful environments (Singh et al. 2004; Hettiarachchi et al. 2005; Horváth et al. 2007).

Besides this, SA thickens the cell walls to block Cd penetration by synthesizing more cell wall polysaccharides and promoting expression of related genes. Salicylic acid also mars the activity of pectin methylesterase, which brings about a reduction in the accumulation of Cd in cell wall regions along with alteration in the partition ratio of Cd (Jia et al. 2021). Salicylic acid mitigates metal toxicity by getting intricately involved in crosstalks with that of ethylene and nitric oxide (NO) (Khan et al. 2021b).

15.2.9 Polyamines

Plants, when subjected to heavy metal toxicity, result in excessive generation of ROS, imbalance in ionic equilibrium, and ROS-mediated disruption of membranal integrity, proteins, lipids, and nucleic acids (DNA/RNA) (Nahar et al. 2016). Under these circumstances, polyamines (PAs) like putrescine (Put), spermidine (Spd), and spermine (Spm) exhibit an active role in alleviating oxidative stress via lowering the production of ROS (Baniasadi et al. 2018; Hassan et al. 2018). Further, redox homeostasis has also been reported to be maintained by PAs with the help of elevated measures of non-enzymatic antioxidants (ascorbic acid plus glutathione), as is the case in Spd-supplemented Cr-stressed radish (*Raphanus sativus*) plants (Choudhary et al. 2012). Enhancement in levels of glutathione by PAs treatment is also quite efficient in chelating heavy metals and compartmentalizing them inside the vacuoles (Nahar et al. 2016; Howladar et al. 2018). Likewise, enzymatic (CAT, APX, SOD, GR, GST, DHAR, MDHAR) antioxidants show a drastic improvement in their performance to mitigate metal stress when PAs are supplemented.

Moreover, PAs have the ability to protect the membranal structure from metal-induced degradation through binding with iron ions and phospholipids. Both the enzymes of the glyoxalase system, i.e., Gly I and Gly II, are stimulated by exogenous treatment of PAs to bring about detoxification of toxic methylglyoxal (Nahar et al. 2016). Besides this, regulation of nutrition and upkeep of a proper ammonium-nitrate ratio by PAs also contributes toward maintaining the cellular structure in response to metal stress as observed in Cd-stressed *Inula crithmoides* (Ghabriche et al. 2017).

Further, Rady et al. in 2016 recorded that Cd stress-exposed wheat seeds, when treated with Put (1 mM), alleviated metal toxicity by making relative water content, membrane integrity, and biosynthesis of photosynthetic pigments much improved in comparison to other polyamines, with a provision of far better osmoprotection. In response to Al toxicity, treatment with Put lowers the enzymatic action of those concerned with the production of H_2O_2 and degradation of the cell wall and plasma membrane (Yu et al. 2018). Polyamines Spm or Spd, when exogenously applied to plants or primed to seedlings, actively promote their tolerance against Cd toxicity as reported in China grass (Boehmeria nivea) plants and Cd + Pb toxicity in wheat seedlings (Rady and Hemida 2015; Gong et al. 2016; Taie et al. 2019).

Additionally, exogenous treatment of PAs upregulated the endogenous measures of free and conjugated (insoluble and soluble) PAs to make the tolerance potential against Cr much better in Kinnow mandarin (*Citrus reticulata*) (Shahid et al. 2018). Enzymes of catabolic (e.g., polyamine oxidase—PAO) and anabolic nature (arginine decarboxylase—ADC, ornithine decarboxylase—ODC, S-adenosylmethionine decarboxylase—SAMDC, etc.) exhibit lower and higher activity levels, respectively, to ameliorate heavy metal stress (Shahid et al. 2018). The negative impact of metal stress in the form of oxidative and cell impairments and their amelioration with the help of phytohormones like MT, SA, BR, and PAs has been summarized in Fig. 15.1.

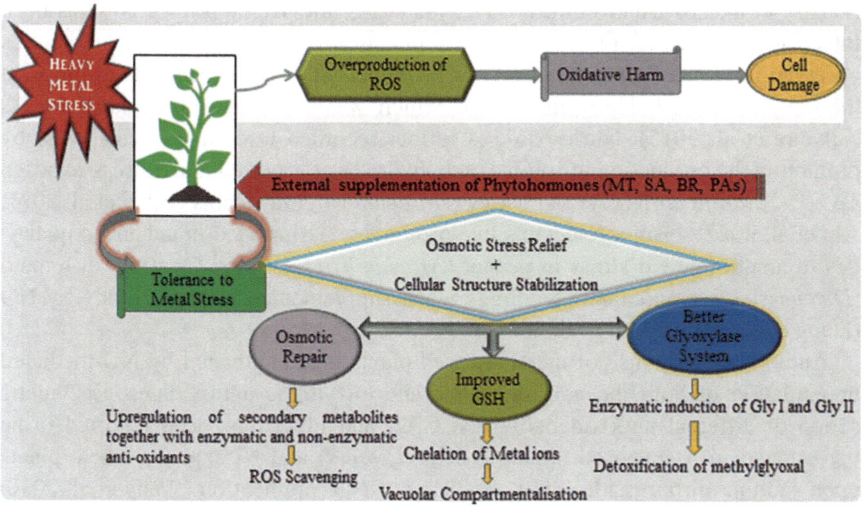

Fig. 15.1 Diagrammatic representation of the effect of metal toxicity in plants and a few major ways of its alleviation by phytohormones

15.3 Role of Other Signaling Molecules in Heavy Metal Stress Tolerance in Plants

15.3.1 Nitric Oxide

Having a major say over a number of signaling pathways, nitric oxide (NO) regulates vital physiological responses of the plants when subjected to hostile and stressful ecological conditions. Its widespread occurrence inside subcellular moieties of plants enables it to control transcriptional plus post-transcriptional modifications so as to regulate various functional aspects of plants (Fancy et al. 2017). One of its significant facets is linked to the counteraction of metal-induced stress in plants. With regard to different parameters of growth in such plants, exogenously supplemented NO can reduce metal uptake and accumulation. Also, notable improvements in germination rates, length, and biomass of seedlings resulted as observed in As stress-exposed rice (*Oryza sativa*) and Ni stress-exposed finger millet (*Eleusine coracana*) seedlings (Kotapati et al. 2017; Praveen and Gupta 2018). Application of NO is further reported to mitigate Cr stress in tall fescue plants through the improvement in various photosynthetic attributes (Huang et al. 2018). In wheat (*Triticum aestivum*) exposed to Zn toxicity, NO can remarkably regulate the photosynthetic efficiency via enhancing Fv/Fm and photochemical quenching (Tripathi et al. 2017).

Injuries to the photosynthetic apparatus, rates of respiration, and non-photochemical quenching are all normalized in Cd-stressed plants upon receiving NO supplementation. Additionally, NO can also upregulate the measures of Rubisco and activity levels of Rubisco activase in response to metal-induced toxic conditions (Khairy et al. 2016; Per et al. 2017). This stress alleviatory role of NO has been linked to its potential to suppress the production of oxidative stress markers as observed in lettuce (*Lactuca sativa*) plants under As stress, where NO diminished the content of both H_2O_2 and MDA, which was otherwise upregulated by As (Silveira et al. 2015). Nitric oxide is further recorded to combat metal stress by promoting the production of soluble carbohydrates as reported in Cd stress-exposed savory (*Satureja hortensis*) and barley (*Hordeum vulgare*) plants (Azizi et al. 2021; Alp et al. 2022). Osmoprotectants like proline are assimilated in enhanced quantities to ameliorate Cd stress in peanut (*Arachis hypogea*) and Cr stress in tomato (*Lycopersicum esculentum*) seedlings when supplemented experimentally by NO (Dong et al. 2016; Khan et al. 2021a).

Antioxidant response of metal-stressed plants is strengthened by NO-mediated upregulation of both the non-enzymatic and enzymatic antioxidants. Enzymatic action of different antioxidants (APX, CAT, and POD) has been reported to be upregulated in As-exposed lettuce (*Lactuca sativa*) and Ni-exposed wheat plants upon sodium nitroprusside (SNP—a source of NO) application (Wang et al. 2010; Farnese et al. 2013). On similar lines, application of NO ameliorated Pb stress in wheat and cowpea (*Vigna unguiculata*) plants via stimulating the activity of different enzymatic antioxidants (APX, CAT, POD, and SOD) (Kaur et al. 2015; Sadeghipour 2016).

Hasanuzzaman and Fujita in 2013 reported that in wheat seedlings subjected to As stress, application of SNP enhanced the functional efficiency of glyoxylase enzymes for the purpose of detoxifying methylglyoxal. In addition to this, NO promotes the measures of glutathione and glutathione homeostasis performs the detoxification of methylglyoxal as well (Mustafiz et al. 2010). Nitric oxide also boosted the expression levels of *HvAOX* gene in barley plants in response to Cd stress (He et al. 2019) and those of DHAR and GR enzymes in wheat plants exposed to Al stress (Sun et al. 2015).

15.3.2 Hydrogen Sulfide

This odorous gas, in its lower concentrations, equips the plants to confront a varied range of abiotic stress-inducing factors like extremes of salt, water, temperature, and heavy metals (Li 2013). Treatment with H_2S in stressed plants is reported to trigger the defense system comprised of enzymatic and non-enzymatic antioxidants for counteracting these stresses. Metallic toxicity in particular is mitigated in stressed plants by elevating endogenous measures of H_2S (He et al. 2018). Furthermore, external supplementation of H_2S can bring about the same results via (a) enhancing various growth and photosynthetic attributes (Ali et al. 2013; Fu et al. 2019); (b) reducing the metal ion uptake while enhancing absorption of essential nutrients (Chen et al. 2013; Ahmad et al. 2019; Zanganeh et al. 2019); (c) triggering activation of antioxidant enzymes to quench ROS for the upkeep of cell structure (Shan et al. 2012; Basharat et al. 2015; Rizwan et al. 2019); (d) upregulating the expression of phytochelatins plus metallothionein associated genes (Valivand et al. 2019); and (e) getting involved in intricate cross talks with other signaling molecules and phytohormones (Qiao et al. 2015; Tian et al. 2017; Kaya et al. 2019). With regard to various metal-afflicted plant species, the alleviatory role of H_2S has been tabulated as follows (Table 15.1):

15.4 Conclusion

Toxicity imposed by heavy metals on the plants has been recognized as one of the detrimental factors for optimum growth and development. In order to ameliorate this abiotic stress, a number of physiological and molecular mechanisms become operational inside the plants, which fall short of their efficacy if this stress is quite severe and a prolonged one. Therefore, external supplementation of these metal-stressed plants with phytoprotectants (signaling molecules plus phytohormones) becomes quite a favorable option. Various studies have highlighted the imperative role of these phytoprotectants in metal-stressed plant species on account of their antagonistic impact on metal-induced toxicity. Exogenously treated stressed plants become more tolerant by these molecules, which could efficiently lead to achieving

Table 15.1 Role of H_2S in alleviation of metal toxicity in different plant species

S.No.	Heavy metal	Plant species	Role of H_2S in metal stress alleviation	References
1.	Cd	(a) *Triticum aestivum*	Promoted rates of seed germination by activating the hydrolytic action of enzymes like amylase and esterase, and reduced oxidative stress with the help of antioxidant enzymes	Huang et al. (2015)
		(b) *Hordeum vulgare*	Lowered cd uptake along with enhancement in photosynthetic activity, amounts of soluble proteins, and the enzymatic action of antioxidants	Fu et al. (2019)
2.	Cr	(a) *Hordeum vulgare*	Reduction in Cr levels inside different plant parts while maintaining the integrity of the photosynthetic apparatus and ameliorated the nuclear damage inside root cells	Ali et al. (2013)
		(b) *Brassica oleracea* var. *botrytis*	Scavenged ROS via stimulating the activity of antioxidant enzymes; lowered Cr accumulation	Ahmad et al. (2019)
3.	Al	*Glycine max*	Upregulation of genes dealing with the reduction in Al accumulation and combat stress caused by Al excess	Wang et al. (2019b)
4.	As	*Pisum sativum*	Enhanced the activity of the ascorbate-glutathione cycle and lowered accumulation of arsenic in stressed pea plants	Singh et al. (2015)
5.	Ni	(a) *Cucurbita pepo*	Made the seedling biomass and photosynthetic attributes better; elevated the activity of the ascorbate-glutathione cycle as well as the expression of phytochelatins genes; reduced H_2O_2 levels and leakage of electrolytes	Valivand et al. (2019)
		(b) *Oryza sativa*	Enhanced enzymatic action plus expression of genes concerned with nitrogen metabolism; improved the growth and photosynthetic parameters	Rizwan et al. (2019)
6.	Pb	(a) *Brassica napus*	Enhanced absorption of essential nutrients; reduction in the MDA and ROS measures by stimulation of the antioxidant defense system; decreased the pb uptake and improved photosynthetic attributes	Ali et al. (2014a, b)
		(b) *Zea mays*	Betterment in growth; more protein plus glutathione levels; decreased pb accumulation; upregulated the activity of nitrate reductase enzyme	Zanganeh et al. (2019)
7.	Zn	(a) *Solanum nigrum*	Reduction in ROS content and uptake of Zn; enhancement in the activity of metallothionines	Liu et al. (2015)
		(b) *Capsicum annuum*	Decreased the uptake of Zn; reduced the content of oxidative stress markers; upregulated the activity of the antioxidant defense system	Kaya et al. (2018)

the goal of sustainable crop production. However, for better comprehension of their mechanistic action, more investigative works are required, which will enable the plant scientists to develop metal stress-tolerant varieties.

References

Ahammed GJ, Wu M, Wang Y, Yan Y, Mao Q, Ren J, Ma R, Liu A, Chen S (2020) Melatonin alleviates iron stress by improving iron homeostasis, antioxidant defense & secondary metabolism in cucumber. Sci Hortic 265:109205

Ahmad P, Alyemeni MN, Wijaya L, Alam P, Ahanger MA, Alamri SA (2017) Jasmonic acid alleviates negative impacts of cadmium stress by modifying osmolytes and antioxidants in faba bean (*Vicia faba* L.). Arch Agron Soil Sci 63(13):1889–1899

Ahmad R, Ali S, Rizwan M, Dawood M, Farid M, Hussain A, Wijaya L, Alyemeni MN, Ahmad P (2019) Hydrogen sulfide alleviates chromium stress on cauliflower by restricting its uptake & enhancing antioxidative system. Physiol Plant 168(2):289–300

Al-Huqail AA, Khan MN, Ali HM, Siddiqui MH, Al-Huqail AA, AlZuaibr FM, Al-Muwayhi MA, Marraiki N, Al-Humaid LA (2020) Exogenous melatonin mitigates boron toxicity in wheat. Ecotoxicol Environ Saf 201:110822

Ali S, Farooq MA, Hussain S, Yasmeen T, Abbasi GH, Zhang G (2013) Alleviation of chromium toxicity by hydrogen sulfide in barley. Environ Toxicol Chem 32(10):2234–2239

Ali B, Mwamba TM, Gill RA, Yang C, Ali S, Daud MK, Wu Y, Zhou W (2014a) Improvement of element uptake & antioxidative defense in *Brassica napus* under lead stress by application of hydrogen sulfide. Plant Growth Regul 74(3):261–273

Ali B, Song WJ, Hu WZ, Luo XN, Gill RA, Wang J, Zhou WJ (2014b) Hydrogen sulfide alleviates lead-induced photosynthetic & ultrastructural changes in oilseed rape. Ecotoxicol Environ Saf 102:25–33

Ali E, Hussain N, Shamsi IH, Jabeen Z, Siddiqui MH, Jiang LX (2018) Role of jasmonic acid in improving tolerance of rapeseed (*Brassica napus* L.) to Cd toxicity. J Zhejiang Univ Sci B 19(2):130

Alp K, Terzi H, Yildiz M (2022) Proteomic & physiological analyses to elucidate nitric oxide-mediated adaptive responses of barley under cadmium stress. Physiol Mol Biol Plants 28(7):1467–1476

Amri B, Khamassi K, Ali MB, da Silva JAT, Kaab LBB (2016) Effects of gibberellic acid on the process of organic reserve mobilization in barley grains germinated in the presence of cadmium and molybdenum. S Afr J Bot 106:35–40

Anuradha S, SSR R (2009) Effect of 24-epibrassinolide on the photosynthetic activity of radish plants under cadmium stress. Photosynthetica 47(2):317–320

Arnao MB, Hernández-Ruiz J (2009) Chemical stress by different agents affects the melatonin content of barley roots. J Pineal Res 46(3):295–299

Arnao MB, Hernández-Ruiz J (2013) Growth conditions determine different melatonin levels in *Lupinus albus* L. J Pineal Res 55(2):149–155

Arshad T, Maqbool N, Javed F, Wahid A, Arshad MU (2017) Enhancing the defensive mechanism of lead affected barley (*Hordeum vulgare* L.) genotypes by exogenously applied salicylic acid. J Agric Sci 9(2):139–146

Ayyaz A, Farooq MA, Kanwal A, Aslam M, Iqbal M, Manzoor A, Khalid A, Umer S, Bano H, Rasool B, Zafar Z (2020) Differential responses of exogenous melatonin on growth, photosynthesis & antioxidant defence system in two *Brassica napus* L. cultivars under chromium stress. Int J Environ Agric Biotechnol 5(2):397–411

Azeez NA, Dash SS, Gummadi SN, Deepa VS (2021) Nano-remediation of toxic heavy metal contamination: Hexavalent chromium [Cr (VI)]. Chemosphere 266:129204

Azizi I, Esmaielpour B, Fatemi H (2021) Exogenous nitric oxide on morphological, biochemical & antioxidant enzyme activity on savory (*Satureja hortensis* L.) plants under cadmium stress. J Saudi Soc Agric Sci 20(6):417–423

Bajguz A (2010) An enhancing effect of exogenous brassinolide on the growth & antioxidant activity in *Chlorella vulgaris* cultures under heavy metals stress. Environ Exp Bot 68(2):175–179

Baniasadi F, Saffari VR, Maghsoudi Moud AAM (2018) Physiological & growth responses of *Calendula officinalis* L. plants to the interaction effects of polyamines & salt stress. Sci Hortic 234:312–317

Basharat A, Ping Q, Rui S, Farooq MA, Gill RA, Jian W, Muhammad A, Weijun Z (2015) Hydrogen sulfide alleviates the aluminum-induced changes in *Brassica napus* as revealed by physiochemical & ultrastructural study of plant. Environ Sci Pollut Res Int 22:3068–3081

Bhardwaj S, Jan S, Sharma D, Kapoor D, Singh R, Bhardwaj R (2022) Heavy metal contamination in plants sources and effects. In: Heavy metals in plants physiological to molecular approach. CRC Press, Boca Raton, pp 50–63

Bruno L, Pacenza M, Forgione I, Lamerton LR, Greco M, Chiappetta A, Bitonti MB (2017) In *Arabidopsis thaliana* cadmium impact on the growth of primary root by altering SCR expression and auxin-cytokinin cross-talk. Front Plant Sci 8:1323

Bücker-Neto L, Paiva ALS, Machado RD, Arenhart RA, Margis-Pinheiro M (2017) Interactions between plant hormones and heavy metals responses. Genet Mol Biol 40:373–386

Byeon Y, Lee HJ, Lee HY, Back K (2016) Cloning & functional characterization of the Arabidopsis N-acetylserotonin O-methyltransferase responsible for melatonin synthesis. J Pineal Res 60(1):65–73

Cao YY, Qi CD, Li S, Wang Z, Wang X, Wang J, Ren S, Li X, Zhang N, Guo YD (2019) Melatonin alleviates copper toxicity via improving copper sequestration & ROS scavenging in cucumber. Plant Cell Physiol 60(3):562–574

Chao Y-Y, Chen C-Y, Huang W-D, Kao CH (2010) Salicylic acid-mediated hydrogen peroxide accumulation & protection against Cd toxicity in rice leaves. Plant Soil 329(1–2):327–337

Chen J, Wang W-H, Wu F-H, You C-Y, Liu T-W (2013) Hydrogen sulfide alleviates aluminum toxicity in barley seedlings. Plant Soil 362:301–318

Choudhary SP, Kanwar M, Bhardwaj R, Yu JQ, Tran LSP (2012) Chromium stress mitigation by polyamine-brassinosteroid application involves phytohormonal & physiological strategies in *Raphanus sativus* L. PLoS One 7(3):e33210

Cortleven A, Ehret S, Schmülling T, Johansson H (2019) Ethylene-independent promotion of photomorphogenesis in the dark by cytokinin requires COP1 and the CDD complex. J Exp Bot 70(1):165–178

Davies PJ (2010) The plant hormones: their nature, occurrence, & functions. In: Plant hormones. Springer, Netherlands, pp 1–15

Dong Y, Chen W, Xu L, Kong J, Liu S, He Z (2016) Nitric oxide can induce tolerance to oxidative stress of peanut seedlings under cadmium toxicity. Plant Growth Regul 79(1):19–28

Elobeid M, Göbel C, Feussner I, Polle A (2012) Cadmium interferes with auxin physiology and lignification in poplar. J Exp Bot 63(3):1413–1421

Emam Verdian A, Ding Y, Mokhberdoran F (2020) The role of salicylic acid and gibberellin signaling in plant responses to abiotic stress with an emphasis on heavy metals. Plant Signal Behav 15(7):1777372

Fancy NN, Bahlmann AK, Loake GJ (2017) Nitric oxide function in plant abiotic stress. Plant Cell Environ 40(4):462–472

Fariduddin Q, Khanam S, Hasan SA, Ali B, Hayat S, Ahmad A (2009) Effect of 28-homobrassinolide on the drought stress-induced changes in photosynthesis & antioxidant system of *Brassica juncea* L. Acta Physiol Plant 31(5):889–897

Farnese FS, de Oliveira JA, Gusman GS, Leão GA, Ribeiro C, Siman LI, Cambraia J (2013) Plant responses to arsenic: the role of nitric oxide. Water Air Soil Pollut 224(9):1–11

Farooq MA, Gill RA, Islam F, Ali B, Liu H, Xu J et al (2016) Methyl jasmonate regulates antioxidant defense and suppresses arsenic uptake in *Brassica napus* L. Front Plant Sci 7:468

Farouk S, Al-Amri SM (2019) Exogenous melatonin-mediated modulation of arsenic tolerance with improved accretion of secondary metabolite production, activating antioxidant capacity & improved chloroplast ultrastructure in rosemary herb. Ecotoxicol Environ Saf 180:333–347

Fattorini L, Ronzan M, Piacentini D, Della Rovere F, De Virgilio C, Sofo A et al (2017) Cadmium and arsenic affect quiescent Centre formation and maintenance in *Arabidopsis thaliana* post-embryonic roots disrupting auxin biosynthesis and transport. Environ Exp Bot 144:37–48

Fu M-M, Dawood M, Wang N-H, Wu F (2019) Exogenous hydrogen sulfide reduces cadmium uptake & alleviates cadmium toxicity in barley. Plant Growth Regul 89(2):227–237

Gangwar S, Singh VP, Srivastava PK, Maurya JN (2011) Modification of chromium (VI) phytotoxicity by exogenous gibberellic acid application in *Pisum sativum* (L.) seedlings. Acta Physiol Plant 33:1385–1397

Gemrotová M, Kulkarni MG, Stirk WA, Strnad M, Van Staden J, Spíchal L (2013) Seedlings of medicinal plants treated with either a cytokinin antagonist (PI-55) or an inhibitor of cytokinin degradation (INCYDE) are protected against the negative effects of cadmium. Plant Growth Regul 71:137–145

Ghabriche R, Ghnaya T, Mnasri M, Zaier H, Baioui R, Vromman D, Abdelly C, Lutts S (2017) Polyamine & tyramine involvement in NaCl-induced improvement of Cd resistance in the halophyte Inula *chrithmoides* L. J Plant Physiol 216:136–144

Ghori NH, Ghori T, Hayat MQ, Imadi SR, Gul A, Altay V, Ozturk M (2019) Heavy metal stress and responses in plants. Int J Environ Sci Technol 16:1807–1828

Gong X, Liu Y, Huang D, Zeng G, Liu S, Tang H, Zhou L, Hu X, Zhou Y, Tan X (2016) Effects of exogenous calcium & spermidine on cadmium stress moderation & metal accumulation in *Boehmeria nivea* (L.) Gaudich. Environ Sci Pollut Res Int 23(9):8699–8708

González-García MP, Vilarrasa-Blasi J, Zhiponova M, Divol F, Mora-García S, Russinova E, Caño-Delgado AI (2011) Brassinosteroids control meristem size by promoting cell cycle progression in Arabidopsis roots. Development 138(5):849–859

Gu Q, Chen Z, Yu X, Cui W, Pan J, Zhao G, Xu S, Wang R, Shen W (2017) Melatonin confers plant tolerance against cadmium stress via the decrease of cadmium accumulation & reestablishment of microRNA-mediated redox homeostasis. Plant Sci 261:28–37

Gu C-S, Yang Y-H, Shao Y-F, Wu K-W, Liu Z-L (2018) The effects of exogenous salicylic acid on alleviating cadmium toxicity in Nymphaea tetragona Georgi. S Afr J Bot 114:267–271

Guedes FRCM, Maia CF, Silva BRSD, Batista BL, Alyemeni MN, Ahmad P, Lobato AKDS (2021) Exogenous 24-Epibrassinolide stimulates root protection, & leaf antioxidant enzymes in lead stressed rice plants: central roles to minimize Pb content & oxidative stress. Environ Pollut 280:116992

Guo J, Zhou R, Ren X, Jia H, Hua L, Xu H, Lv X, Zhao J, Wei T (2018) Effects of salicylic acid, Epi-brassinolide & calcium on stress alleviation & Cd accumulation in tomato plants. Ecotoxicol Environ Saf 157:491–496

Hasanuzzaman M, Fujita M (2013) Exogenous sodium nitroprusside alleviates arsenic-induced oxidative stress in wheat (*Triticum aestivum* L.) seedlings by enhancing antioxidant defense & glyoxalase system. Ecotoxicology 22(3):584–596

Hassan FAS, Ali EF, Alamer KH (2018) Exogenous application of polyamines alleviates water stress-induced oxidative stress of *Rosa damascena* M. Var. trigintipetala Dieck. S Afr J Bot 116:96–102

He H, Li Y, He LF (2018) The central role of hydrogen sulfide in plant responses to toxic metal stress. Ecotoxicol Environ Saf 157:403–408

He L, Wang X, Feng R, He Q, Wang S, Liang C et al (2019) Alternative pathway is involved in nitric oxide-enhanced tolerance to cadmium stress in barley roots. Plants 8(12):557

He J, Zhuang X, Zhou J, Sun L, Wan H, Li H, Lyu D (2020) Exogenous melatonin alleviates cadmium uptake & toxicity in apple rootstocks. Tree Physiol 40(6):746–761

Hernández-García J, Briones-Moreno A, Blázquez MA (2021, January) Origin and evolution of gibberellin signaling and metabolism in plants. In: Seminars in cell & developmental biology, vol 109. Academic Press, pp 46–54

Hettiarachchi GH, Reddy MK, Sopory SK, Chattopadhyay S (2005) Regulation of TOP2 by various abiotic stresses including cold & salinity in pea & transgenic tobacco plants. Plant Cell Physiol 46(7):1154–1160

Horváth E, Szalai G, Janda T (2007) Induction of abiotic stress tolerance by salicylic acid signaling. J Plant Growth Regul 26(3):290–300

Howladar SM, Al-Robai SA, Al-Zahrani FS, Howladar MM, Aldhebiani AY (2018) Silicon & its application method effects on modulation of cadmium stress responses in *Triticum aestivum* (L.) through improving the antioxidative defense system & polyamine gene expression. Ecotoxicol Environ Saf 159:143–152

Hu Z, Fu Q, Zheng J, Zhang A, Wang H (2020) Transcriptomic & metabolomic analyses reveal that melatonin promotes melon root development under copper stress by inhibiting jasmonic acid biosynthesis. Hortic Res 7(1):79

Huang TL, Nguyen QTT, Fu SF, Lin CY, Chen YC, Huang HJ (2012) Transcriptomic changes and signalling pathways induced by arsenic stress in rice roots. Plant Mol Biol 80:587–608

Huang ZQ, Ye SC, Hu LY, Hu KD, Yan H, Li WJ, Jiao H, Zhang H (2015) Hydrogen sulfide promotes wheat grain germination under cadmium stress. Proc Natl Acad Sci India Sect B Biol Sci 86:887–895

Huang M, Ai H, Xu X, Chen K, Niu H, Zhu H, Sun J, Du D, Chen L (2018) Nitric oxide alleviates toxicity of hexavalent chromium on tall fescue & improves performance of photosystem II. Ecotoxicol Environ Saf 164:32–40

Hussain B, Ashraf MN, Abbas A, Li J, Farooq M (2021) Cadmium stress in paddy fields: effects of soil conditions and remediation strategies. Sci Total Environ 754:142188

Jahan MS, Guo S, Baloch AR, Sun J, Shu S, Wang Y, Ahammed GJ, Kabir K, Roy R (2020) Melatonin alleviates nickel phytotoxicity by improving photosynthesis, secondary metabolism & oxidative stress tolerance in tomato seedlings. Ecotoxicol Environ Saf 197:110593

Jamla M, Khare T, Joshi S, Patil S, Penna S, Kumar V (2021) Omics approaches for understanding heavy metal responses and tolerance in plants. Curr Plant Biol 27:100213

Jan S, Noman A, Kaya C, Ashraf M, Alyemeni MN, Ahmad P (2020) 24-epibrassinolide alleviates the injurious effects of Cr(VI) toxicity in tomato plants: insights into growth, physio-biochemical attributes, antioxidant activity & regulation of ascorbate–glutathione & glyoxalase cycles. J Plant Growth Regul 39(4):1587–1604

Janeczko A, Koscielniak J, Pilipowicz M, Szarek-Lukaszewska G, Skoczowski A (2005) Protection of winter rape photosystem 2 by 24- epibrassinolide under cadmium stress. Photosynthetica 43(2):293–298

Javed T, Ali MM, Shabbir R, Anwar R, Afzal I, Mauro RP (2021) Alleviation of copper-induced stress in pea (*Pisum sativum* L.) through foliar application of gibberellic acid. Biology 10(2):120

Jia H, Wang X, Wei T, Wang M, Liu X, Hua L, Ren X, Guo J, Li J (2021) Exogenous salicylic acid regulates cell wall polysaccharides synthesis & pectin methylation to reduce Cd accumulation of tomato. Ecotoxicol Environ Saf 207:111550

Kamran M, Danish M, Saleem MH, Malik Z, Parveen A, Abbasi GH et al (2021) Application of abscisic acid and 6-benzylaminopurine modulated morpho-physiological and antioxidative defense responses of tomato (*Solanum lycopersicum* L.) by minimizing cobalt uptake. Chemosphere 263:128169

Kaur G, Singh HP, Batish DR, Mahajan P, Kohli RK, Rishi V (2015) Exogenous nitric oxide (NO) interferes with lead (Pb)-induced toxicity by detoxifying reactive oxygen species in hydroponically grown wheat (*Triticum aestivum*) roots. PLoS One 10(9):e0138713

Kaya C, Ashraf M, Akram NA (2018) *Hydrogen sulfide regulates the levels of key metabolites & antioxidant defense system to counteract oxidative stress in pepper (Capsicum annuum* L.) plants exposed to high zinc regime. Environ Sci Pollut Res Int 25:1–7

Kaya C, Ashraf M, Alyemeni MN, Ahmad P (2019) Responses of nitric oxide & hydrogen sulfide in regulating oxidative defence system in wheat plants grown under cadmium stress. Physiol Plant 168(2):345–360

Kaya C, Higgs D, Ashraf M, Alyemeni MN, Ahmad P (2020) Integrative roles of nitric oxide & hydrogen sulfide in melatonin-induced tolerance of pepper (*Capsicum annuum* L.) plants to iron deficiency & salt stress alone or in combination. Physiol Plant 168(2):256–277

Keunen E, Schellingen K, Vangronsveld J, Cuypers A (2016) Ethylene and metal stress: small molecule, big impact. Front Plant Sci 7:23

Khairy AIH, Oh MJ, Lee SM, Kim DS, Roh KS (2016) Nitric oxide overcomes Cd & Cu toxicity in in vitro-grown tobacco plants through increasing contents & activities of RuBisCO & RuBisCO Activase. Biochimie Open 2:41–51

Khalil R, Haroun S, Bassyoini F, Nagah A, Yusuf M (2021) Salicylic acid in combination with kinetin or calcium ameliorates heavy metal stress in *Phaseolus vulgaris* plant. J Agric Food Res 5:100182

Khan MIR, Jahan B, Alajmi MF, Rehman MT, Khan NA (2019) Exogenously-sourced ethylene modulates defense mechanisms and promotes tolerance to zinc stress in mustard (*Brassica juncea* L.). Plants 8(12):540

Khan MN, Alamri S, Al-Amri AA, Alsubaie QD, Al-Munqedi B, Ali HM, Singh VP, Siddiqui MH (2021a) Effect of nitric oxide on seed germination & seedling development of tomato under chromium toxicity. J Plant Growth Regul 40(6):2358–2370

Khan MIR, Jahan B, AlAjmi MF, Rehman MT, Iqbal N, Irfan M, Sehar Z, Khan NA (2021b) Crosstalk of plant growth regulators protects photosynthetic performance from arsenic damage by modulating defense systems in rice. Ecotoxicol Environ Saf 222:112535

Kobylińska A, Posmyk MM (2016) Melatonin restricts Pb-induced PCD by enhancing BI-1 expression in tobacco suspension cells. Biometals 29(6):1059–1074

Kohli A, Sreenivasulu N, Lakshmanan P, Kumar PP (2013) The phytohormone crosstalk paradigm takes center stage in understanding how plants respond to abiotic stresses. Plant Cell Rep 32(7):945–957

Kohli SK, Handa N, Gautam V, Bali S, Sharma A, Khanna K et al (2017) ROS signaling in plants under heavy metal stress. In: Reactive oxygen species and antioxidant systems in plants: role and regulation under abiotic stress. Springer Singapore, Singapore, pp 185–214

Korver RA, Koevoets IT, Testerink C (2018) Out of shape during stress: a key role for auxin. Trends Plant Sci 23(9):783–793

Kotapati KV, Palaka BK, Ampasala DR (2017) Alleviation of nickel toxicity in finger millet (*Eleusine coracana* L.) germinating seedlings by exogenous application of salicylic acid & nitric oxide. Crop J 5(3):240–250

Krishnamurthy A, Rathinasabapathi B (2013) Auxin and its transport play a role in plant tolerance to arsenite-induced oxidative stress in *Arabidopsis thaliana*. Plant Cell Environ 36(10):1838–1849

Lee HY, Back K (2017) Cadmium disrupts subcellular organelles, including chloroplasts, resulting in melatonin induction in plants. Molecules 22(10):1791

Li ZG (2013) Hydrogen sulfide: a multifunctional gaseous molecule in plants. Russ J Plant Physiol 60(6):733–740

Li MQ, Hasan MK, Li CX, Ahammed GJ, Xia XJ, Shi K, Zhou YH, Reiter RJ, Yu JQ, Xu MX, Zhou J (2016) Melatonin mediates selenium-induced tolerance to cadmium stress in tomato plants. J Pineal Res 61(3):291–302

Li BB, Fu YS, Li XX, Yin HN, Xi ZM (2022) Brassinosteroids alleviate cadmium phytotoxicity by minimizing oxidative stress in grape seedlings: toward regulating the ascorbate-glutathione cycle. Sci Hortic 299:111002

Liu X, Chen J, Wang GH, Wang WH, Shen ZJ, Luo MR, Gao GF, Simon M, Ghoto K, Zheng HL (2015) Hydrogen sulfide alleviates zinc toxicity by reducing zinc uptake & regulating genes expression of antioxidative enzymes & metallothioneins in roots of the cadmium/zinc hyperaccumulator *Solanum nigrum* L. Plant Soil 400:177–192

Lu Q, Zhang T, Zhang W, Su C, Yang Y, Hu D, Xu Q (2018) Alleviation of cadmium toxicity in Lemna minor by exogenous salicylic acid. Ecotoxicol Environ Saf 147:500–508

Mansour MM, Kamel EAR (2005) Interactive effect of heavy metals and gibberellic acid on mitotic activity and some metabolic changes of *Vicia faba* L. plants. Cytologia 70(3):275–282
Montero Palmero MB, Martín Barranco A, Escobar C, Hernández LE (2014) Early transcriptional responses to mercury: a role for ethylene in mercury induced stress. New Phytol 201(1):116–130
Morkunas I, Woźniak A, Mai VC, Rucińska-Sobkowiak R, Jeandet P (2018) The role of heavy metals in plant response to biotic stress. Molecules 23(9):2320
Moussa HR, El-Gamal SM (2010) Effect of salicylic acid pretreatment on cadmium toxicity in wheat. Biol Plant 54(2):315–320
Mustafiz A, Sahoo KK, Singla-Pareek SL, Sopory SK (2010) Metabolic engineering of glyoxalase pathway for enhancing stress tolerance in plants. In: Sunkar R (ed) Plant stress tolerance: methods & protocols. Springer, New York, pp 95–118
Nahar K, Hasanuzzaman M, Alam MM, Rahman A, Suzuki T, Fujita M (2016) Polyamine & nitric oxide crosstalk: antagonistic effects on cadmium toxicity in mung bean plants through upregulating the metal detoxification, antioxidant defense & methylglyoxal detoxification systems. Ecotoxicol Environ Saf 126:245–255
Nazir F, Hussain A, Fariduddin Q (2019) Interactive role of epibrassinolide & hydrogen peroxide in regulating stomatal physiology, root morphology, photosynthetic & growth traits in *Solanum lycopersicum* L. under nickel stress. Environ Exp Bot 162:479–495
Parashar A, Yusuf M, Fariduddin Q, Ahmad A (2014) Salicylic acid enhances antioxidant system in Brassica juncea grown under different levels of manganese. Int J Biol Macromol 70:551–558
Per TS, Khan NA, Masood A, Fatma M (2016) Methyl jasmonate alleviates cadmium-induced photosynthetic damages through increased S-assimilation and glutathione production in mustard. Front Plant Sci 7:1933
Per TS, Masood A, Khan NA (2017) Nitric oxide improves S-assimilation & GSH production to prevent inhibitory effects of cadmium stress on photosynthesis in mustard (*Brassica juncea* L.). Nitric Oxide Biol Chem 68:111–124
Piotrowska-Niczyporuk A, Bajguz A, Zambrzycka E, Godlewska-Żyłkiewicz B (2012) Phytohormones as regulators of heavy metal biosorption and toxicity in green alga Chlorella vulgaris (Chlorophyceae). Plant Physiol Biochem 52:52–65
Praveen A, Gupta M (2018) Nitric oxide confronts arsenic stimulated oxidative stress & root architecture through distinct gene expression of auxin transporters, nutrient related genes & modulates biochemical responses in *Oryza sativa* L. Environ Pollut 240:950–962
Qiao Z, Jing T, Liu Z, Zhang L, Jin Z, Liu D, Pei Y (2015) H_2S acting as a downstream signaling molecule of SA regulates Cd tolerance in Arabidopsis. Plant Soil 393(1):137–146
Rady MM, Hemida KA (2015) Modulation of cadmium toxicity & enhancing cadmium-tolerance in wheat seedlings by exogenous application of polyamines. Ecotoxicol Environ Saf 119:178–185
Rady MM, El-Yazal MAS, Taie HAA, Ahmed SMA (2016) Response of wheat growth & productivity to exogenous polyamines under lead stress. J Crop Sci Biotechnol 19(5):363–371
Ramakrishna B, Rao SSR (2015) Foliar application of brassinosteroids alleviates adverse effects of zinc toxicity in radish (*Raphanus sativus* L.) plants. Protoplasma 252(2):665–677
Rao X, Dixon RA (2017) Brassinosteroid mediated cell wall remodeling in grasses under abiotic stress. Front Plant Sci 8:806
Ren Y, Li X, Liang J, Wang S, Wang Z, Chen H, Tang M (2023) Brassinosteroids & gibberellic acid actively regulate the zinc detoxification mechanism of *Medicago sativa* L. seedlings. BMC Plant Biol 23(1):75
Rhaman MS, Imran S, Rauf F, Khatun M, Baskin CC, Murata Y, Hasanuzzaman M (2020) Seed priming with phytohormones: an effective approach for the mitigation of abiotic stress. Plants 10(1):37
Riaz M, Kamran M, Fang Y, Wang Q, Cao H, Yang G et al (2021) Arbuscular mycorrhizal fungi-induced mitigation of heavy metal phytotoxicity in metal contaminated soils: a critical review. J Hazard Mater 402:123919

Rivas-San Vicente M, Plasencia J (2011) Salicylic acid beyond defence: its role in plant growth & development. J Exp Bot 62(10):3321–3338

Rizwan M, Mostofa MG, Ahmad MZ, Zhou Y, Adeel M, Mehmood S, Ahmad MA, Javed R, Imtiaz M, Aziz O, Aziz O (2019) Hydrogen sulfide enhances rice tolerance to nickel through the prevention of chloroplast damage & the improvement of nitrogen metabolism under excessive nickel. Plant Physiol Biochem 138:100–111

Ronzan M, Piacentini D, Fattorini L, Della Rovere F, Eiche E, Riemann M et al (2018) Cadmium and arsenic affect root development in *Oryza sativa* L. negatively interacting with auxin. Environ Exp Bot 151:64–75

Rostami S, Azhdarpoor A (2019) The application of plant growth regulators to improve phytoremediation of contaminated soils: a review. Chemosphere 220:818–827

Roy M, McDonald LM (2015) Metal uptake in plants and health risk assessments in metal-contaminated smelter soils. Land Degrad Dev 26(8):785–792

Sabagh AE, Mbarki S, Hossain A, Iqbal MA, Islam MS, Raza A et al (2021) Potential role of plant growth regulators in administering crucial processes against abiotic stresses. Front Agron 3:648694

Sabagh AE, Islam MS, Hossain A, Iqbal MA, Mubeen M, Waleed M et al (2022) Phytohormones as growth regulators during abiotic stress tolerance in plants. Front Agron 4:765068

Sadeghipour O (2016) Pretreatment with nitric oxide reduces lead toxicity in cowpea (*Vigna unguiculata* [L.] walp.). Arch Biol Sci 68(1):165–175

Saini S, Sharma I, Pati PK (2017) Integrating the knowledge of auxin homeostasis with stress tolerance in plants. Mech Plant Horm Signal Stress 1:53–70

Saini S, Kaur N, Pati PK (2021) Phytohormones: key players in the modulation of heavy metal stress tolerance in plants. Ecotoxicol Environ Saf 223:112578

Sanjari S, Keramat B, Nadernejad N, Mozafari H (2019) Ameliorative effects of 24-epibrassinolide & thiamine on excess cadmium-induced oxidative stress in Canola (*Brassica napus* L.) plants. J Plant Interact 14(1):359–368

Schellingen K, Van Der Straeten D, Vandenbussche F, Prinsen E, Remans T, Vangronsveld J, Cuypers A (2014) Cadmium-induced ethylene production and responses in Arabidopsis thaliana rely on ACS2 and ACS6 gene expression. BMC Plant Biol 14:1–14

Schopfer P (1996) Hydrogen peroxide-mediated cell-wall stiffening in vitro in maize coleoptiles. Planta 199(1):43–49

Shahid MA, Balal RM, Khan N, Rossi L, Rathinasabapathi B, Liu G, Khan J, Cámara-Zapata JM, Martínez-Nicolas JJ, Garcia-Sanchez F (2018) Polyamines provide new insights into the biochemical basis of Cr-tolerance in Kinnow mandarin grafted on diploid & double-diploid rootstocks. Environ Exp Bot 156:248–260

Shan CJ, Dai HP, Sun YF (2012) Hydrogen sulfide protects wheat seedlings against copper stress by regulating the ascorbate & glutathione metabolism in leaves. Aust J Crop Sci 6(2):248–254

Sharaf AEMM, Farghal II, Sofy MR (2009) Role of gibberellic acid in abolishing the detrimental effects of Cd and Pb on broad bean and lupin plants. Res J Agric Biol Sci 5:668–673

Sharma P, Bhardwaj R (2007) Effect of 24-epibrassinolide on seed germination, seedling growth & heavy metal uptake in *Brassica juncea* L. Gen Appl Plant Physiol 33(1–2):59–73

Siddiqi KS, Husen A (2019) Plant response to jasmonates: current developments and their role in changing environment. Bull Natl Res Cent 43(1):1–11

Silveira NM, de Oliveira JA, Ribeiro C, Canatto RA, Siman L, Cambraia J, Farnese F (2015) Nitric oxide attenuates oxidative stress induced by arsenic in lettuce (*Lactuca sativa*) leaves. Water Air Soil Pollut 226(11):1–9

Singh BN, Mishra RN, Agarwal PK, Goswami M, Nair S, Sopory SK, Reddy MK (2004) A pea chloroplast translation elongation factor that is regulated by abiotic factors. Biochem Biophys Res Commun 320(2):523–530

Singh VP, Singh S, Kumar J, Prasad SM (2015) Hydrogen sulfide alleviates toxic effects of arsenate in pea seedlings through up-regulation of the ascorbate–glutathione cycle: possible involvement of nitric oxide. J Plant Physiol 181:20–29

Soares C, de Sousa A, Pinto A, Azenha M, Teixeira J, Azevedo RA, Fidalgo F (2016) Effect of 24-epibrassinolide on ROS content, antioxidant system, lipid peroxidation & Ni uptake in *Solanum nigrum* L. under Ni stress. Environ Exp Bot 122:115–125

Srivastava S, Srivastava AK, Suprasanna P, D'souza SF (2013) Identification and profiling of arsenic stress-induced microRNAs in *Brassica juncea*. J Exp Bot 64(1):303–315

Sun C, Liu L, Yu Y, Liu W, Lu L, Jin C, Lin X (2015) Nitric oxide alleviates aluminum-induced oxidative damage through regulating the ascorbate-glutathione cycle in roots of wheat. J Integr Plant Biol 57(6):550–561

Surgun-Acar Y, Zemheri-Navruz F (2022) Exogenous application of 24-epibrassinolide improves manganese tolerance in *Arabidopsis thaliana* L. via the modulation of antioxidant system. J Plant Growth Regul 41(2):546–557

Sytar O, Kumari P, Yadav S, Brestic M, Rastogi A (2019) Phytohormone priming: regulator for heavy metal stress in plants. J Plant Growth Regul 38:739–752

Tadaiesky LBA, da Silva BRS, Batista BL, Lobato AKDS (2021) Brassinosteroids trigger tolerance to iron toxicity in rice. Physiol Plant 171(3):371–387

Taie HAA, Seif El-Yazal MA, Ahmed SMA, Rady MM (2019) Polyamines modulate growth, antioxidant activity, & genomic DNA in heavy metal–stressed wheat plant. Environ Sci Pollut Res Int 26(22):22338–22350

Tian B, Zhang Y, Jin Z, Liu Z, Pei Y (2017) Role of hydrogen sulfide in the methyl jasmonate response to cadmium stress in foxtail millet. Front Biosci 22(3):530–538

Tripathi DK, Mishra RK, Singh S, Singh S, Vishwakarma K, Sharma S (2017) Nitric oxide ameliorates zinc oxide nanoparticles phytotoxicity in wheat seedlings: implication of the ascorbate–glutathione cycle. Front Plant Sci 8:1

Ur Rahman S, Xuebin Q, Yasin G, Cheng H, Mehmood F, Zain M et al (2021) Role of silicon on root morphological characters of wheat (*Triticum aestivum* L.) plants grown under cd-contaminated nutrient solution. Acta Physiol Plant 43:1–13

Ur Rahman S, Li Y, Hussain S, Hussain B, Riaz L, Ashraf MN, Khaliq MA, Du Z, Cheng H (2023) Role of phytohormones in heavy metal tolerance in plants: a review. Ecol Indic 146:109844

Valivand M, Amooaghaie R, Ahadi A (2019) Seed priming with H_2S & Ca^{2+} trigger signal memory that induces crossadaptation against nickel stress in zucchini seedlings. Plant Physiol Biochem 143:286–298

Ventrella A, Catucci L, Piletska E, Piletsky S, Agostiano A (2009) Interactions between heavy metals & photosynthetic materials studied by optical techniques. Bioelectrochemistry 77(1):19–25

Veselov D, Kudoyarova G, Symonyan M, Veselov S (2003) Effect of cadmium on ion uptake, transpiration and cytokinin content in wheat seedlings. Bulg J Plant Physiol 29(3–4):353–359

Vitti A, Nuzzaci M, Scopa A, Tataranni G, Remans T, Vangronsveld J, Sofo A (2013) Auxin and cytokinin metabolism and root morphological modifications in *Arabidopsis thaliana* seedlings infected with Cucumber mosaic virus (CMV) or exposed to cadmium. Int J Mol Sci 14(4):6889–6902

Wang SH, Zhang H, Jiang SJ, Zhang L, He QY, He HQ (2010) Effects of the nitric oxide donor sodium nitroprusside on antioxidant enzymes in wheat seedling roots under nickel stress. Russ J Plant Physiol 57(6):833–839

Wang M, Duan S, Zhou Z, Chen S, Wang D (2019a) Foliar spraying of melatonin confers cadmium tolerance in *Nicotiana tabacum* L. Ecotoxicol Environ Saf 170:68–76

Wang H, Ji F, Zhang Y, Hou J, Liu W, Huang J, Liang W (2019b) Interactions between hydrogen sulphide & nitric oxide regulate two soybean citrate transporters during the alleviation of aluminium toxicity. Plant Cell Environ 42(8):2340–2356

Wu X, Li H, Luo H, Lin D, and Zhou X (2019) Effect of exogenous melatonin on photosynthetic characteristics in cucumber seedlings under cadmium stress: a rapid detection method for the cadmium resistance. In: E3S Web of conferences. EDP Sciences, p 136

Xiang C, Oliver DJ (1998) Glutathione metabolic genes coordinately respond to heavy metals and jasmonic acid in Arabidopsis. Plant Cell 10(9):1539–1550

Xu LL, Fan ZY, Dong YJ, Kong J, Bai XY (2015) Effects of exogenous salicylic acid & nitric oxide on physiological characteristics of two peanut cultivars under cadmium stress. Biol Plant 59(1):171–182

Yadav P, Kaur R, Kohli SK, Sirhindi G, Bhardwaj R (2016) Castasterone assisted accumulation of polyphenols & antioxidant to increase tolerance of *B. juncea* plants towards copper toxicity. Cogent Food Agric 2(1):1276821

Yasir MW, Siddique MBA, Shabbir Z, Ullah H, Riaz L, Shah AA (2021) Biotreatment potential of co-contaminants hexavalent chromium and polychlorinated biphenyls in industrial wastewater: individual and simultaneous prospects. Sci Total Environ 779:146345

Yu Y, Zhou W, Zhou K, Liu W, Liang X, Chen Y, Sun D, Lin X (2018) Polyamines modulate aluminum-induced oxidative stress differently by inducing or reducing H_2O_2 production in wheat. Chemosphere 212:645–653

Yuan HM, Huang X (2016) Inhibition of root meristem growth by cadmium involves nitric oxide-mediated repression of auxin accumulation and signalling in Arabidopsis. Plant Cell Environ 39(1):120–135

Zanganeh R, Jamei R, Rahmani F (2019) Role of salicylic acid & hydrogen sulfide in promoting lead stress tolerance & regulating free amino acid composition in *Zea mays* L. Acta Physiol Plant 41:94–99

Zhang H, Dang Z, Zheng LC, Yi XY (2009) Remediation of soil co-contaminated with pyrene & cadmium by growing maize (*Zea mays* L.). Int J Environ Sci Technol 6(2):249–258

Zhang Y, Xu S, Yang S, Chen Y (2015) Salicylic acid alleviates cadmium-induced inhibition of growth & photosynthesis through upregulating antioxidant defense system in two melon cultivars (*Cucumis melo* L.). Protoplasma 252:911–924

Zhang N, Sun Q, Li H, Li X, Cao Y, Zhang H, Li S, Zhang L, Qi Y, Ren S, Guo YD (2016) Melatonin improved anthocyanin accumulation by regulating gene expressions & resulted in high reactive oxygen species scavenging capacity in cabbage. Front Plant Sci 7:197

Zhang J, Zeng B, Mao Y, Kong X, Wang X, Yang Y, Zhang J, Xu J, Rengel Z, Chen Q (2017) Melatonin alleviates aluminium toxicity through modulating antioxidative enzymes & enhancing organic acid anion exudation in soybean. Funct Plant Biol 44(10):961–968

Zhao FY, Hu F, Zhang SY, Wang K, Zhang CR, Liu T (2013) MAPKs regulate root growth by influencing auxin signaling and cell cycle-related gene expression in cadmium-stressed rice. Environ Sci Pollut Res 20:5449–5460

Zhong W, Xie C, Hu D, Pu S, Xiong X, Ma J, Sun L, Huang Z, Jiang M, Li X (2020) Effect of 24-epibrassinolide on reactive oxygen species & antioxidative defense systems in tall fescue plants under lead stress. Ecotoxicol Environ Saf 187:109831

Zhou M, Ghnaya T, Dailly H, Cui G, Vanpee B, Han R, Lutts S (2019) The cytokinin trans-zeatine riboside increased resistance to heavy metals in the halophyte plant species *Kosteletzkya pentacarpos* in the absence but not in the presence of NaCl. Chemosphere 233:954–965

Zhu XF, Jiang T, Wang ZW, Lei GJ, Shi YZ, Li GX, Zheng SJ (2012) Gibberellic acid alleviates cadmium toxicity by reducing nitric oxide accumulation and expression of IRT1 in *Arabidopsis thaliana*. J Hazard Mater 239:302–307

Zouari M, Ahmed CB, Elloumi N, Bellassoued K, Delmail D, Labrousse P, Abdallah FB, Rouina BB (2016) Impact of proline application on cadmium accumulation, mineral nutrition & enzymatic antioxidant defense system of *Olea europaea* L. cv Chemlali exposed to cadmium stress. Ecotoxicol Environ Saf 128:195–205

GPSR Compliance

The European Union's (EU) General Product Safety Regulation (GPSR) is a set of rules that requires consumer products to be safe and our obligations to ensure this.

If you have any concerns about our products, you can contact us on ProductSafety@springernature.com

In case Publisher is established outside the EU, the EU authorized representative is:

Springer Nature Customer Service Center GmbH
Europaplatz 3
69115 Heidelberg, Germany

Batch number: 10725825

Printed by Printforce, the Netherlands